T0332720

Stochastic Differential Equations

Mathematics and Its Applications (*East European Series*)

Volume 40

Stochastic Differential Equations

With Applications to Physics and Engineering

by

Kazimierz Sobczyk

Institute of Fundamental Technological Research,
Polish Academy of Sciences,
Warsaw, Poland

KLUWER ACADEMIC PUBLISHERS

DORDRECHT / BOSTON / LONDON

Library of Congress Cataloging-in-Publication Data

Sobczyk, Kazimierz.
Stochastic differential equations : with applications to physics and engineering / by Kazimierz Sobczyk.
p. cm. -- (Mathematics and its applications. East European series)
Includes index.
Bibliography: p.
ISBN 0-7923-0339-3
1. Stochastic differential equations. I. Title. II. Series: Mathematics and its applications (Kluwer Academic Publishers). East European series.
QA274.23.S55 1990
519.2--dc20 89-15287

Published by Kluwer Academic Publishers,
P.O. Box 17, 3300 AA Dordrecht, The Netherlands.

Kluwer Academic Publishers incorporates
the publishing programmes of
D. Reidel, Martinus Nijhoff, Dr W. Junk and MTP Press.

Sold and distributed in the U.S.A. and Canada
by Kluwer Academic Publishers,
101 Philip Drive, Norwell, MA 02061, U.S.A.

In all other countries, sold and distributed
by Kluwer Academic Publishers Group,
P.O. Box 322, 3300 AH Dordrecht, The Netherlands.

Printed on acid-free paper

To
my wife Anna
and children
Joanna, Paweł, Jacek, Marcin

SERIES EDITOR'S PREFACE

'Et moi, ..., si j'avait su comment en revenir, je n'y serais point allé.'

Jules Verne

The series is divergent; therefore we may be able to do something with it.

O. Heaviside

One service mathematics has rendered the human race. It has put common sense back where it belongs, on the topmost shelf next to the dusty canister labelled 'discarded nonsense'.

Eric T. Bell

Mathematics is a tool for thought. A highly necessary tool in a world where both feedback and nonlinearities abound. Similarly, all kinds of parts of mathematics serve as tools for other parts and for other sciences.

Applying a simple rewriting rule to the quote on the right above one finds such statements as: 'One service topology has rendered mathematical physics ...'; 'One service logic has rendered computer science ...'; 'One service category theory has rendered mathematics ...'. All arguably true. And all statements obtainable this way form part of the raison d'être of this series.

This series, *Mathematics and Its Applications*, started in 1977. Now that over one hundred volumes have appeared it seems opportune to reexamine its scope. At the time I wrote

> "Growing specialization and diversification have brought a host of monographs and textbooks on increasingly specialized topics. However, the 'tree' of knowledge of mathematics and related fields does not grow only by putting forth new branches. It also happens, quite often in fact, that branches which were thought to be completely disparate are suddenly seen to be related. Further, the kind and level of sophistication of mathematics applied in various sciences has changed drastically in recent years: measure theory is used (non-trivially) in regional and theoretical economics; algebraic geometry interacts with physics; the Minkowsky lemma, coding theory and the structure of water meet one another in packing and covering theory; quantum fields, crystal defects and mathematical programming profit from homotopy theory; Lie algebras are relevant to filtering; and prediction and electrical engineering can use Stein spaces. And in addition to this there are such new emerging subdisciplines as 'experimental mathematics', 'CFD', 'completely integrable systems', 'chaos, synergetics and large-scale order', which are almost impossible to fit into the existing classification schemes. They draw upon widely different sections of mathematics."

By and large, all this still applies today. It is still true that at first sight mathematics seems rather fragmented and that to find, see, and exploit the deeper underlying interrelations more effort is needed and so are books that can help mathematicians and scientists do so. Accordingly MIA will continue to try to make such books available.

If anything, the description I gave in 1977 is now an understatement. To the examples of interaction areas one should add string theory where Riemann surfaces, algebraic geometry, modular functions, knots, quantum field theory, Kac-Moody algebras, monstrous moonshine (and more) all come together. And to the examples of things which can be usefully applied let me add the topic 'finite geometry'; a combination of words which sounds like it might not even exist, let alone be applicable. And yet it is being applied: to statistics via designs, to radar/sonar detection arrays (via finite projective planes), and to bus connections of VLSI chips (via difference sets). There seems to be no part of (so-called pure) mathematics that is not in immediate danger of being applied. And, accordingly, the applied mathematician needs to be aware of much more. Besides analysis and numerics, the traditional workhorses, he may need all kinds of combinatorics, algebra, probability, and so on.

In addition, the applied scientist needs to cope increasingly with the nonlinear world and the extra mathematical sophistication that this requires. For that is where the rewards are. Linear models are honest and a bit sad and depressing: proportional efforts and results. It is in the non-linear world that infinitesimal inputs may result in macroscopic outputs (or vice versa). To appreciate what I am hinting at: if electronics were linear we would have no fun with transistors and computers; we would have no TV; in fact you would not be reading these lines.

There is also no safety in ignoring such outlandish things as nonstandard analysis, superspace and anticommuting integration, p-adic and ultrametric space. All three have applications in both electrical engineering and physics. Once, complex numbers were equally outlandish, but they frequently proved the shortest path between 'real' results. Similarly, the first two topics named have already provided a number of 'wormhole' paths. There is no telling where all this is leading - fortunately.

Thus the original scope of the series, which for various (sound) reasons now comprises five subseries: white (Japan), yellow (China), red (USSR), blue (Eastern Europe), and green (everything else), still applies. It has been enlarged a bit to include books treating of the tools from one subdiscipline which are used in others. Thus the series still aims at books dealing with:

- a central concept which plays an important role in several different mathematical and/or scientific specialization areas;
- new applications of the results and ideas from one area of scientific endeavour into another;
- influences which the results, problems and concepts of one field of enquiry have, and have had, on the development of another.

Many phenomena in nature have a stochastic component or, for a variety of reasons, can be well modelled in terms of equations with stochastic aspects. So the topic of stochastic differential equations should be a flourishing one in mathematics, with many applications. And so it is, and, indeed the applications range from oil platforms in the north sea (stochastic mechanics) to biology and from physics to computer networks. The topic is also, mathematically speaking, a deep and beautiful one.

There are several good books on stochastic differential equations. But a book that combines a thorough, selfcontained treatment of the topic with actual real life applications; a book that also really tells the reader what all this beautiful theory is good for and, moreover, discusses how to deal with these things numerically for simulation purposes, is rare. This is such a book and it is a pleasure to welcome it in this series.

The shortest path between two truths in the real domain passes through the complex domain.

J. Hadamard

La physique ne nous donne pas seulement l'occasion de résoudre des problèmes ... elle nous fait pressentir la solution.

H. Poincaré

Never lend books, for no one ever returns them; the only books I have in my library are books that other folk have lent me.

Anatole France

The function of an expert is not to be more right than other people, but to be wrong for more sophisticated reasons.

David Butler

Amsterdam, December 1990

Michiel Hazewinkel

CONTENTS

"Nothing would be done at all if a man waited till he could do it so well that no one could find fault with it."
John Henry Newman
(1801–1890)

Preface

This book is the outgrowth of my lectures given in the first quarter of 1985 for the research staff of the Technical University of Denmark, and constitutes an extension of the lecture notes: "Stochastical Differential Equations for Applications" published in 1985 by this University. Various parts of the material presented here were also lectured in the Department of Technical Physics and Applied Mathematics of the Warsaw Technical University as well as within a special series of lectures in "mathematics for applications" organized by the Mathematical Institute of the Polish Academy of Science.

The primary objective of this book is to provide a uniform and concise presentation of the basic knowledge on stochastic differential equations and their applications to various engineering disciplines. My special concern is focused on such presentation which gives an appropriate insight into the mathematical structure of the contemporary theory of stochastic equations as well as into the most efficient practical methods.

The book is intended for applied mathematicians, physicists and engineers who are interested in studying the behaviour of dynamical systems subjected to random excitation. It can be used both as a textbook at the graduate level and as a reference book by researchers. The reader is only assumed to be familiar with basic probability theory and with the most common methods of applied mathematics. Some elementary background in functional analysis would undoubtedly help but is not considered necessary.

The organization of the book is as follows. The text starts (Introduction) from a short description of the physical and mathematical origins of stochastic differential equations. Then I give a systematic summary of the basic concepts and results of the stochastic process theory (Chapter I) including the concepts, interpretations and constructions of the stochastic calculus with special attention to continuity differentiation and integration of stochastic processes (Chapter II). This knowledge forms the basis for the formulation, interpretation and analysis of stochastic equations.

In Chapter III the theory of stochastic differential equations is outlined with an attempt to show its mathematical structure and recent advances. Chapter IV, the largest one, presents a systematic discussion of the most effective analytical methods and solutions along with illustrative and applicatory examples elaborated in stochastic analysis of real dynamical systems. Chapter V gives a systematic presentation of numerical methods for stochastic differential equations; an effort has been made to reflect the existing developments in this very fresh field. Chapter VI is devoted to applications of stochastic differential equations in analysis of practical problems; the material presented can be regarded as a methodical introduction to stochastic dynamics of engineering systems.

A distinctive feature of this book is that it provides a unified treatment of both regular (often called, random) and Itô stochastic differential equations together with the methods and solutions elaborated very recently. In addition, an effort has been made to guide the reader through a large variety of research work devoted to both theoretical and applicatory aspects of stochastical differential equations. The bibliography attached is directly connected with the material presented in the content of the book. Although the references listed do not cover all existing and interesting contributions to the subject, I hope they will be found to be representative of the present state of the art.

In closing I wish to express my sincere thanks to all those who by their direct or indirect contribution have made the appearance of this book possible.

Above all, I owe a great debt of gratitude to Professor Ove Ditlevsen from the Technical University of Denmark for his kind invitation extended to me to visit his Department and for all arrangements associated with my lectures and publishing the lecture notes. I would also like to express my thanks to Dr. Peter Bjerager for his kindness in undertaking the arduous task of reading the manuscript of the lecture notes and for making many valuable improvements. Finally, it is my great pleasure to acknowledge the help I received during the preparation of the manuscript of the book. In particular, I wish to express my thanks to Mrs. Anna Gutweter for her excellent typing, and to Mr. Jerzy Trębicki for his painstacking work in bringing the typescript to its final form.

K. Sobczyk
Warszawa, March 1989

INTRODUCTION: ORIGIN OF STOCHASTIC DIFFERENTIAL EQUATIONS

Differential equations play a central role in applications of mathematics to natural and engineering sciences. However, differential equations governing real processes always contain some elements (e.g. coefficients, inhomogeneous part) which characterize physical features of the phenomenon and environment and are experimentally determined. Due to errors in the measurements and inherent randomness of a phenomenon these elements cannot most often be expressed by one uniquely defined function $f(x)$, but they have to be characterized by a family of functions $f_\gamma(x)$ depending on a certain parameter γ. Usually, we are, however, not able to foresee which of these functions shall be observed. Hence, in modelling of majority of real phenomena function $f(\cdot)$ have to be replaced by a random function $f_\gamma(\cdot)$; parameter γ is then interpreted as element of the space of elementary events Γ on which a probability is defined.

A similar situation occurs when we think about initial and boundary conditions. In the theory they are assumed to be given. In reality, however, these conditions are known only within certain ranges of values. Hence a more realistic formulation of initial-value and boundary value problems should involve treating the data as random.

In view of the above we can say in general that more realistic formulation of the differential equations arising in applied sciences (taking into account uncertainties and random noises associated with the process considered) should involve **stochastic differential equations**, that is—differential equations for stochastic processes. In other words we can say that a stochastic differential equation is a differential equation which contains random elements*. These elements may be random constants (random variables) or some random processes, but their statistical properties are assumed to be given. Accordingly, the solution of the equation will be a stochastic process and the basic problem consists in finding its probabilistic properties.

The genesis of stochastic differential equations is connected with the problems of physics. We wish to mention first the **Gibbsian ensemble method** (described by J.W. Gibbs in 1903) in the classical statistical mechanics where the problem of a time evolution of a large number of material particles (described by the Hamilton differential equations) is treated probabilistically. Though this model accounts only for the statistical nature of the initial state of a system (governing Hamiltonian equations are deterministic) it deserves, nevertheless, to be regarded as a first fruitful attempt in joining differential equations and probability.

The basic idea is as follows. The Gibbsian method characterizes a state of a

* We do not make the distinction of some authors between "stochastic" and "random" differential equations.

system of N material particles by the ensemble of dynamical (or canonical) variables, that is by generalized coordinates and conjugate momenta $(q_1, \ldots, q_N, p_1, \ldots, p_N)$ of the considered system of particles. The Hamiltonian equations

$$\begin{aligned} \frac{dq_k}{dt} &= \frac{\partial H}{\partial p_k}, \\ \frac{dp_k}{dt} &= -\frac{\partial H}{\partial q_k}, \end{aligned} \qquad H = \sum_{k=1}^{3N} p_k \dot{q}_k - L, \tag{0.1}$$

(L is the difference of kinetic and potential kinetic energy of a system) constitute a system of $6N$ differential equations of the first order defining a state of the system at time t provided the state of the system at $t = t_0$ is determined. This makes it possible to introduce a geometrical interpretation of the system evolution using the concept of the phase space. In the situation considered, this is a $6N$-dimensional space, whose points represent the system's states in each fixed instant of time. As the p's and q's vary in time, the phase point moves and traces out a path in the phase space.

Because of the difficulties in exact determination of the dynamical variables and due to the fact that the systems considered in statistical physics include a very large number of particles, a statistical description of the motion of a system has been adopted. This is usually done by introducing the so-called phase probability density $f(\mathbf{x}, t)$ where $\mathbf{x}$ denotes a point of the phase space: $\mathbf{x} = (x_1, x_2, \ldots, x_{6N})$. Since the motion of the system is governed by deterministic laws (the Hamilton equations) it is shown in statistical mechanics that the phase probability density $f(\mathbf{x}, t)$ satisfies the following Liouville's equation

$$\begin{aligned} &\frac{\partial f}{\partial t} + \sum_{k=1}^{6N} \frac{\partial}{\partial x_k} (\dot{x}_k f) = 0, \\ &f(\mathbf{x}, t_o) = f_o(\mathbf{x}), \end{aligned} \tag{0.2}$$

where $f_o(\mathbf{x})$ denotes the probability density of the initial state of the system.

On the other hand, it is clear that the problem of classical statistical mechanics described above may be formulated in the language of stochastical processes. We , simply, deal with the motion of a dynamical system governed by a set of Hamiltonian equations with random initial conditions, that is symbolically

$$\begin{aligned} &\frac{d\mathbf{X}}{dt} = \mathbf{F}(\mathbf{X}, t) \\ &\mathbf{X}(t_o) = \mathbf{X}^o(\gamma) \end{aligned} \tag{0.3}$$

where $\mathbf{X}^o(\gamma)$ is a $6N$-dimensional random vector with a given probability density $f_o(\mathbf{x}^o)$.

It is clear that differential equations with random initial conditions occur naturally in many other fields (cf. chemical kinetics, structural mechanics, heat conduction etc.). In this context it is worth noting that one of the first stochastic problems for partial differential equations was concerned with heat conduction in an infinite beam when its initial temperature is charaterized by a random function (J. Kampe de Fériét, 1955).

If the system is open (which means that it interacts with an external field) then the Hamiltonian is modified to include the externally applied excitations. Since the external fields should often be treated as randomly fluctuating in time we come directly to the system of equations

$$\begin{aligned} \frac{d\mathbf{X}}{dt} &= \mathbf{F}(\mathbf{X},t) + \mathbf{G}(\mathbf{X},t)\eta(t,\gamma), \\ \mathbf{X}(t_o) &= \mathbf{X}^o(\gamma), \end{aligned} \tag{0.4}$$

where the term $\mathbf{G}(\mathbf{X},t)\eta(t,\gamma)$ describes random flutuations (with state dependent intensity) in an external field interacting with the sytem considered. The equations of the form (0.4) constitute adequate models for numerous natural and engineering systems subjected to time-varying random excitation.

One of the simplest examples of the above type of description is the one-dimensional **Langevin equation** for the **Brownian motion** (of a small particle suspended in a liquid). The phenomenon of the Brownian motion and the Langevin equation are commonly regarded as the origin of the stochastic differential equations. Due to this fact it is worthy to look more carefully at the physics of the Langevin equation.

The phenomenon of Brownian motion—discovered experimentally by the Scottish botanist R. Brown in 1827—is one of the most interesting examples of random physical processes. Mathematical modelling of this phenomenon took the attention of many physicists and mathematicians. The first fruitful results are connected with the names of Einstein and Smoluchowski, who derived (1905–1906) a partial differential equation for the probability density of the displacement of the Brownian particle. One of the significant conclusions of this early achievement is that the mean square of the displacement of the particle grows linearly in time—the result (as we know today) has serious implications in the mathematical theory of Brownian motion.

Nearly in the same time (in 1908) Paul Langevin formulated a phenomenological description of the erratic motion of a heavy "Brownian" particle of mass m immersed in a liquid. He assumed that this erratic behaviour could be described by means of the Newtonian equation for the particle. The interaction of the surrounding fluid with the Brownian particle gives rise to two distinct forces: a dissipative force (due to dynamic friction in the course of a motion of a particle through a viscous fluid) and a fluctuating force (arising from the molecular collisions). So, Langevin wrote down a differential equation of the motion of the particle

$$m\frac{dv}{dt} = f(t) \tag{0.5}$$

where v is the component of velocity of particle along the x axis and $f(t)$ is the force caused by a surrounding medium (a sum of dissipative and fluctuating forces). Langevin assumed that

$$f(t) = -\beta v + \eta(t) \tag{0.6}$$

where $-\beta v$ characterizes the systematic (regular) dissipative part of the influence of the surrounding medium, and $\eta(t)$ represents the irregular force exerted on the particle by the collisions. Equations (0.5) and (0.6) give (let $\alpha = \beta/m, \xi(t) = \frac{1}{m}\eta(t)$)

$$\frac{dv}{dt} + \alpha v = \xi(t) \tag{0.7}$$

or, equivalently

$$\frac{d^2x}{dt^2} + \alpha\frac{dx}{dt} = \xi(t). \tag{0.8}$$

Equation (0.7) is commonly termed as the **Langevin equation**. Comparing (0.7) with general equation (0.4) it is seen that the classical Langevin equation is really a simple case, when: $\mathbf{F} = -\alpha v, \mathrm{G}\eta = \frac{1}{m}\eta(t) = \xi(t)$. If we assume that a motion of a spherical Brownian particle with radius a in a liquid with viscosity ν is governed by the Stokes law, then the constant $\alpha = \frac{6\pi a\nu}{m}$.

Random impulses acting on the particle are very small in strength and occur "infinitely" frequently (in normal conditions the particle undergoes about 10^{21} molecular collisions per second from all directions). In addition, the process of collisions has normally very small correlation time, but in each instant of time the fluctuating force has a finite value which is comparable with the dissipative force (in conditions of statistical equilibrium). The above physical features of the molecular collisions are usually formalized in physics as follows

$$\begin{aligned} &E[\xi(t)] = 0, \\ &E[\xi(t_1)\xi(t_2)] = D\delta(t_2 - t_1), \quad D > 0, \end{aligned} \tag{0.9}$$

where $\delta(t_2 - t_1)$ is the Dirac function, symbol $E[\cdot]$ denotes averaging over the ensemble of collision processes and D is a constant.

Therefore, $\xi(t)$ in Langevin equation (0.7) is a very peculiar process. In stochastic theory it is called a "white noise". Of course, such a stochastic process does not exist in the conventional sense, since it would have to have uncorrelated values at all points $t_1 \neq t_2$ and infinite variance. Nevertheless, the white noise is a useful mathematical idealization for describing and analysis of rapidly varying random fluctuations which are practically uncorrelated for different instants of time.

Although equation (0.7) can be formally integrated (using the properties of Dirac function) and statistics of the solution process can be explicitly determined—it constitutes only a formal relation and its mathematically rigorous meaning requires an adequate theory. Such a theory is especially desirable in the case of general equation (0.4)—accounting for state dependent fluctuations—in which in the place of $\eta(t,\gamma)$ appears a white noise $\xi(t)$; it will be presented in Chapter III of this book.

In mathematics the theory of stochastic differential equations was initiated as a method of construction of diffusion Markov processes on the basis of the Brownian motion process. Such an appraoch, historically associated with the work of Bernstein (in thirties), has been systematically formulated by Itô (in early forties) and developed by many others in our days and now is known as a theory of the **Itô stochastic differential equations**. This theory also constitutes a base for rigorous treatment of the Langevin-type equations in which a random element is idealized as the so-called white noise.

Recently (roughly, since beginning of sixties) stochastic differential equations have been widely accepted as an important mathematical tool in modelling and analysis of numerous processes in engineering, especially in control and mechanical systems. Needs for the investigation of performance of various engineering systems subjected to environmental random excitations has created stimuli for the systematic analysis of various types of stochastic differential equations (see Chapter IV and VI of the book.).

Today the theory of stochastic differential equations—although still in its formative stage—has a very extensive literature dealing both with mathematical bases as well as with applications. It is the purpose of this book to give a systematic and concise exposition of this new subject with a special emphasis on the applicatory aspects of the theory.

Chapter I

STOCHASTIC PROCESSES—SHORT RÉSUMÉ

1. INTRODUCTORY REMARKS

In the observation of reality we often encounter phenomena which cannot be uniquely characterized. Dealing with such phenomena we observe that under the same conditions, the outcomes of a repeated experiment do not coincide; between the results there exist some unpredictable, random variations. Such phenomena are commonly regarded as random phenomena. Of course, the concept of "randomness" comes from our experience with a real world and is mostly used in its intuitive sense. Most often an event is regarded to be "random" if we do not know all causes and conditions of its realization.

In order to investigate quantitatively the regularities of random phenomena one has to introduce first a mathematical model of randomness together with an appropriate measure of the possibility of the occurence of various "uncertain" outcomes of an experiment. The notions which should be rigorously defined in constructing such a general model of random phenomena are first of all: a **sample space** (i.e. a collection of all possible outcomes of an experiment), **random event** and **probability**. The mathematical discipline which constructs and investigates models of random phenomena is widely known as the probability theory. Although the questions concerning the true relationship between real phenomena and the language of probability theory constitute, in general, very essential and often, controversial, problem of the methodical nature (cf. different attempts to interpretation and formalization of randomness [12]), the contemporary probability theory has proven to be the most solid basis for studying various complicated and uncertain phenomena including processes governed by differential equations.

This theory defines its basic concepts in the language of measure theory. In this way the specific terminology of the probability theory (associated with its intuitive origins) and various situations considered within its framework are expressible in terms of the measure spaces, measurable functions and their transformations (cf. [23]). This feature of modern probability theory dominates the recent developments in the stochastic process theory and in the theory of stochastic differential equations.

In this chapter we shall provide a brief résumé of the concepts and results of the probability theory which are of prime importance for formulation and analysis of stochastical differential equations.

2. PROBABILITY AND RANDOM VARIABLES

2.1. Basic concepts

In the theoretical model of random phenomena a basic role is played by the so-called probability space $(\Gamma, \mathcal{F}, P)$. The set Γ — called the space of elementary events or the *sample space* — represents all possible, the simplest (or elementary) outcomes of a trial associated with the considered random phenomenon. $\mathcal{F}$ is a σ-algebra of subsets of Γ; that is $\mathcal{F}$ is a non-empty class of subsets of Γ satisfying the following conditions: (1): $\Gamma \in \mathcal{F}$, (2): if $A \in \mathcal{F}$ and $B \in \mathcal{F}$ then $A - B \in \mathcal{F}$, and (3): $A_i \in \mathcal{F}$, $i = 1, 2, \ldots$ then a countable sum of A_i also belongs to $\mathcal{F}$. The elements of $\mathcal{F}$ are called *random events*. P is a probability , or probability measure defined on all $\mathcal{F}$; that is P is a set function whose arguments are random events (elements of $\mathcal{F}$) such that (axioms of probability):

1. $0 \leq P(A) \leq 1 \quad , \qquad A \in \mathcal{F}$
2. $P(\Gamma) = 1,$
3. for a countable collection of mutually disjoint
$$A_1, A_2, \ldots \text{in } \mathcal{F}: \quad P\{\cup_j A_j\} = \sum_j P(A_j)$$

Hence, $P(A)$ is a countably additive, non-negative set function such that $P(A) \in [0, 1]$ for all $A \in \mathcal{F}$ with $P(\Gamma) = 1$.

It is clear, that in experiment on random phenomenon various outcomes (elementary events) can occur. These outcomes are in most situations represented by real numbers. Other types of outcomes, while not originally numerical can be made to correspond to numbers by a suitable choice. Therefore, it is justified to assume that in general (in all experiments) one can assign a real number $X(\gamma)$ for each elementary event $\gamma \in \Gamma$. This leads to the concept of a random variable. The definition is the following

Definition 1.1. *A* random variable *is a real function $X(\gamma), \gamma \in \Gamma$ defined on a space Γ and measurable with respect to P, that is for every real number x, the set $\{\gamma : X(\gamma) < x\}$ is an event in $\mathcal{F}$.*

The probabilistic behaviour of a random variable $X(\gamma)$ is completely and uniquely specified by its distribution function $F(x)$ which is defined as follows

$$F(x) = P\{X(\gamma) < x\} \tag{1.1}$$

By the definition, the distribution function always exists and it is non-negative and non-decreasing function of real variable x. Furthermore, it is continuous from the left.

A random variable $X = X(\gamma)$ is a *discrete* random variable if its distribution function takes the form of a staircase with a finite or countably finite jumps. The jumps x_i and its probabilities $p_i(\sum_i p_i = 1)$ characterizes a discrete random variable. A random variable $X(\gamma)$ is called a *continuous random variable* if its distribution function is absolutely continuous; that is, there exists such non-negative and integrable function $f(x)$ that for each x

$$F(x) = \int_{-\infty}^{x} f(z)dz \tag{1.2}$$

In this case exists derivative $\frac{dF}{dx} = f(x)$ which is called the *density function* of random variable $X(\gamma)$ and has the properties:

$$f(x) \geq 0, \qquad \int_a^b f(x)dx = F(b) - F(a), \qquad \int_{-\infty}^{+\infty} f(x)dx = 1. \tag{1.3}$$

It is worth noticing that the density function of a discrete random variable does not exist in the ordinary sense. It can be, however, constructed by use of the Dirac delta.

The next important concept of probability theory is the expectation or mean value of random variable.

Definition 1.2. *A mean or average value of a random variable $x = X(\gamma)$ is the following Stieltjes integral (if exists)*

$$m_X = E(X) = \int_\Gamma X(\gamma)dP(\gamma) = \int_{-\infty}^{+\infty} xdF(x). \tag{1.4}$$

From the properties of the Stieltjes integrals follows that

$$m_X = \begin{cases} \sum_i x_i p_i & \text{for discrete r.v.} \\ \int_{-\infty}^{+\infty} xf(x)dx\ , & \text{for continuous r.v.} \end{cases} \tag{1.5}$$

An important class of expectations is constituted by the averages of various powers of random variable. These average values are called the *moments*. Exactly, the k-th order moment of a random variable $X(\gamma)$ is defined by

$$m_k = E[X^k] = \begin{cases} \sum_i x_i^k p_i & \text{for discrete r.v.} \\ \int_{-\infty}^{+\infty} x^k f(x)dx & \text{for continuous r.v.} \end{cases} \tag{1.6}$$

The *central moment* of order k of a random variable $X(\gamma)$ is the quantity: $\mu_k = E[(X(\gamma) - m_X)^k]$. In particular

$$\sigma_X^2 = \mu_2 = \begin{cases} \sum_i (x_i - m_X)^2 p_i & \text{for discrete r.v.} \\ \int_{-\infty}^{+\infty} (x - m_X)^2 f(x)dx, & \text{for continuous r.v.} \end{cases} \tag{1.7}$$

is known as the *variance* of $X(\gamma)$. The variance measures the spread or dispersion of the random variable $X(\gamma)$ about the mean.

The positive square root of the variance, namely σ_X is called the *standard deviation* of $X(\gamma)$. For $k > 0$ and any n such that $E|X(\gamma)|^n$ is finite, the following useful inequalities (due to Markov and Chebyshev, respectively) hold:

$$P\{|X(\gamma)| \geq k\} \leq \frac{E|X(\gamma)|^n}{k^n},$$

$$P\{|X(\gamma) - m_X| > \epsilon\} \leq \frac{\sigma_X^2}{\epsilon^2}. \tag{1.8}$$

Another type of expectation—playing a role in theory and applications—is the average value of an exponential function of a real random variable.

Definition 1.3. *A complex function $\phi(\lambda)$ of the real parameter λ defined as follows*

$$\phi(\lambda) = E\left[e^{i\lambda X(\gamma)}\right] = \begin{cases} \sum_k p_k e^{i\lambda x_k}, & \text{for discrete r.v.} \\ \int_{-\infty}^{+\infty} e^{i\lambda x} f(x)dx, & \text{for continuous r.v.} \end{cases} \tag{1.9}$$

is called the characteristic function of $X(\gamma)$.

It is seen that the characteristic function $\phi(\gamma)$ and the density function $f(x)$ form a Fourier transform pair. Since for every λ and x, $\left|e^{i\lambda x}\right| = 1$, the characteristic function always exists. Here are the basic properties of the characteristic function:

1) $\phi(0) = 1$,
2) $|\phi(\lambda)| \leq 1$, for each λ,
3) $\phi(\lambda) = \overline{\phi(-\lambda)}$ where the bar denotes the complex conjugate,
4) $\phi(\lambda)$ is uniformly continuous over the entire real line.

The moments of a random variable $X(\gamma)$ can be computed from the characteristic function. A rigorous statement of this property is as follows: if the r-moment of a random variable $X(\gamma)$ exists, then its characteristic function has derivative of the order r and for $0 \leq k \leq r$

$$m_k = \frac{1}{i_k} \frac{d^k \phi(\lambda)}{d\lambda^k}\bigg|_{\lambda=0} \tag{1.10}$$

Using the characteristic function we can define important quantities which are called the *cumulants* (or *semi-invariants*) of $X(\gamma)$. The cumulants can be defined by the characteristic function as follows

$$\kappa_k = \frac{1}{i^k} \frac{d^k}{d\lambda^k} ln\phi(\lambda)\bigg|_{\lambda=0}. \tag{1.11}$$

The following relationships between moments and cumulants are valid:

$$\kappa_1 = m_1, \qquad \kappa_2 = m_2 - m_1^2 = \mu_2, \qquad \kappa_3 = m_3 - 3m_1m_2 - 2m_1^3 = \mu_3,$$
$$\kappa_4 = \mu_4 - 3\mu_2^2, \qquad \kappa_5 = \mu_5 - 10\mu_3\mu_2.$$

Very often there is a need for the investigation of the joint behaviour of two or more variables. In such cases it is convenient to deal with a multidimensional random variable or random vector. The *n-dimensional random vector*

$$\mathbf{X}(\gamma) = [X_1(\gamma), X_2(\gamma), \ldots, X_n(\gamma)]$$

is a function (measurable with respect to P) which takes every element $\gamma \in \Gamma$ into a point $x \in R_n$. For each random vector the probability distribution of its values is well defined, and—analoguously to the case of one random variable–is characterized conveniently by the joint distribution function

$$F(x_1, x_2, \ldots, x_n) = P\{X_1(\gamma) < x_1, X_2(\gamma) < x_2, \ldots, X_n(\gamma) < x_n\} \tag{1.12}$$

The function $F(x_1, x_2, \ldots, x_n)$ is non-negative, non-decreasing and continuous to the left with respect to each of the variables. It also satisfies some simple relations, which in the case $n = 2$ take the form

$$F(-\infty, +\infty) = F(-\infty, x_2) = F(x_1, -\infty) = 0$$
$$F(+\infty, +\infty) = 1 \quad , \quad F(x_1, +\infty) = F_{X_1}(x_1) \quad , \quad F(+\infty, x_2) = F_{X_2}(x_2)$$

In the above, the functions $F_{X_1}(x_1)$, $F_{X_2}(x_2)$ are the distribution functions of $X_1(\gamma)$ and $X_2(\gamma)$, respectively. These distribution functions are commonly called the *marginal* distribution functions (they specify the probability distribution of one random variable when the second takes arbitrary values). In the case of n-dimensional random variable ($n > 2$) there exists $\binom{n}{k}$ k-dimensional marginal distributions for $k = 1, 2, \ldots, n-1$.

If

$$F(x_1, x_2, \ldots, x_n) = F_1(x_1)F_2(x_2)\ldots F_n(x_n) \tag{1.13}$$

the random variables $X_1, X_2, \ldots, X_n$ are called *independent*.

If the distribution function $F(x_1, x_2, \ldots, x_n)$ is differentiable with respect to each variable then its mixed derivative

$$f(x_1, x_2, \ldots, x_n) = \frac{\partial^n F(x_1, x_2, \ldots, x_n)}{\partial x_1 \partial x_2 \ldots \partial x_n} \tag{1.14}$$

is called the *joint probability density function* of $X_1, X_2, \ldots, X_n$.

For independent random variables, $f(x_1, x_2, \ldots, x_n) = f(x_1)f(x_2)\ldots f(x_n)$. The density function is non-negative, integrable and it has some simple properties which in the case $n = 2$ can be written in the form

$$\int_{-\infty}^{x_1}\int_{-\infty}^{x_2} f(z_1, z_2,)dz_1dz_2 = F(x_1, x_2),$$

$$\int_{-\infty}^{+\infty}\int_{-\infty}^{+\infty} f(x_1, x_2)dx_1dx_2 = 1,$$

(1.15)

$$\int_{-\infty}^{+\infty} f(x_1, x_2)dx_2 = f_{X_1}(x_1), \qquad \int_{-\infty}^{+\infty} f(x_1, x_2)dx_1 = f_{X_2}(x_2)$$

The functions $f_{X_1}(x_1), f_{X_2}(x_2)$ are the *marginal* density functions of $X_1(\gamma)$ and $X_2(\gamma)$, respectively.

The moments of two or more random variables are defined in a similar way as in the case of one random variable. The joint central moment μ_{ln} of the order $l + n$ is defined by

$$\mu_{ln} = E\Big\{ \big[X_1(\gamma) - m_{X_1}\big]^l \big[X_2(\gamma) - m_{X_2}\big]^n \Big\} \tag{1.16}$$

The central moment μ_{11} is called the *covariance* of X_1 and X_2. It is often convenient to deal with a normalized covariance

$$\rho_{X_1X_2} = \frac{\mu_{11}}{\sigma_{X_1}\sigma_{X_2}} \tag{1.17}$$

called the *correlation coefficient* of X_1 and X_2. It is obvious that $-1 \le \rho_{X_1X_2} \le 1$. If $\rho_{X_1X_2} = 0$ the random variables X_1 and X_2 are said to be *uncorrelated.* The correlation coefficient is a measure of linear dependence between X_1 and X_2; its value is a measure of accuracy with which one random variable can be approximated by a linear function of the other. If $\rho_{X_1X_2} = 1$, then $X_2 = aX_1 + b$.

2.2. Some probability distributions

As we have already said, every random variable is specified by its probability distribution. The concrete probability distributions described analytically by distribution functions $F(x)$ can take different forms. It turns out, however, that not all distributions are of the same importance. Some of them proved to be extremely useful both in theory and applications; for example, uniform distribution, Gaussian (or normal) distribution, gamma distribution and others. Here, we shall characterize briefly only Poisson, uniform and normal distributions.

A discrete random variable $X(\gamma)$ taking non-negative and integer values $k = 0, 1, 2, \ldots,$ is said to have a *Poisson distribution* if

$$P\{X(\gamma) = k\} = \frac{\nu^k}{k!} e^{-\nu}, \qquad \nu\text{-positive constant} \tag{1.18}$$

The basic characteristics of the Poisson distribution are

$$m_X = \nu, \qquad \sigma_X^2 = \nu, \qquad \phi(\lambda) = e^{\nu(e^{i\nu}-1)}. \tag{1.19}$$

One of the simplest distributions of continuous random variable is the *uniform distribution*. A random variable $X(\gamma)$ has the uniform distribution in the interval $[a, b]$ if

$$f(x) = \begin{cases} \frac{1}{b-a}, & a \leq x \leq b \\ 0, & x < a, x > b \end{cases} \tag{1.20}$$

Simple calculations lead to the following formulae

$$m_X = \frac{a+b}{2}, \qquad \sigma_X^2 = \frac{(b-a)^2}{12}, \qquad \phi(\lambda) = \frac{e^{i\lambda b} - e^{i\lambda a}}{i\lambda(b-a)}. \tag{1.21}$$

A random variable $X(\gamma)$ has the *Gaussian* or *normal distribution* if its density function can be written in the form

$$f(x) = \frac{1}{\sigma_X \sqrt{2\pi}} exp\left[\frac{-(x - m_X)^2}{2\sigma_X^2}\right], \tag{1.22}$$

where m_X is the mean value, and σ_X denotes the standard deviation. It is seen that a Gaussian random variable is completely characterized by the two parameters, that is by its mean and variance. Since distribution (1.22) is symmetric with respect to its mean value m_X, all odd central moments are equal to zero. In general, it can be shown that

$$\begin{aligned} &\mu_{2k+1} = 0, \\ &\mu_{2k} = 1 \cdot 3 \cdot \ldots \cdot (2k-1)\sigma_X^{2k} \end{aligned} \tag{1.23}$$

The characteristic function corresponding to distribution (1.22) is of the form

$$\phi(\lambda) = e^{i\lambda m_X - \frac{1}{2}\sigma_X^2 \lambda^2}. \tag{1.24}$$

A random variable $[X_1(\gamma), \ldots, X_n(\gamma)]$ has an *n-dimensional Gaussian distribution* if its probability density function is the following

$$f(x_1, \ldots, x_n) = \frac{1}{(2\pi)^{n/2}\sqrt{|\mathbf{K}|}} exp\left[-\frac{1}{|\mathbf{K}|} \sum_{i,j=1}^{n} |K_{ij}|(x_i - m_i)(x_j - m_j)\right], \tag{1.25}$$

where $|\mathbf{K}| \neq 0$ is the determinant of the matrix of covariances, that is $\mathbf{K} = \{k_{ij}\}$, $i, j, = 1, \dots, n$; $k_{ij} = E\left[(X_i - m_{x_i})(X_j - m_{x_j})\right]$; $|K_{ij}|$ is the cofactor of the element k_{ij} of matrix $\{k_{ij}\}$ and $m_i = E\left[X_i(\gamma)\right]$, $i = 1, 2, \dots, n$.

The characteristic function of an n-dimensional Gaussian distribution is

$$\begin{aligned} \phi(\lambda_1, \dots, \lambda_n) &= E\left\{ exp\left[i \sum_{k=1}^{n} \lambda_k X_k(\gamma) \right] \right\} \\ &= exp\left(i \sum_{k=1}^{n} m_k \lambda_k - \frac{1}{2} \sum_{k,l=1}^{n} k_{kl} \lambda_k \lambda_l \right) \\ &= exp\left[i(\mathbf{m}, \boldsymbol{\lambda}) - \frac{1}{2}(\mathbf{K}\boldsymbol{\lambda}, \boldsymbol{\lambda}) \right] \end{aligned} \tag{1.26}$$

where

$$\mathbf{m} = [m_1, \dots, m_n], \quad \boldsymbol{\lambda} = [\lambda_1, \quad \dots, \lambda_n], \quad \mathbf{K} = \{k_{ij}\}$$

The Gaussian distribution has many simple and interesting properties (for example, it is invariant under linear transformation). This distribution appears in various applications very frequently. A reason for the enormous popularity of the Gaussian distribution stems from the so-called central limit theorem of the probability theory, which asserts that under quite general conditions, the sum of independent random components, when their number tends to infinity is approximately normally distributed (see—next subsection).

2.3. Convergence of sequences of random variables

Let us consider now a sequence of random variables $X_1(\gamma), X_2(\gamma), \dots$ defined on the same probability space $(\Gamma, \mathcal{F}, P)$. Let $X(\gamma)$ be another random variable defined on the same probability space.

A sequence of random variables $\{X_n(\gamma)\}, n = 1, 2, \dots$ is said to be *convergent* to random variable $X(\gamma)$ *with probability one* or *almost surely*, if

$$P\left\{ \lim_{n \to \infty} X_n(\gamma) = X(\gamma) \right\} = 1 \tag{1.27}$$

This kind of convergence is often abbreviated in notation as:

$$X_n \xrightarrow{\text{a.s.}} X.$$

A sequence $\{X_n(\gamma)\}$, $n = 1, 2, \dots$ is said to be convergent to $X(\gamma)$ *in probability* ($X_n \xrightarrow{\text{in pr.}} X$), if for every $\varepsilon > 0$

$$\lim_{n \to \infty} P\left\{ |X_n(\gamma) - X(\gamma)| > \varepsilon \right\} = 0 \tag{1.28}$$

Convergence in probability is sometimes called stochastic convergence or convergence in measure.

If a sequence $\{X_n(\gamma)\}$ is convergent to $X(\gamma)$ almost surely, then it is also convergent to $X(\gamma)$ in probability. An opposite assertion is, in general, not valid.

A sequence of random variables $\{X_n(\gamma)\}$ with distribution functions $F_1(x)$, $F_2(x), \ldots$ is *convergent in distribution* to random variable $X(\gamma)$ with the distribution $F(x)$ if a sequence $\{F_n(x)\}$ converges to $F(x)$ in every continuity point of the distribution function $F(x)$.

A necessary and sufficient condition for $F_x(x) \longrightarrow F(x)$ in all continuity points of $F(x)$, is that the corresponding sequence of characteristic functions $\phi_n(\lambda)$ is convergent to $\phi(\lambda)$ for every real λ, where $\phi(\lambda)$ is any function continuous at $\lambda = 0$. If this holds, $\phi(\lambda)$ is continuous for all λ and is the characteristic function corresponding to the limiting distribution $F(x)$.

It can be shown (cf.[23]) that convergence in probability implies convergence in distribution, but the opposite assertion does not hold.

The above three modes of convergence are concerned with arbitrary sequences of random variables. Now, we shall define a concept of convergence which is concerned only with random variables possessing finite second moments.

A sequence of random variables $\{X_n(\gamma)\}$ such that $E\left\{|X_n(\gamma)|^2\right\} < \infty, n = 1, 2, \ldots$ *converges in the mean square* (in m.s.) to the random variable $X(\gamma)$ such that $E\left\{|X(\gamma)|^2\right\} < \infty$ if*

$$\lim_{n \to \infty} E\left\{|X_n(\gamma) - X(\gamma)|^2\right\} = 0 \tag{1.29}$$

It can be shown (cf.[23]) that the sequence $\{X_n(\gamma)\}$ converges in m.s. if and only if

$$E\left[X_m \overline{X_n}\right] \longrightarrow \text{finite limit } c \tag{1.30}$$

as m and n tend independently to infinity, where $X_n(\gamma)$ can be, in general, complex-valued random variables, and $\overline{X_n}$ denotes the complex conjugate.

It can also be shown [23] that convergence in the mean square implies convergence in probability. If the sequence $\{X_n(\gamma)\}$ converges according to any of three modes (a.s.,in pr-ty, m.s.) the limiting random variables are equal with probability one. Summing up, in general:

$$\begin{matrix} \text{a.s.} \searrow & & \\ & \text{in pr.} \longrightarrow \text{in distr.} \\ \text{m.s.} \nearrow & & \end{matrix}$$

* The mean-square convergence is often denoted: l.i.m..

The following **assertion** holds: if for some $k > 0$

$$\sum_{n=1}^{\infty} E\left|X_n(\gamma) - X(\gamma)\right|^k < \infty \tag{1.31}$$

then $X_n \xrightarrow{\text{a.s.}} X$.

Let us assume that $\{X_n(\gamma)\}$ is a sequence of Gaussian random variables with parameters (m_n, σ_n). A sequence $\{X_n(\gamma)\}$ is convergent in distribution to a certain limiting random variable $X(\gamma)$ if and only if (cf. [16])

$$m_n \longrightarrow m, \qquad \sigma_n \longrightarrow \sigma, \qquad as\ n \longrightarrow \infty;$$

in this case the limiting distribution is Gaussian with parameters (m, σ).

This theorem is easily extendable to the case of random vectors. As we have already mentioned, under some fairly general conditions, the sum of a large number of independent random variables is approximately normally distributed. A rigorous statement gives the following **central limit theorem (Lapunov theorem)**:

Let $\{X_k(\gamma)\}$ be a sequence of independent random variables with mean values m_k, standard deviations $\sigma_k \neq 0$ and finite absolute central moments $b_k = E\big|X_k - m_k\big|^3$. Let us denote

$$B_n = \left(\sum_{k=1}^{n} b_k\right)^{1/3} \qquad C_n = \left(\sum_{k=1}^{n} \sigma_k^2\right)^{1/2}$$

If

$$\lim_{n\to\infty} \frac{B_n}{C_n} = 0$$

then the sequence $\{F_k(z)\}$ of the distribution functions of random variables $Z_n(\gamma)$ defined as

$$Z_n(\gamma) = \frac{\sum_{k=1}^{n}(X_k - m_k)}{C_n}$$

tends (as $n \to \infty$) to the Gaussian distribution function $N(0,1)$, that is

$$\lim_{n\to\infty} F_n(z) = \frac{1}{\sqrt{2\pi}} \int_{-\infty}^{z} e^{-\frac{y^2}{2}}\,dy$$

The above limit theorem has been generalized in many important directions. The formulation given above is of the *integral type* (convergence of distribution functions); in the literature there exist appropriate *local limit theorems* (convergence of probability densities)—cf. [13]. The central limit theorem holds also for multidimensional random variables. Furthermore, it is extendable to the case of "weakly" dependent random variables (cf. [30]).

2.4. Entropy and information of random variables

A result of an experiment associated with a random phenomenon can not be predicted before observation; a characteristic feature of any random phenomenon is its uncertainty. This uncertainty (or indeterminacy) can often be estimated qualitatively comparing uncertainties of different situations. However, in the analysis of real random phenomena (especially, those in engineering systems) it is desirable to evaluate an uncertainty quantitavely. A convenient measure of uncertainty is a certain function of probabilities (of the outcome of the experiment) which is called *entropy*. The entropy of a random variable is a measure of indeterminacy (or randomness) of its values—a priori, before experiment. As we shall see below, a value of entropy of a discrete random variable can be interpreted as an average number of "bits" necessary to differentiate its possible values. Very close to entropy is the concept of *information*. It turns out that the amount of information contained in a random variable or the amount of mutual information between two random variables can be defined in terms of entropy.

Let $X(\gamma)$ be a discrete random variable assuming its values x_i with probabilities $P\{X(\gamma) = x_i\} = p_i \qquad i = 1, 2, \ldots, n.$

Definition 1.4. *The quantity $H(X)$ defined*

$$H(X) = -\sum_{i=1}^{n} p_i \log p_i \tag{1.32}$$

is called the entropy *of a discrete random variable $X(\gamma)$.*

Sign "−" is taken to make entropy non-negative. The logarithm in (1.32) can be taken with arbitrary base $a > 1$. When logarithm to base 2 is used, the unit of entropy is called a "bit" (binary digit), and when natural logarithm to base e is used, the unit is called a "nit".

Let us consider a random variable $X(\gamma)$ which assumes two values x_1, x_2 with equal probabilities $\frac{1}{2}$. In the case of logarithm base 2 we have

$$H(X) = -\left(\frac{1}{2}\log\frac{1}{2} + \frac{1}{2}\log\frac{1}{2}\right) = 1\text{bit}$$

Let us notice that entropy of a discrete random variable depends only on the number of values and their probabilities and does not depend on the values themselves.

Let (X, Y) be a two-dimensional random variable assuming values (x_i, y_i) with probabilities p_{ik}, $i = 1, 2, \ldots, n$; $k = 1, 2, \ldots, m$. Entropy of random vector (X, Y) is defined as

$$H(X, Y) = -\sum_{i=1}^{n}\sum_{k=1}^{m} p_{ik} \log p_{ik}. \tag{1.33}$$

Definition 1.4'. *Entropy of a continuous random variable $X(\gamma)$ with probability density $f(x)$ is the quantity*

$$H(X) = -\int_{-\infty}^{+\infty} f(x)\log f(x)dx$$
$$= E\left[-\log f(X)\right]. \tag{1.34}$$

It can be easily calculated that entropy of the uniform distribution on interval $[a, b]$ is

$$H(X) = \log(b-a) \tag{1.35}$$

whereas the entropy of Gaussian random variable is equal to

$$H(X) = \frac{1}{2}\log(2\pi e\sigma_X^2). \tag{1.36}$$

It is seen that the entropy of uniform and Gaussian distribution is a monotonic function of the interval $a-b$ and variance σ^2, respectively. It implies that the entropy of a continuous random variable can assume both positive and negative values.

Entropy of n-dimensional random vector $\mathbf{X} = \left[X_1,\ldots,X_n\right]$ with the probability density $f(x_1,\ldots,x_n)$ is defined as

$$H(\mathbf{X}) = -\int_{-\infty}^{+\infty}\cdots\int_{-\infty} f(x_1,\ldots,x_n)\log f(x_1,\ldots,x_n)dx_1\ldots dx_n$$
$$= E\left[-\log f(\mathbf{X})\right] \tag{1.37}$$

The *conditional entropy* of a random variable $X(\gamma)$ with respect to random variable $Y(\gamma)$ is the quantity

$$H(X|Y) = -\int_{-\infty}^{+\infty} f(x|y)\log f(x|y)dx \tag{1.38}$$

The above quantity depends on the values of the random variable $Y(\gamma)$, so it is a random variable.

The *average conditional entropy* of $X(\gamma)$ with respect to $Y(\gamma)$ is defined

$$\overline{H_Y}(X) = E\left[H(X|Y)\right] = \int_{-\infty}^{+\infty} H(X|Y)f(y)dy$$
$$= -\int_{-\infty}^{+\infty} f(x,y)\log f(x|y)dxdy \tag{1.39}$$

An analoguous definition of conditional entropy holds for discrete random variables. For both types of random variables

$$H(X,Y) = H(X) + \overline{H}_X(Y) = H(Y) + \overline{H}_Y(X),$$
$$H(X,Y) \leq H(X) + H(Y) \tag{1.40}$$

where equality holds if and only if the random variables are independent.

Entropy of an n-dimensional Gaussian vector $\mathbf{X}$ is given by the formula

$$H(\mathbf{X}) = \log \sqrt{(2\pi e)^n \, |\mathbf{K}|} \tag{1.41}$$

Definition 1.5. *The* mutual information $I(X,Y)$ *between random variables* $X(\gamma)$ *and* $Y(\gamma)$ *is defined as*

$$I(X,Y) = H(X) - \overline{H}_Y(X) \tag{1.42}$$

From formula (1.40) one obtains

$$I(X,Y) = H(X) + H(Y) - H(X,Y). \tag{1.43}$$

For continuous random variables

$$I(X,Y) = \int_{-\infty}^{+\infty} \int_{-\infty}^{+\infty} f(x,y) \log \frac{f(x,y)}{f(x)f(y)} dx dy \tag{1.44}$$

Here are the basic properties of the mutual information:

a) $I(X,Y) \geq 0$, equality holds if and only if $X(\gamma)$ and $Y(\gamma)$ are independent

b) $I(X,Y) = I(Y,X)$;

c) if $Z = g(X)$ is one-to-one mapping, then $I(X,Y) = I(Z,Y)$

3. STOCHASTIC PROCESSES - BASIC CONCEPTS

A.

In many physical applications one has to deal with random quantities depending upon a parameter. One of the commonly known examples of such phenomena is the Brownian motion. Each coordinate of the Brownian particle is a random variable that depends on time. Just as often one encounters phenomena in which the dependence of random variables on the non-temporal parameters is observed. For example, the velocity of a fluid particle in turbulent motion is a random variable which depends on the spatial coordinates. To create a mathematical model of phenomena similar to these mentioned above the concept of random function has been introduced.

Let T be a set of real numbers or its subset, and $t \in T$. Here T has the meaning of time.

Definition 1.6. *A stochastic process $X(t)$ is a family of random variables $\{X_t(\gamma), t \in T, \gamma \in \Gamma\}$ depending upon the parameter t and defined on the probability space $(\Gamma, \mathcal{F}, P)$.*

One can also say that a stochastic process $X(t)$ is a function which maps the index set T into the space S of random variables defined on $(\Gamma, \mathcal{F}, P)$.

Since a stochastic process is a **family of** random variablesits specification is analoguous to that of random vectors; the differences are concerned with the fact that now the number of random variables may be (countably or uncountably) infinite.

For every fixed t we have a random variable $X(t)$ with probability distribution $F_t(x) = P\{X(t) < x\}$. For an arbitray finite set of t-values, say $\{t_1, t_2, \ldots, t_n\}$ the random variables $X(t_1), X(t_2), \ldots, X(t_n)$ have a joint n-dimensional distribution with the distribution function

$$F_{t_1,t_2,\ldots,t_n}(x_1, x_2, \ldots, x_n) = P\{X(t_1) < x_1, \ldots, X(t_n) < x_n\} \tag{1.45}$$

The family of all these joint probability distributions for $n = 1, 2, \ldots$ and all possible values of the t_j constitutes the *familiy of finite-dimensional distributions* associated with the process $X(t)$. Most of the essential properties of a stochastic process can be expressed in terms of the corresponding finite-dimensional distributions. This family of distribution functions satisfies the following conditions:

a) the *symmetry condition*: if $\{i_1, \ldots, i_n\}$ is a permutation of numbers $1, \ldots, n$, then for arbitrary $n \geq 1$

$$F_{t_{i_1},\ldots,t_{i_n}}(x_{i_n}, \ldots, x_{i_n}) = F_{t_1,\ldots,t_n}(x_1, \ldots, x_n) \tag{1.46}$$

b) the *consistency condition*: for $m < n$ and arbitrary $t_{m+1}, \ldots, t_n \in T$

$$F_{t_1,\ldots,t_m,t_{m+1},\ldots,t_n}(x_1, \ldots, x_m, \infty, \ldots, \infty) = F_{t_1,\ldots,t_m}(x_1, \ldots, x_m) \tag{1.47}$$

Theorem 1.1. (Kolmogorov theorem) *For every given family of finite-dimensional distributions satisfying the conditions a), b), there exists a probability space $(\Gamma, \mathcal{F}, P)$ and a stochastic process $X_t(\gamma), t \in T$ defined on it that possesses the given distributions as finite-dimensional distributions.*

For the proof of this fundamental theorem—see, e.g. to [6], [16].

A stochastic process $X(t)$ can be equivalently specified by a family of its finite-dimensional characteristic functions defined as follows

$$\varphi_{t_1,\ldots t_k}(\lambda_1 \ldots, \lambda_n) = E\left\{exp\left[i\sum_{k=1}^{n} \lambda_k X(t_k)\right]\right\} =$$

$$= \int \cdots \int exp\left[i\sum_{k=1}^{n} \lambda_k x_k\right] dF_{t_1,\ldots,t_n}(x_1, \ldots, x_n) \tag{1.48}$$

A stochastic process can be also defined in another way. For every fixed elementary event γ in the given probability space, $X(t)$ or in the more explicit notation

$X(t, \gamma)$ is a function of t, defined for all $t \in T$. This function $x(t) = X(t, \gamma)$ with a fixed $\gamma \in \Gamma$ describes a fixed realization of the stochastic process; it is called a *realization*, or *trajectory* or a *sample function* of the process. The sample function $x(t)$ may be regarded as a "point" in the space $\mathcal{X}$ of all finite and real-valued functions of $t \in T$; $\mathcal{X}$ is called the *sample space* of the process $X(t)$. Thus, the stochastic process can be defined as follows.

Definition 1.6'. *A stochastic process $X(t, \gamma)$ is a measurable function which maps the space of elementary events Γ into the sample space $\mathcal{X}$ (cf. definition 1.27 in Sec. 11).*

Analogously as in the case of one random variable the function $X(t, \gamma)$ or $X_\gamma(t)$ induces a probability measure in $\mathcal{X}$ and a stochastic process is specified if the probability measure on the sample function space $\mathcal{X}$ is characterized. A possible way of characterization of the probability in the sample function space is associated with the concept of a characteristic functional of the process—cf. [26].

If a non-random function $\lambda(t)$ and the process $X(t)$ are such that the integral

$$\int_T \lambda(t) X(t) dt$$

exists for almost all realizations of $X(t)$, then the functional defined as follows

$$\Phi[\lambda] = E\left\{exp\left[i \int_T \lambda(t) X(t) dt\right]\right\} \tag{1.49}$$

is called the *characteristic functional* of the process $X(t)$.

The characteristic functional $\Phi[\lambda]$ has the properties analogous to those of characteristic functions. Namely,

$$\Phi[\lambda]_{\lambda(t) \equiv 0} = 1$$

$$\sum_{k=1}^{n} \sum_{l=1}^{n} \Phi[\lambda_k(t) - \lambda_l(t)] c_k \overline{c_l} \geq 0 \tag{1.50}$$

for arbitrary functions $\lambda_1(t), \ldots, \lambda_n(t)$ and arbitrary complex numbers $c_1, \ldots, c_n$. Inequality (1.50) asserts that the functional $\Phi[\lambda]$ is non-negative definite. It should be emphasized that (unlike for characteristic functions) conditions (1.50) are only necessary ones; a statement that a continuous (in a suitable sense) characteristic functional possessing properties (1.50) defines certain stochastic process is not valid in general (cf. [26],[27]).

Here are other properties of the characteristic functional.

1) If Φ_1 and Φ_2 are characteristic functionals and $p_1 + p_2 = 1$ where p_1, p_2 are non-negative numbers, then $p_1\Phi_1 + p_2\Phi_2$ is also a characteristic functional.

2) If Φ_1 and Φ_2 are characteristic functional, then $\Phi_1\Phi_2$ is also a characteristic functional.
3) If $\Phi_\alpha[\lambda]$ is a family of characteristic functionals depending upon random parameter α, and $F(\alpha)$ is the distribution function of α, then

$$\int_A \Phi_\alpha[\lambda]dF(\alpha) \tag{1.51}$$

is also a characteristic functional.
4) Let $\Phi_X[\lambda]$ be a characteristic functional of the process $X(t)$, and $\Phi_Y[\lambda]$ — a characteristic functional of another process $Y(t)$ being a linear transformation of $X(t)$.
   a) If $Y(t) = X(t) + a(t)$, where $a(t)$ — nonrandom function, then

$$\Phi_Y[\lambda] = \Phi_X[\lambda]exp\left\{i\int_T a(t)\lambda(t)dt\right\} \tag{1.52}$$

   b) If A is an integral linear operator of the form

$$Y(t) = AX(t) = \int_T G(s,t)X(s)ds \tag{1.53}$$

   where $G(s,t)$ is a suitable non-random function, then

$$\Phi_Y[\lambda] = \Phi_X[A^*\lambda] = \Phi_X\left[\int_T G(t,s)\lambda(s)ds\right] \tag{1.54}$$

   where A^* is a conjugate to A.

Having the characteristical functional of process $X(t)$ we can obtain the moments. A counterpart of formula (1.10) is

$$m_k(t) = E\left[X^k(t)\right] = \frac{1}{i^k}\frac{\delta^k\Phi[\lambda]}{\delta\lambda^k(t)}\bigg|_{\lambda(t)=0} \tag{1.55}$$

where the symbol $\frac{\delta}{\delta\lambda}$ denotes the functional (or, Volterra) derivative.

A stochastic process can also be understood as follows.

Definition 1.6″. A stochastic process *is a function $X(t,\gamma)$ of two variables $t \in T, \gamma \in \Gamma$ which assumes real values and which is P-measurable as a function of γ, for each fixed $t \in T$.*

The above defintion asserts that a stochastic process is a function $X(t,\gamma)$ which maps $T \times \Gamma$ into R and all its "intersections" $X_t(\gamma), t \in T$ are random variables.

If function $X(t,\gamma)$ which maps $T \times \Gamma$ into R is measurable with respect to a product measure $\mu \times P$ where μ is the Lebesgue measure on T, we say that $X(t,\gamma)$ is a **measurable stochastic process** .

Let us notice that if $X(t,\gamma)$ is a measurable stochastic process then due to the Fubini theorem, $X(t,\gamma)$ is μ-measurable as a function of t, for almost all $\gamma \in \Gamma$. This means that sample functions of $X(t,\gamma)$ are measurable with probability one.

In theory and applications it is often important to know the probability of events in a sample function space; for example, we may want to know the probability that a sample functions of a given process have specified properties, that is, they belong to some set A in the space $\mathcal{X}$ (e.g. sample functions are continuous). But most often we know the finite-dimensional distributions of a process and the question which arises is — how much can we deduce (from the given distributions) about the corresponding probability distribution in a sample function space (cf. Exercise 1).

If t is a real valued parameter (t may take any value in a given interval $T \subset R_1$) there exists many such sets A belonging to $\mathcal{X}$ which are not measurable and it can be shown by simple examples (cf. [9]) that their probabilities have not sense; if even the probabilities can be properly defined they may not be uniquely determined by the finite-dimensional distributions of the process. To get over this difficulty Doob has introduced a concept of separability. A stochastic process is **separable** if, roughly speaking, its all probabilistic properties are determined by its values on countable sets $\{t_1, t_2, \ldots\}$ of instants (dense in the interval T). For a separable process, probabilities of the kind mentioned above are uniquely determined by the finite-dimensional distributions. Doob has shown (cf. [9]) that, to any given process $X(t,\gamma)$ always exists an equivalent separable process $\tilde{X}(t,\gamma)$.

We say that two stochastic process $X(t,\gamma)$ and $Y(t,\gamma)$ are **equivalent** if, for every fixed $t \in T$, $X(t,\gamma)$ and $Y(t,\gamma)$ are equal with probability one. Two equivalent stochastic processes have the same family of finite dimensional distributions.

In the analysis of stochastic processes, in addition to the basic σ-algebra $\mathcal{F}$ of subsets of Γ a family $\mathcal{F}_t$ of sub-algebras plays an important role. This is for each $t \geq 0$ a sub-algebra (of $\mathcal{F}$) of events observed from the past until time t.

We say that the process $X(t)$ defined for $t \geq 0$ is **adapted to a family** $\mathcal{F}_t$, if for all $t \geq 0$ a quantity $X(t)$ is $\mathcal{F}_t$-measurable; This means that $X(t)$ is , for every t, completely determined by $\mathcal{F}_t$. Usually, the family $\mathcal{F}_t$ is considered to be simply a σ-algebra generated by events $\{X(s) \in B\}, s \leq t$, where B is a measurable subset of R_1 (or R_n — in the case of vector processes).

B.

Like random variables stochastic processes can be conveniently characterized by the moments; the simplest characteristic is the **mean** or **average value** $m_X(t)$. It is defined as a function which for each t is equal to the mean value of the corresponding random variable, that is

$$m_X(t) = E\{X_t(\gamma)\} \tag{1.56}$$

The average value of a real stochastic process $X(t)$ is expressed by the one-dimensional density function of the process $f_t(x)$ as follows

(1.57) $$m_X(t) = \int_{-\infty}^{+\infty} x dF_t(x) = \int_{-\infty}^{+\infty} x f_t(x) dx$$

The **variance** of the process $X(t)$ is defined by

(1.58) $$\sigma_X^2(t) = E\left\{|X(t) - m_X(t)|^2\right\}$$

One of the basic concepts in the theory and practice is the covariance function of the stochastic process. In general, when process $X(t)$ is complex-valued, i.e. $X(t) = X_1(t) + iX_2(t)$ the **covariance function** of the process $X(t)$ is defined as

(1.59) $$K_X(t_1, t_2) = E\left\{[X(t_1) - m_X(t_1)]\,\overline{[X(t_2) - m_X(t_2)]}\right\}$$

where "—" denotes the complex conjugate.

It is clear that

(1.59′) $$K_X(t_1, t_2) = \tilde{K}(t_1, t_2) - m_X(t_1)\overline{m_X(t_2)}$$

where $\tilde{K}_X(t_1, t_2) = E\left[X(t_1)\overline{X(t_2)}\right]$ is often called the *correlation function* of the process.

The covariance function $K_X(t_1, t_2)$ has the following properties:

1) $K_X(t, t) = \sigma_X^2(t) \geq 0$,
2) $K_X(t_1, t_2)$ is a symmetric function, i.e.

$$K_X(t_2, t_1) = \overline{K_X(t_1, t_2)};$$

in the case of a real stochastic process

$$K_X(t_2, t_1) = K_X(t_1, t_2);$$

3) $\left|K_X(t_1, t_2)\right|^2 \leq K_X(t_1, t_1) \cdot K_X(t_2, t_2)$.

Every covariance function has the fundamental property of being of *non-negative definite type*, that is for arbitrary n, arbitrary finite sets of time points $\{t_1, \ldots, t_n\}$ and arbitrary complex numbers $z_1, \ldots, z_n$

(1.60) $$\sum_{j,k=1}^{n} K_X(t_j, t_k) z_j z_k \geq 0$$

The above property is, in fact, a characteristic property of the class of all covariance functions.

Often, the simultaneous behaviour of n stochastic processes $X_1(t), X_2(t) \ldots,$ $X_n(t)$ is of interest. Then one introduces an n-dimensional **stochastic vector process** $\mathbf{X}(t) = [X_1(t), \ldots, X_n(t)]$. The mean value of such a process is defined as: $m_{\mathbf{X}}(t) = [m_{X_1}(t), \ldots, m_{X_n}(t)]$. The component processes $X_i(t)$ may be statistically related to each other; the mutual dependence is conveniently characterized by the covariance matrix $\mathbf{R}(t_1, t_2) = \{R_{X_i X_j}(t_1, t_2)\}, i, j, = 1, 2, \ldots, n$, where

$$R_{X_i X_j}(t_1, t_2) = E\left\{[X_i(t_1 - m_{X_i}(t_1)] \overline{[X_j(t_2) - m_{X_j}(t_2)]}\right\} \tag{1.61}$$

is said to be a **cross-covariance function** of the processes $X_i(t)$ and $X_j(t)$. Obviously, $R_{X_i X_i}(t_1, t_2) = K_{X_i}(t_1, t_2)$.

4. GAUSSIAN PROCESSES

An important class of stochastic processes is formed by processes with Gaussian distributions. Such processes are extremely useful and most often encountered in applications.

Definition 1.7. *A stochastic process $X(t), t \in T$ is said to be Gaussian or normal process if for any integer n and any subset $\{t_1, t_2, \ldots, t_n\}$ of T the random variables $X(t_1), X(t_2), \ldots, X_n(t)$ have the join Gaussian distribution, that is, their characteristic function is given for any real numbers $\lambda_1, \lambda_2, \ldots, \lambda_n$, by*

$$\varphi_{t_1, \ldots, t_n}(\lambda_1, \ldots, \lambda_n) = E\left\{exp\left[i \sum_{j=1}^{n} \lambda_j X(t_j)\right]\right\} =$$

$$= exp\left\{i \sum_{j=1}^{n} \lambda_j m_X(t_j) - \frac{1}{2} \sum_{j,k=1}^{n} \lambda_j \lambda_k K_X(t_j, t_k)\right\} \tag{1.62}$$

where $m_X(t)$ and $K_X(t_1, t_2)$ are the mean and covariance function of a process $X(t)$. If the covariance matrix $K_X(t_i, t_j), i, j = 1, 2, \ldots, n$ is not singular, then the joint probability density function of $X(t)$ has the form

$$f_{t_1, \ldots, t_n}(x_1, x_2, \ldots, x_n) =$$

$$= \frac{1}{(2\pi)^{\frac{n}{2}} \sqrt{|\mathbf{K}|}} exp\left\{-\frac{1}{2|\mathbf{K}|} \sum_{i,j=1}^{n} |K_{ij}| [x_i - m_X(t_i)] [x_j - m_X(t_j)]\right\} \tag{1.63}$$

where $x_i = X(t_i)$, $|\mathbf{K}|$ — is the determinant of the covariance matrix $K_X(t_i, t_j)$, $|K_{ij}|$ — is the cofactor of the element in the i-the row and the j-th column of $\{K_X(t_i, t_j)\}$.

An alternative definition of a Gaussian process can be given by the following necessary and sufficient condition (cf. [39]): a stochastic process $X(t), t \in T$ is Gaussian if and only if every finite linear combination of the form

$$Z = \sum_{i=1}^{n} \alpha_i X(t_i) \tag{1.64}$$

is a Gaussian random variable.

The obvious necessary condition for process $X(t)$ to be a Gaussian process is that $X(t)$ should be, for each $t \in T$, a Gaussian random variable. The following theorem is valid (cf. [9]).

Theorem 1.2. *Let T be a set of real numbers. Let $m(t)$ be an arbitrary function of $t \in T$ and $K(s,t)$ — an arbitrary function of two variables $s, t \in T$, such that*

a) $K(s,t) = \overline{K(s,t)}$;

b) *if $\{t_1, t_2, \ldots, t_n\}$ is an arbitrary finite set of values of T, then matrix $K(t_i, t_j)$ is non-negative definite type;*

then there exists a Gaussian stochastic process $X(t)$, for which

$$E[X(t)] = m(t)$$

$$E\left[X(s)\overline{X(t)}\right] - m(s)\overline{m(t)} = K(s,t)$$

Hence, for an arbitrary stochastic process with finite moments of first and second order there exists a Gaussian process $X(t)$ having the same mean and covariance function.

Here are the basic properties of a Gaussian stochastic process.

1) A Gaussian process is entirely determined by its mean $m_X(t)$ and correlation function $K_X(t_1, t_2)$.
2) If $X(t)$ is a Gaussian process then all processes obtained by linear transformations of $X(t)$ are also Gaussian (normality is invariant under linear transformation).
3) If $X(t)$ is a Gaussian process then all its moments of odd order are equal to zero, that is for each finite n (it is assumed that $m_X(t) = 0$)

$$E\{X(t_1)X(t_2)\ldots X(t_{2n+1})\} = 0 \tag{1.65}$$

and all even order moments may be expressed in terms of the second order moments by the following formula

$$E\{X(t_1), \ldots, X(t_{2n})\} = \\ = \sum E\{X(t_{i_1})X(t_{i_2})\} \ldots E\{X(t_{i_{2n-1}})X(t_{i_{2n}})\} \tag{1.66}$$

where summation is over all possible ways of dividing the $2n$ points into n combinations of pairs $(i_1, i_2), (i_3, i_4), \ldots, (i_{2n-1}, i_{2n})$. The number of terms in the summation is:

$$\frac{(2n)!}{2^n n!} = 1 \cdot 3 \cdot 5 \cdot \ldots \cdot (2n-1).$$

For example

$$E\{X(t_1)X(t_2)X(t_3)X(t_4)\} = K_X(t_1,t_2)K_X(t_3,t_4)+$$
$$+K_X(t_1,t_3)K_X(t_2,t_4) + K_X(t_1,t_4)K_X(t_2,t_3),$$
$$E\{X(t_1)X(t_2X^2(t_3))\} = K_X(t_1,t_2)E\{X^2(t_3)\} + 2K_X(t_1,t_3)K_X(t_2,t_3)$$

The first property follows directly from the definition of a Gaussian process and the second one expresses a known fact from the theory of multidimensional random variables. The third property can be shown by differentiation of characteristic functions of process $X(t)$ and appropriate manipulations (cf. [16]).

5. STATIONARY PROCESSES

Another class of stochastic process playing a very significant role in theory and applications is constituted by processes whose probabilistic structure is invariant under a translation of the time parameter.

Definition 1.8. *A stochastic process $X(t), t \in T$, is said to be* **strictly stationary** *(or stationary in a narrow sense) if for any n, any subset $\{t_1, t_2, \ldots, t_n\}$ of T and for an arbitrary h such that $t_i + h \in T, (i = 1, 2, \ldots, n)$, the following equality holds*

$$F_{t_1,t_2,\ldots,t_n}(x_1, x_2, \ldots, x_n) = F_{t_1+h,t_2+h,\ldots,t_n+h}(x_1, x_2, \ldots, x_n) \tag{1.67}$$

The one-dimensional distribution of a strictly stationary process is the same for each instant of time t, whereas two-dimensional distribution depends only on the difference of time instants. Many of the most important properties of a stationary process can be expressed in term of these two first distributions. So, it is natural to introduce a wider class of stationary processes.

Definiton 1.9. *A stochastic process $X(t)$ such that $E\left\{|X(t)|^2\right\} < \infty, t \in T$ is called a* **weakly stationary** *(or stationary in a wide sense) if*

$$m_X(t) = \text{const.}, \qquad K_X(t_2 - t_1) = K_X(\tau), \quad \tau = t_2 - t_1 \tag{1.68}$$

The basic properties of the covariance function stated in Sec.3 are the following

1) $K_X(0) = \sigma^2 = \text{const.}$,
2) $K_X(\tau) = \overline{K_X(-\tau)}$; for real process $K_X(\tau) = K_X(-\tau)$
3) $|K_X(\tau)| \leq K_X(0)$

Stationary processes are of great practical importance. This is mainly due to their nice properties. The estimation of the statistical characteristics of such processes from experimental data is much simpler than in the case of non-stationary processes. Furthermore, for stationary processes there exists the apparatus of spectral analysis analogous to the harmonic analysis of deterministic functions. The spectral analysis originates in the following theorem (on spectral representation of a stationary process).

Theorem 1.3. *For weakly stationary stochastic process $X(t)$ which has continuous correlation function there exists process $\Phi(\omega)$ with orthogonal increments, such that for each t (in the mean square sense, cf. Sec. 14.4)*

$$X(t) = \int_{-\infty}^{+\infty} e^{i\omega t} d\Phi(\omega) \tag{1.69}$$

and conversely, an arbitrary process of the form (1.69) is weakly stationary and its correlation function has the representation

$$K_X(\tau) = \int_{-\infty}^{+\infty} e^{i\omega\tau} dG(\omega) \tag{1.70}$$

where $G(\omega)$ is real, non-decreasing, bounded and continuous from the left (the spectral distribution function) and

$$E\left[\Phi(\omega)\right] = 0, \qquad E\left|\Phi(\omega)\right|^2 = G(\omega), \qquad E\left|d\Phi(\omega)\right|^2 = dG(\omega).$$

If $G(\omega)$ is absolutely continuous, i.e. there exists a function $g(\omega)$ such that

$$G(\omega) = \int_{-\infty}^{\omega} g(\nu)d\nu, \qquad g(\omega) = \frac{dG(\omega)}{d\omega} \tag{1.71}$$

then function $g(\omega)$ is called the **spectral density** of the process $X(t,\gamma)$. In this case formula (1.70) takes the form

$$K_X(\tau) = \int_{-\infty}^{+\infty} e^{i\omega\tau} g_x(\omega)d\omega \tag{1.72'}$$

and if $\displaystyle\int_{-\infty}^{+\infty} |K_X(\tau)|\, d\tau < \infty\,,$ then

$$g_x(\omega) = \frac{1}{2\pi}\int_{-\infty}^{+\infty} e^{-i\omega\tau} K_X(\tau)d\tau. \tag{1.72''}$$

It is easily seen that:

a) $g_X(\omega) \geq 0$

b) $\sigma_X^2 = K_X(0) = \int_{-\infty}^{\infty} g_X(\omega)d\omega$,

c) for real processes $g_X(\omega) = g_X(-\omega)$.

It is clear that a strictly stationary process whose second moment is finite is also weaky stationary. The converse statement is not true in general. An exception is the Gaussain process since it is completely characterized by its moments of first and second order. Therefore, a weakly stationary Gaussain process is also strictly stationary.

Definition 1.10. *A stochastic vector process* $\mathbf{X}(t) = [X_1(t), X_2(t), \ldots X_n(t)]$ *is weakly stationary if its mean is constant and all elements of its covariance matrix (1.61) depend only on difference* $\tau = t_2 - t_1$.

From this definition follows that stationarity of a vector stochastic process is not assured by the stationarity of its components; the component-processes have to be also stationarily correlated.

In analysis of various practical problems it is important to know what is the range of frequences which carry the most significant values of a power of random signal. This leads to the concepts of a **wide-band** and **narrow-band** processes. Though the concept of bandwidth of a stationary process has mostly been used by engineers in the intuitive sense it is possible to define it rigorously.

If the spectral density $g_X(\omega)$ has only one local maximum at, say, $\omega = \omega_o$ then the **bandwidth** Λ_X of the process $X(t)$ can be defined by the relation (cf. also [21])

$$\int_{-\infty}^{+\infty} g_X(\omega)d\omega = \Lambda_X g_X(\omega_o) \tag{1.73}$$

The above definition can be extended to the case where $g_X(\omega)$ has several local maxima and then the bandwidth Λ_X will consist of several disjoint intervals.

If process $X(t)$ plays the role of an excitation of dynamical system and if Λ_X is small (large) in comparison with the frequency band of the system in question then the process $X(t)$ is said to be a narrow-band (wide-band) excitation.

Remark.

Analysis of stricly stationary processes is often performed in the framework of measure preserving transformations (cf. [9]). Such an analysis relies on the fact that if X_o is a random variable and T_t, $0 \leq t < \infty$ is a translation semi-group of measure preserving transformations, then stochastic process $X(t) = T_t X_o$ is strictly stationary. Similarly, weakly-stationary processes can be represented in the form: $X(t) = U_t X_o$, where U_t is a semi-group of isometric operators defined on the linear manifold of the Hilbert space generated by the considered process.

6. MARKOV PROCESSES

6.1. Basic definitions

An important class of stochastic processes is characterized by the property that the "future" is independent of the "past" when the "present" is known. Such processes are named after A.A. Markov who initiated in 1906 the study of random variables connected in a chain without so called "after-effects".

Let $X(t), t \in [t_0, \infty)$ be a stochastic process defined on the probability space $(\Gamma, \mathcal{F}, P)$ with R_1 as the set of its possible values (state space). Let us denote by $\mathcal{B}$ a family of Borel sets in R_1.

Definition 1.11. *A stochastic process $X(t)$ is called a* **Markov process** *if for every n, any arbitrary finite subset $\{t_1 < t_2 <, \ldots, < t_n\}$ of the index set $[t_0, \infty)$ and for any real numbers $x_1, x_2, \ldots, x_n$*

$$\begin{aligned} &P\left\{X(t_n) < x_n \middle| X(t_{n-1}) = x_{n-1}, X(t_{n-2}) = x_{n-2}, \ldots, X(t_1) = x_1\right\} = \\ &= P\left\{X(t_n) < x_n \middle| X(t_{n-1}) = x_{n-1}\right\} \end{aligned} \tag{1.74}$$

that is, the conditional distribution of $X(t_n)$ for given values of $X(t_1), X(t_2), \ldots, X(t_{n-1})$ depends only on the most recent known value $X(t_{n-1})$.

The above definition implies that the basic characteristics of a Markov process is a function

$$P(s, x; t, B) = P\left\{X(t) \in B \middle| X(s) = x\right\}, \quad s < t \tag{1.75'}$$

or

$$F(s, x; t, y) = F\left\{X(t) < y \middle| X(s) = x\right\} \tag{1.75''}$$

where B is an arbitrary Borel set in R_1, i.e. $B \in \mathcal{B}$.

The function $P(s, x; t, B)$ is called a *transition probability* (or transition function). It gives the probability of a transition of a process (or system) from a state x at time s to one of the states belonging to B at time t. The function $P(s, x; t, B)$ associated with a Markov process $X(t)$ has the following properties:

a) for fixed s, t, x the function $P(s, x; t, -)$ is a probability measure on $\mathcal{B}$; $P(s, x; t, R_1) = 1$;

b) for fixed $s, t, B \in \mathcal{B}$ the function $P(s, -; t, B)$ is $\mathcal{B}$ — measurable;

c)

$$P(s, x; s, B) = \begin{cases} 1 & \text{for} \quad x \in B \\ 0 & \text{for} \quad x \notin B; \end{cases}$$

d) for $s < u < t$ such that $X(u) = \xi$ the following Chapman-Kolmogorov equation holds

$$P(s,x;t,B) = \int_{R_1} P(u,\xi;t,B)dP(s,x;u,d\xi) \tag{1.76}$$

The properties stated above can serve as another definition of a Markov process (cf. [16]).

The significance of the transition probabilities for Markov process and the Chapman-Kolmogorov equation lies in the fact that all finite-dimensional distributions of the process can be constructed from them and from the initial distribution at time t_0.

Theorem 1.4. *If $X(t)$ is a Markov process, $t \in [t_0, \infty)$ with its transition probability $P(s,x;t,B)$, and if $P_{t_0}(A)$ is the distribution of $X(t_0)$, that is $P_{t_0}(A) = P\{X(t_0) \in A\}$, $A \in \mathcal{B}$, then*

$$P\{X(t_1) \in B_1, X(t_2) \in B_2, \ldots X(t_n) \in B_n\} =$$

$$= \int_{R_1} P_{t_0}(dx_0) \prod_{k=1}^{n-1} \int_{B_k} P(t_{k-1}, x_{k-1}; t_k, dx_k) P(t_{n-1}, x_{n-1}; t_n, B_n) \tag{1.77}$$

where $t_0 < t_1 < t_2 < \cdots < t_n$ and $B_k \in \mathcal{B}$ for all k.

In particular

$$P\{X(t) \in B\} = \int_{R_1} P(t_0, x; t, B) P_{t_0}(dx) \tag{1.78}$$

If the probability $P(s,x;t,B)$ has a density, that is

$$P(s,x;t,B) = \int_B p(s,x;t,y)dy \tag{1.79}$$

then the Chapman-Kolmogorov equation (1.76) takes the form

$$p(s,x;t,y) = \int_{R_1} p(s,x;u,\xi)p(u,\xi;t,y)d\xi \tag{1.80}$$

All finite-dimensional probability densities are determined in the following way

$$f(x_1,t_1;x_2,t_2;\ldots,x_n,t_n) =$$
$$= f(x_n,t_n|x_1,t_1;\ldots;x_{n-1},t_{n-1})f(x_1,t_1;\ldots;x_{n-1},t_{n-1})$$
$$= f(x_n,t_n|x_{n-1},t_{n-1})f(x_1,t_1;\ldots;x_{n-1},t_{n-1}).$$

By subsequent application of the above formula one easily obtains

$$f(x_1,t_1;x_2,t_2;\ldots;x_n,t_n) = f(x_1,t_1) \prod_{k=2}^{n} f(x_k,t_k|x_{k-1},t_{k-1}) \tag{1.81}$$

what constitutes a common counterpart of (1.77).

In the analysis of Markov processes and in modelling of real random phenomena we frequently deal with so called stationary transition probabilities characterizing a homogeneous Markov process.

Definition 1.11′. *A Markov process $X(t), t \in [t_0, \infty)$ is said to be* **homogeneous** *(with respect to time) if for arbitrary $s \in [t_0, \infty)$ and $t \in [t_0, \infty)$, $s < t$, the transition probability depends only on the difference $t - s = \tau$, that is if*

$$P(s, x; t, B) = P(x, \tau, B)$$

$$F(s, x; t, y) = F(x, \tau, y) \tag{1.82}$$

The function $P(x, \tau, B)$ determines the probability of transition from x to B in time τ; this probability does not depend on the actual position of the interval of length τ on the time axis. The Chapman-Kolmogorov equation in this case takes the form

$$P(x, \tau_1 + \tau_2, B) = \int_{R_1} P(\xi, \tau_2, B) dP(x, \tau_1, d\xi) \tag{1.83}$$

It is worth noting that a Markov process with a stationary transition probability (1.82) is not a stationary process. For example, the Wiener process — as we shall see it in the next section — has a stationary transition probability but it is not a stationary process.

The Chapman-Kolmogorov equation (1.83) has important implications in the theory of Markov processes; among others, as it can be shown (cf. [38]), it implies the existence of a semi-group of operators associated with a homogeneous Markov process.

Indeed, let $X(t)$, $t \in [t_0, \infty]$ denotes a homogeneous Markov process with transition probability $P(t, x, B) = P\{X(t) \in B | X(t_0) = x\}$. Let $f(x)$ be a bounded measurable function defined on the state space R_1 of the process. Let us define the operator T_t acting on the functions f as follows

$$T_t f(x) = E_X f(X_t) = \int_{R_1} f(y) dP_y(t, x, y) \tag{1.84}$$

Operators T_t are linear, positive and continuous.

The Chapman-Kolmogorov equation (1.83) implies that they have a semi-group property

$$T_{t+s} = T_t T_s = T_s T_t \qquad s, t > 0 \tag{1.85}$$

in the sense that for every bounded measurable function f we have

$$T_{t+s} f = T_t(T_s f) = T_s(T_t f).$$

We say that T_t is a one-parameter **semi-group of Markov transition operators.**

It is natural to introduce an operator A analogous to a derivative and defined as

$$Af(x) = \lim_{t \to 0} \frac{T_t f(x) - f(x)}{t} \tag{1.86}$$

where the domain of the definition consists of all functions for which the limit (in the appropriate sense of operator convergence) exists. Operator A is the **infinitesimal operator** of $T_t, t \geq t_o$ and characterizes the average infinitesimal rate of change of $f(X_t)$.

If the transition probabilities of $X(t)$ are continuous, what in term of operators T_t can be expressed as

$$\lim_{t \to 0} T_t f(x) = f(x) \tag{1.87}$$

then operators T_t, and the transition probabilities $P(t, x, B)$ of the homogeneous stationary process can be uniquely defined by A. For example,. Markov process with continuous sample functions has continuous transition probability. It can be shown (cf. [38]) that when A is well defined, the average

$$u(t, x) = E_X f(X_t) = T_t f(x) \tag{1.88}$$

can be determined as the unique (bounded) solution of the equation

$$\frac{\partial u(t, x)}{\partial t} = Au(t, x) \qquad t > 0 \tag{1.89}$$

with the initial condition: $u(t_o, x) = f(x)$.

The idea presented above can be extended to the non-homogeneous case, where X_t denotes an arbitrary Markov process with transition probability $P(s, x, t, B)$.

Depending on whether the index set or the state space of the process are countable or not various classes of Markov stochastic processes have been distinguished (e.g. processes discrete in time and space, processes with continuous time and discrete in space, processes with continuous time and space).Among continuous parameter Markov processes with a continuous state space a special role is played by the **diffusion processes**.

6.2. Diffusion processes

6.2.1, Definition and Kolgomorov equations

There are two basic approaches to the characterization and analysis of diffusion Markov processes. The first approach (which is also historically earlier) defines these processes in terms of the conditions on the transition probabilities $P(s, x; t, B)$. In the second approach the basic object is the process $X(t, \gamma)$ itself and its variations with respect to time; this strictly stochastic method leads to stochastic differential equations for $X(t, \gamma)$ and will be considered in the next chapter.

Definition 1.12. *A Markov process $X(t), t \in T$ with values in R_1 is called a* **diffusion process** *if its transition probability function $F(s,x;t,y)$ satisfies the following three conditions*

(i)

$$\lim_{\Delta t \to 0} \frac{1}{\Delta t} \int_{|y-x|>\epsilon} d_y F(s,x;s+\Delta t,y) = 0 \tag{1.90}$$

(ii) there exists a function $a(s,x)$ such that

$$\lim_{\Delta t \to 0} \int_{|y-x|\leq\epsilon} (y-x) dF_y(s,x;s+\Delta t,y) = a(s,x) \tag{1.91}$$

(iii) there exists a function $b(s,x) > 0$ such that

$$\lim_{\Delta t \to 0} \int_{|y-x|\leq\epsilon} (y-x)^2 dF_y(s,x;s+\Delta t,y) = b(s,x) \tag{1.92}$$

and the convergence in (1.91), (1.92) is uniform in x.

The function $a(s,x)$ and $b(s,x)$ are called the **drift** and **diffusion coefficients** of a diffusion process, respectively.

It is seen that in conditions (ii), (iii) truncated moments are used. The existence of the above limits does not necessarily imply that $F(s,x;t,y)$ has first and second order moments (on the entire state space R_1). The above limits characterize the infinitesimal mean and variance of the change in $X(t)$, respectively. Let us notice that condition (i) means that any large change in $X(t)$ over a short period of time Δt is improbable when $\Delta t \to 0$. Indeed,

$$P\{[|X(s+\Delta t) - X(s)| > \epsilon] | X(s) = x\} = o(\Delta t).$$

The processes defined above have very important property: it turns out that the coefficients $a(s,x)$ and $b(s,x)$, under certain conditions, determine uniquely the transition probability $P(s,x;t,B)$. The explicit form of this fact is expressed by the Kolgomorov theorems. They can be stated as follows (for proofs — see [1], [16]).

Theorem 1.5. *If $X(t), t \in T$ is a diffusion Markov process and if there exist (for every s, x, y and $t > s$ continuous derivatives*

$$\frac{\partial F(s,x;t,y)}{\partial x}, \qquad \frac{\partial^2 F(s,x;t,y)}{\partial x^2}$$

then the function F — as a function of s and x — satisfies the equation

$$\frac{\partial F}{\partial s} = -a(s,x)\frac{\partial F}{\partial x} - \frac{1}{2}b(s,x)\frac{\partial^2 F}{\partial x^2} \tag{1.93}$$

This equation is called the **backward Kolgomorov equation**. If we assume that there exists density

$$p(s,x;t,y) = \frac{\partial}{\partial y} F(s,x;t,y)$$

then direct differentiation gives the following backward equation for density:

$$\frac{\partial p}{\partial s} + a(s,x)\frac{\partial p(s,x;t,y)}{\partial x} + \frac{1}{2}b(s,x)\frac{\partial^2 p(s,x;t,y)}{\partial x^2} = 0 \tag{1.93'}$$

Theorem 1.6. *If $X(t), t \in T$ is a diffusion Markov process and if there exist continuous derivatives*

$$\frac{\partial p(s,x;t,y)}{\partial t}, \frac{\partial}{\partial y}\left[a(t,y)p(s,x;t,y)\right], \frac{\partial^2}{\partial y^2}\left[b(t,y)p(s,x;t,y)\right]$$

then $p(s,x;t,y)$ — as a function of t and y — satisfies the equation:

$$\frac{\partial p}{\partial t} + \frac{\partial}{\partial y}\left[a(t,y)p(s,x;t,y)\right] - \frac{1}{2}\frac{\partial^2}{\partial y^2}\left[b(t,y)p(s,x;t,y)\right] = 0 \tag{1.94}$$

The above equation is called the **forward** Kolmogorov equation or the **Fokker-Planck-Kolmogorov** equation.

The equations presented above play a very important role in the theory and applications; they allow us to determine the probabilistic structure of a diffusion process by solving classical partial differential equations.

The equations (1.93′) and (1.94) should be supplemented by the appropriate initial and boundary conditions as well as by the conditions associated with the probabilistic nature of unknown functions. These probabilistic conditions are in the case of equation (1.94) the following:

$$p(s,x;t,y) \geq 0, \qquad \int_{-\infty}^{+\infty} p(s,x;t,y)dy = 1$$

$$\lim_{t \to s} \frac{1}{t-s} \int_{|y-x|>\epsilon} p(s,x;t,y)dy = 0 \tag{1.95}$$

From the point of view of the theory of partial differential equations, equation (1.93′) means that the density of the transition probability is a fundamental solution of parabolic equation

$$\frac{\partial u}{\partial s} + Lu = 0, \qquad L \equiv \frac{1}{2}b(s,x)\frac{\partial^2}{\partial x^2} + a(s,x)\frac{\partial}{\partial x} \tag{1.96}$$

Indeed, a fundamental solution is just function $p(s,x;t,y), s < t$ which satisfies, for each fixed y equation (1.93′) and

(1.97) $$p(s,x;t,y) \xrightarrow[s \to t]{} \delta(y-x)$$

Condition (1.97) means that

(1.98) $$\int p(s,x;t,y)f(y)dy \longrightarrow f(x) \qquad \text{as} \quad s \to t$$

for any finite integrable function f.

In other words, a unique bounded solution of the Cauchy problem for equation (1.96) with "initial" condition $u(t_o, x) = f(x)$ is represented as

(1.99) $$u(s,x) = \int p(s,x;t,y)f(y)dy \qquad s < t.$$

It should be noticed that equation (1.96), as equation with respect to s, is solved in backward direction, hence, the condition $f(x)$ is not initial in usual sense.

Equation (1.94) means that the transition probability density is a fundamental solution of the equation

(1.100) $$\frac{\partial u}{\partial t} = L^* u$$

If the distribution of process $X(t)$ at $t = t_o$ is given, then by virtue of (1.78) the one-dimensional probability density $f(x,t)$ of the process is

(1.101) $$f(x,t) = \int_{-\infty}^{+\infty} p(t_o, x_o; t, x) f(x_o, t_o) dx_o$$

The one-dimensional distribution of $X(t)$ can, however, be determined directly as a solution of the Fokker-Planck-Kolmogorov equation for $f(x,t)$. Indeed, multiplying equation (1.94) written for $p(t_o, x_o; t, x)$ by $f(x_o, t_o)$ and integrating with respect to x_o gives

(1.102) $$\frac{\partial f(x,t)}{\partial t} + \frac{\partial}{\partial x}[a(x,t)f(x,t)] - \frac{1}{2}\frac{\partial^2}{\partial x^2}[b(x,t)f(x,t)] = 0$$

The associated initial condition is: $f(x,t_o) = f_o(x)$.

If the initial condition $f_o(x)$ is concentrated on one value (Dirac — delta distribution) then the one-dimensional density $f(x,t)$ is simply the transition density and equations (1.94) and (1.102) coincide.

The first existence theorem for Kolmogorov equation (1.96) has been proved by Feller in paper [10]. Assuming that coefficients are sufficiently smooth and bounded and function $b(s,x)$ is not uniformly degenerated Feller showed the existence of fundamental solution $p(s,x;t,y)$ $s < t$ of equation (1.96). The solution constructed was continuous, non-negative and had all the other properties of the transition probability (e.g. (1.95)). The assumptions which Feller posed on $a(s,x)$ and $b(s,x)$ are rather strong, but the contemporary theory of partial differential equations generates the existence of fundamental solution of the Kolmogorov equation under more relaxed conditions (cf. [14]).

6.2.2. Boundary conditions

If a state space of the diffusion process is restricted to interval $I = (r_1, r_2)$ which is finite or semi-finite then it is necessary to specify the appropriate **boundary conditions** associated with the Kolmogorov equation. These conditons may differ depending on the behaviour of the process at the boundaries; this behaviour is associated with the physical features of the particular process. Feller (cf. [11]) gave the classification of boundaries associated with one-dimensional diffusion processes (later, the extension to the multi-dimensional processes with boundaries was provided by others (cf. [37], [40]). This classification enables us to formulate correctly the boundary value problems for equations (1.93′), (1,94) and (1.102).

The point is that depending on the infinitesimal characteristics $a(x,t)$ and $b(x,t)$ the process can exhibit various types of behaviour at the boundaries of the interval. First, the coefficients may be such that the process $X(t)$ never reaches the boundaries; in these cases no boundary conditions have to be imposed, and the boundaries are called **natural**. Also, the coefficients may be such that $X(t)$ has positive probability of taking the value r_1 or r_2. For such processes, however, there are two possible cases: (i) the drift towards the boundaries can be such that the boundaries "automatically" act as absorbing barriers and no boundary conditions can be imposed (in this case the boundaries are called **exit** boundaries), (ii) the process may behave like the classical diffusion process on a finite interval, and various boundary conditions can be imposed (in this case the boundaries are termed **regular**). Boundaries which are exit or regular are called **accessible**. Other boundaries are called inaccessible. For example, the first-passage time problems can be considered only for processes admitting accessible boundaries.

The classification of boundaries depends on the Lebesgue integrability of the functions:

$$l_1(x) = exp\left\{-\int_{x'}^{x} \frac{a(y)}{b(y)} dy\right\} \tag{1.103}$$

$$l_2(x) = \frac{1}{b(x) l_1(x)}, \qquad l_3(x) = l_1(x) \int_{x'}^{x} l_2(z) dz \tag{1.104}$$

where $x' \in (r_1, r_2)$ is fixed. The function $g(x)$ is Lebesgue integrable on interval $I_i (i.e. g(x) \in L(I_i))$ if

$$\int_{I_i} g(x)dx < \infty.$$

Feller's classification of boundaries, asserts that

1) the boundary r_i is an exit boundary if

$$l_2(x) \notin L(I_i) \quad \text{and} \quad l_3(x) \in L(I_i) \tag{1.105}$$

2) the boundary r_i is regular if

$$l_1(x) \in L(I_i) \qquad l_2(x) \in L(I_i) \tag{1.106}$$

where $I_i = I_1$ or I_2, and $I_1 = (r_1, x'), I_2 = (x', r_2)$.

The existence and uniqueness of the solution of equations (1.93′), (1.94) and (1.102) for the process with boundaries depends on the kind of boundaries and, of course, on the imposed conditions. By use of the theory of semi-groups of operators Feller in his paper [11] obtained results which relate the existence and uniqueness problem to that of classifying the boundaries. Other relevant references are [19], [29].

In formulating of boundary conditions for equations (1.93′), (1.94) and especially for equation (1.102) the following physical interpretation originating from Stratonovich [35] may be useful. Let us treat the probability as a kind of substance filling the considered spatial domain. Then, the probability density $f(x,t)$ can be interpreted as a concentration or density of such a "probabilistic mass" at point x and at time t. The flux Φ along the axis x is composed of the systematic part: $a(x,t)f(x,t)$ and the random (diffusional) part: $-\frac{1}{2}\frac{\partial}{\partial x}[b(x,t)f(x,t)]$ i.e.

$$\Phi(x,t) = a(x,t)f(x,t) - \frac{1}{2}\frac{\partial}{\partial x}[b(x,t)f(x,t)] \tag{1.107}$$

Making use of (1.107) we conclude that the Fokker-Planck-Kolmogorov equation (1.102) represents the continuity equation (or the equation of conservation of probability)

$$\frac{\partial}{\partial t}f(x,t) + \frac{\partial}{\partial x}\Phi(x,t) = 0 \tag{1.108}$$

characterizing the preservation of the "probabilistic mass" during the time evolution. For infinitesimal increments Δx and Δt equation (1.108) can be written as

$$\frac{f(x,t+\Delta t) - f(x,t)}{\Delta t} + \frac{\Phi(x+\Delta x,t) - \Phi(x,t)}{\Delta x} = 0$$

or

$$[f(x,t+\Delta t)-f(x,t)]\,\Delta x=[\Phi(x,t)-\Phi(x+\Delta x,t)]\,\Delta t \tag{1.109}$$

The last equation shows that the increment of probability within a small interval of time Δt on the element Δx of the state space is equal to the difference of fluxes in this time interval Δt. Formula (1.109) can be interpreted as a principle of the preservation of probability and it enables us to formulate the boundary conditions which are in accord with the physics of the problems.

If a considered stochastic process $X(t)$ can assume all possible values from $-\infty$ to ∞, then the equation (1.108) is valid on entire real line. As the boundary conditions we take in this case the conditions at $\pm\infty$. Integrating (1.108) with respect to x from $-\infty$ to ∞ and accounting for the normalization conditions gives

$$\Phi(-\infty,t)=\Phi(\infty,t) \tag{1.110}$$

However, in spite of this equality usually stronger conditions are required, namely

$$f(-\infty,t)=f(+\infty,t)=0 \tag{1.111}$$

If a stochastic process takes its values only from the bounded interval $[r_1,r_2]$ then the Fokker-Planck-Kolmogorov equation (1.102) should be considered only in this interval. In this case the conditions

$$\Phi(r_1,t)=0, \qquad \Phi(r_2,t)=0 \tag{1.112}$$

mean that a flow of "probability mass" through boundaries r_1 and r_2 is not allowed. We can say that in these points the *reflecting boundaries* exist. So, condition (1.112) can be interpreted as expressing a reflection on boundaries.

If at the boundaries r_1 and r_2 the *absorbing barriers* are placed, a particle which reached one of the barriers is absorbed, i.e. it remains there for ever. In such situation the probability $f(x,t)$ should vanish, i.e.

$$f(r_1,t)=0, \qquad f(r_2,t)=0 \tag{1.113}$$

More complex boundary conditions should be imposed if in points r_1 and r_2 there are elastic barriers; this is a situation when a part of particles is reflected while, the remaining are absorbed. In such case the boundary conditions take the form of combination of absorbtion and reflection.

Remark: The solution of the Fokker-Planck-Kolmogorov equation (1.102) with absorbing or elastic barriers have the following property. During the time in which a particle did not touch the barrier, the process is described by equation (1.102), but when the time increases more and more particles remain at the barrier and when

$t \to \infty$ practically all particles will be absorbed, that is within the interval and on its boundary we have: $\lim_{t\to\infty} f(x,t) = 0$. This implies that in this case the condition

$$\int_{-\infty}^{+\infty} f(x,t)dx = 1$$

has not to be satisfied.

Of course, other boundary conditions are also possible. For example, at point r_1 there may be posed an absorbing barrier, and at r_2 a reflecting one, etc. Sometimes, a condition of periodicity of the form: $f(x + 2\pi, t) = f(x,t)$ may be required.

6.2.3. Stationary diffusion processes

An important question is concerned with the existence of stationary Markov processes, and more specifically—with the existence of stationary diffusion processes. When does a Markov process $X(t)$ show the properties of a stationary process ?

First, it is necessary that for arbitrary random events $A, B \in \mathcal{F}$ the probabilities of events $\{X(t) \in A\}$ and $\{X(t+h) \in B\}$ be independent of t. Representing these probabilities in terms of the transition probability we come to the conclusion that the transition probability of a stationary Markov process should necessarily be homogeneous (with respect to time) and the initial distribution $P_o(A) = P\{X(0) \in A\}$ be such that for arbitrary $\tau > 0$ and all $A \in \mathcal{F}$:

$$P_o(A) = \int P(x,\tau,A)P_o(dx) \tag{1.114}$$

The homogeneity of the transition function and condition (1.114) imply that all finite-dimensional distributions

$$P\{X(t+\tau_1) \in B_1, X(t+\tau_2) \in B_2, \ldots, X(t+\tau_n) \in B_n\}$$

are independent of t.

The above leads to the **assertion**: a necessary and sufficient conditions for stationarity of the Markov process $X(t)$ are:

(i) $X(t)$ is homogeneous in time Markov process,

(ii) there exists an invariant probability distribution P_o in all the state space, that is such a distribution that for all $A \in \mathcal{F}$ and all $\tau > 0$, the formula (1.114) holds. If this invariant distribution P_o is taken as initial distribution of the process , then $X(t)$ is stationary .

However, not all homogeneous transition probabilities admit existence of invariant distributions P_o. If $P(x,\tau,A)$ is such that for some x belonging to the state space and for $U_R = \{x : |x| > R\}$

$$\lim_{R\to\infty} \lim_{T\to\infty} \frac{1}{T}\int_0^T P(x,\tau,U_R)d\tau = 0 \tag{1.115}$$

then there exists an invariant measure P_o, that is the process considered is stationary (see: Khasminski [103]).

As far as diffusion processes are concerned, their probabilistic properties are expressed by the drift and diffusion coefficients. So, it is necessary, they be independent of time, that is $a(x,t) = a(x), b(x,t) = b(x)$. Analytical condition for the existence of stationary diffusion process (provided in [27]) has the form: on the considered finite or infinite interval in the state space

$$\int_{r_1}^{r_2} l_2(x)dx < \infty \tag{1.116'}$$

where

$$l_2(x) = \frac{1}{b(x)l_1(x)} \quad , \quad l_1(x) = exp\left\{-\int_{x'}^{x} \frac{a(\xi)}{b(\xi)}d\xi\right\} \tag{1.116''}$$

For example, let the state space be the entire real axis ($r_1 = -\infty, r_2 = +\infty$) and let the coefficients $a(x)$ and $b(x)$ have the form

$$a(x) = a_o + a_1 x \quad , \quad b(x) = b_o + b_1 x + b_2 x^2$$

where a_o, a_1 and b_o, b_1, b_2 are constant ($a_1 < 0, b_2 > 0$). In this case there exists a stationary probability measure with density $f(x)$ which satisfies the "time-independent" Fokker-Planck-Kolmogorov equation ($\frac{\partial f}{\partial t} = 0$). Solution of this equation for different combinations of the constants gives the family of Pearson distributions.

Therefore, the one-dimensional stationary probability density $f_{st}(x) = \lim_{t\to\infty} f(x,t)$, if it exists, does not depend on time t and on the initial density $f_o(x)$. The Fokker-Planck-Kolmogorov equation takes the form

$$\frac{\partial}{\partial x}\left[a(x)f_{st}(x)\right] - \frac{1}{2}\frac{\partial^2}{\partial x^2}\left[b(x)f_{st}(x)\right] = 0 \tag{1.117}$$

or, after integration with respect to x

$$\frac{d}{dx}\left[b(x)f_{st}(x)\right] - 2a(x)f_{st}(x) = -2\Phi_o$$

where the flux Φ defined by (1.107) is constant, i.e. $\Phi(x,t) = \Phi_o =$ const.. The last equation is a first order inhomogeneous equation for function $h(x) = b(x)f_{st}(x)$:

$$\frac{dh(x)}{dx} - 2\frac{a(x)}{b(x)}h(x) = -2\Phi_o$$

Hence, the general solution for $f_{st}(x)$ is

$$f_{st}(x) = \frac{C}{b(x)}exp\left\{2\int_{x'}^{x}\frac{a(\xi)}{b(\xi)}d\xi\right\} - \frac{2\Phi_o}{b(x)}\int_{x'}^{x} exp\left\{2\int_{x'}^{y}\frac{a(\xi)}{b(\xi)}d\xi\right\}dy \tag{1.118}$$

where the constant C is determined from the normalization condition, whereas the constant value of the flux Φ_0 is found from the boundary conditions; as lower integration limit x' one can take an arbitrary point from the state space of the process.

If $\Phi = 0$, equation (1.117) gives

$$\frac{d}{dx}[b(x)f_{st}(x)] - 2a(x)f_{st}(x) = 0 \tag{1.119}$$

with the following general solution

$$f_{st}(x) = \frac{C}{b(x)} exp\left\{2\int_{x'}^{x}\frac{a(\xi)}{b(\xi)}d\xi\right\} \tag{1.120}$$

The above formula is often used in applications.

6.3. Methods of solving the Kolmogorov equation

a. The simplest case; application of the Laplace transform

Let us consider first the simplest diffusion process , when $a(x,t) = 0, b(x,t) = 1$. Equation (1.102) takes the form

$$\frac{\partial f}{\partial t} = \frac{1}{2}\frac{\partial^2 f}{\partial x^2} \tag{1.121}$$

Let us assume the following initial condition

$$f(x,0) = 0 \tag{1.122}$$

and the "boundary" condition at infinity

$$f(-\infty,t) = f(\infty,t) = 0 \tag{1.123}$$

Let $\tilde{f}(x,s)$ be the Laplace transform of $f(x,t)$. Then $\tilde{f}(x,s)$ satisfies the equation

$$\frac{1}{2}\frac{d^2\tilde{f}}{dx^2} - s\tilde{f} = 0 \tag{1.124}$$

since $\tilde{f}(x,0) = s\tilde{f}$. Two linearly independent solutions of (1.124) are

$$\tilde{f}_1(x,s) = e^{-\sqrt{2s}x} \quad , \quad \tilde{f}_2(x,s) = e^{\sqrt{2s}x}$$

Because of (1.123) we take

$$\tilde{f}(x,s) = \tilde{f}_1(x,s) = e^{-\sqrt{2s}x} \tag{1.125}$$

By inverting we obtain

$$f(x,t) = \frac{1}{[2\pi t]^{1/2}} exp\left\{-\frac{x^2}{2t}\right\} \tag{1.126}$$

If we consider the transition $p(t_\circ, x_\circ; t, x)$ instead of $f(x,t)$, we have

$$p(t_\circ, x_\circ; t, x) = \frac{1}{[2\pi(t - t_\circ)]^{1/2}} exp\left\{-\frac{(x - x_\circ)^2}{2(t - t_\circ)}\right\} \tag{1.126$'$}$$

If the infinitesimal variance of the process is $b(x,t) = \sigma^2 > 0$, then instead of (1.126) we obtain

$$p(t_\circ, x_\circ; t, x) = \frac{1}{\sigma\,[2\pi(t - t_\circ)]^{1/2}} exp\left\{\frac{x - x_\circ)^2}{2(t - t_\circ)\sigma^2}\right\} \tag{1.127}$$

The Laplace transform can also be applied to more general cases.

It is clear that in the case of equation ($b(x,t) = \sigma^2 > 0, \quad a(x,t) = \mu$)

$$\frac{\partial f}{\partial t} = \frac{1}{2}\sigma^2 \frac{\partial^2 f}{\partial x^2} - \mu \frac{\partial f}{\partial x} \tag{1.128}$$

application of the Laplace transform gives for conditions (1.122), (1.123) the following solution for the transition density

$$p(t_\circ, x_\circ; t, x) = \frac{1}{\sigma\,[2\pi(t - t_\circ]^{1/2}} exp\left\{-\frac{(x - x_\circ - \mu t)^2}{2(t - t_\circ)\sigma^2}\right\} \tag{1.129}$$

The above formulae can be used in constructing the solutions of Kolmogorov equation for processes with boundary conditions (cf. Exercise 9).

b. Method of separation of variables

Let us assume that the infinitesimal coefficients do not depend on time (the process is homogeneous with respect to time). The Fokker-Planck-Kolmogorov equation (1.102) takes the form

$$\frac{\partial f}{\partial t} + \frac{\partial}{\partial x}[a(x)f(x,t)] - \frac{1}{2}\frac{\partial^2}{\partial x^2}[b(x)f(x,t)] = 0 \tag{1.130}$$

where the coefficients $a(x)$ and $b(x)$ are assumed to be given. Using the standard methods of separation of variables we assume that

$$f(x,t) = \Lambda(x)T(t) \tag{1.131}$$

where Λ and T are functions of x and t, respectively. Substitution of (1.131) in (1.130) gives the two ordinary differential equation

$$\frac{1}{T(t)}\frac{dT}{dt} = -\lambda^2 \tag{1.132}$$

$$\frac{1}{2}\frac{d^2}{dx^2}\left[b(x)\Lambda(x)\right] - \frac{d}{dx}\left[a(x)\Lambda(x)\right] + \lambda^2\Lambda(x) = 0 \tag{1.133}$$

Equation (1.132) is a simple first order equation with solution

$$T(t) = e^{-\lambda^2 t} \tag{1.134}$$

while equation (1.133) is a second order equation with variable coefficients. In general, the solution $\Lambda(x, A, B, \lambda)$ depends on two arbitrary constants A and B. If the eigenvalues of equation (1.133) are discrete and distinct, then by virtue of linearity of the problem the solution can be represented symbolically in the form

$$f(x,t) = \sum_{n=0}^{\infty} \Lambda(x; A_n, B_n, \lambda_n)e^{-\lambda_n^2 t} \tag{1.135}$$

where A_n, B_n and λ_n are determined from the initial and boundary conditions which must be specified for any particular diffusion process.

In general case of variable coefficients $a(x)$ and $b(x)$ the eigenfunction–eigenvalue problem (1.133) is not easy to solve. Only in restricted numbers of cases (special form of $a(x)$ and $b(x)$) a solution can be determined in a closed form—in terms of special functions (cf. example 2 in Sec. 27.1.1.). The solution of the Fokker-Planck-Kolmogorov equation (1.130) can be then expressed as an expansion in terms of the eigenfunctions $\Lambda_n(x)$.

For example, if equation (1.130) is supplemented by the absorbing boundary conditions $f(\pm a, t) = 0$ and the initial condition is $f(x, t_o) = \delta(x - x_o)$ then the solution is represented as (cf. [35])

$$f(x,t) = p(t_o, x_o; t, x) = \sum_{n=0}^{\infty} \frac{1}{f_{st}(x)}\Lambda_n(x)\Lambda_n(x_o)e^{-\lambda_n^2(t-t_o)} \tag{1.136}$$

where $f_{st}(x)$ is the stationary solution (1.120) of equation (1.130) and the functions $\Lambda_n(x)$ are orthonormal with the weight function $\frac{1}{f_{st}(x)}$, i.e.

$$\int_{-a}^{+a} \frac{1}{f_{st}(x)}\Lambda_m(x)\Lambda_n(x)dx = \begin{cases} 1, & n = m \\ 0, & n \neq m \end{cases} \tag{1.137}$$

The above form of solution is often used in various applications.

c. Transformations of variables

In many cases the Fokker-Planck-Kolmogorov equation can be siginifantly simplified by introducing appropriate new variables; sometimes it can be even reduced to the simplest diffusion equation (1.121).

Let

$$z = \Psi(x) \tag{1.138}$$

be a one to one transformation of the state space of a given diffusion process $X(t)$. As a result we obtain another diffusion process $Z(t)$ with probability density $\tilde{f}(z,t)$ such that

$$f(x,t) = \tilde{f}(z,t)\left|\frac{d\Psi(x)}{dx}\right| \tag{1.139}$$

In the case when $a(x,t) \equiv a(x)$, $b(x,t) \equiv b(x)$ the above transformation leads to the process $Z(t)$ with the drift and diffusion coefficients $a(z)$ and $b(z)$ which in terms of x are expressed by the formulae

$$\begin{aligned} \tilde{a}(x) &= a(x)\Psi'(x) + \frac{1}{2}b(x)\Psi''(x) \\ \tilde{b}(x) &= b(x)\left[\Psi'(x)\right]^2 \end{aligned} \tag{1.140}$$

where $\Psi'(x)$ denotes the derivative with respect to x. The above formulae imply that if we want to obtain a process with drift equal to zero i.e. $\tilde{a}(x) \equiv 0$, then the transformation given by (1.138) should be such that

$$\Psi'(x) = exp\left\{-2\int_{x_0}^{x}\frac{a(\xi)}{b(\xi)}d\xi\right\} \tag{1.141}$$

If transformation (1.138) is such that

$$\Psi'(x) = [b(x)]^{-\frac{1}{2}} \tag{1.142}$$

we obtain the process $Z(t)$ with the diffusion coefficient equal to one, i.e. $\tilde{b}(x) = 1$.

Let us consider now the general Fokker-Planck-Kolmogorov equation (1.102) with time-dependent coefficients. Let us introduce the following one-to-one transformation

$$z = \Psi(x,t) \quad , \quad \tau = \varphi(t) \tag{1.143}$$

The associated transformation of the probability density function is

$$f(x,t) = \tilde{f}(z,t)\left|\frac{d\Psi(x,t)}{dx}\right| \tag{1.144}$$

As a result of transformation (1.143), the probability density $\tilde{f}(z,\tau)$ is governed by the equation

$$\frac{\partial \tilde{f}}{\partial \tau} + \frac{\partial}{\partial z}\left[\tilde{a}(z,\tau)\tilde{f}(z,\tau)\right] - \frac{1}{2}\frac{\partial^2}{\partial z^2}\left[\tilde{b}(z,\tau)\tilde{f}(z,\tau)\right] = 0 \tag{1.145}$$

where

$$\begin{aligned}\tilde{a}(z,\tau) &= \frac{1}{\varphi'(t)}\left[\frac{1}{2}b(x,t)\frac{\partial^2\Psi(x,t)}{\partial x^2} + a(x,t)\frac{\partial\Psi(x,t)}{\partial x} + \frac{\partial\Psi(x,t)}{\partial t}\right]\\ \tilde{b}(z,\tau) &= \frac{1}{\varphi'(t)}b(x,t)\left[\frac{\partial\Psi(x,t)}{\partial x}\right]^2\end{aligned} \tag{1.146}$$

and where x and t on the right hand side of (1.146) are assumed to be expressed by z and τ.

Depending on the specific form of $a(x,t)$ and $b(x,t)$ the use of transformation (1.143) can lead to various simplifications. It is important to emphasize that Cherkasov ([4], cf. also [28]) proposed a transformation which reduces the general equation (1.102) to the simplest diffusion equation (1.121).

Making use of the Cherkasov theorem one can show, for example:

a) the process $X(t)$ with coefficients

$$a(x,t) = \frac{2t}{(\cos x + 2t)^3} - \frac{2x}{\cos x + 2t}, \qquad b(x,t) = \frac{2t}{(\cos x + 2t)^2}$$

can be reduced to the simplest diffusion process governed by (1.121) by the transformation

$$\begin{aligned}\Psi(x,t) &= \left(\frac{k_1}{2t_o}\right)^{1/2}\left[\sin x + 2t(x - x_1) - \sin x_1\right] + k_2\\ \varphi(t) &= \frac{k_1}{2t_o}(t^2 - t_1^2) + k_3\end{aligned}$$

where x_1 is an arbitrary point in the state space, t_1—an arbitrary time instant greater then t_o, and k_1, k_2, k_3 are arbitrary constants;

b) the process $X(t)$ with coefficients

$$a(x,t) = \alpha(t)\left[x - c_o\right] \quad , \qquad b(x,t) = \beta(t)\left[x - c_o\right]^2$$

can be reduced to the simplest diffusion process by the transformation

$$\begin{aligned}\Psi(x,t) &= ln(x - c_o) + \int\left[\beta(t) - \alpha(t)\right]dt\\ \varphi(t) &= \int \beta(t)dt\end{aligned}$$

In some cases, equation (1.102) can be reduced to the simpler one by a change of the unknown function $f(x,t)$; for example, $g(x,t) = \ln f(x,t)$. Reduction of equation (1.102) in which coefficients depend on time only to the simplest equation (1.121) is shown in Example 13.

6.4. Vector diffusion processes

Let $\mathbf{X}(t) = [X_1(t), X_2(t), \ldots, X_n(t)]$ be an n-dimensional vector Markov process. States of this process are vectors $\mathbf{x} = [x_1, \ldots, x_n]$. The probabilistic structure is characterized by the initial distribution and the transition probability

$$P\{\mathbf{X}(t) < \mathbf{y} | \mathbf{X}(s) = \mathbf{x}\} = F(s, \mathbf{x}; t, \mathbf{y}) \tag{1.147}$$

The density of the transition probability is defined as

$$\begin{aligned} p(s,\mathbf{x};t,\mathbf{y}) &= p(s; x_1, x_2, \ldots, x_n; t; y_1, y_2, \ldots, y_n) \\ &= \frac{\partial^n F(s; x_1, x_2, \ldots, x_n; t, y_1, y_2, \ldots, y_n)}{\partial y_1, \partial y_2, \ldots, \partial y_n} \end{aligned} \tag{1.148}$$

A diffusion vector Markov process is defined in the same way as the one-dimensional process with the difference that integrals occuring in the conditions (i), (ii), (iii) are n-fold integrals. The functions $a(t,x)$ and $b(t,x)$ are now vector and matrix valued, respectively. They can be defined as

$$\lim_{\Delta t \to 0} \frac{1}{\Delta t} E\{[X_i(t+\Delta t) - X_i(t)] | \mathbf{X}(t) = \mathbf{x}\} = a_i(\mathbf{x}, t) \tag{1.149}$$

(1.150)
$$\lim_{\Delta t \to 0} \frac{1}{\Delta t} E\left\{[X_i(t+\Delta t) - X_i(t)][X_j(t+\Delta t) - X_j(t)] | \mathbf{X}(t) = \mathbf{x}\right\} = b_{ij}(\mathbf{x}, t)$$

The corresponding Fokker-Planck-Kolmogorov equation has the form

$$\begin{aligned} &\frac{\partial p(s,\mathbf{x};t,\mathbf{y})}{\partial t} + \sum_{i=1}^{n} \frac{\partial}{\partial y_i} [a_i(\mathbf{y},t) p(s,\mathbf{x};t,\mathbf{y})] \\ &- \frac{1}{2} \sum_{i,j=1}^{n} \frac{\partial^2}{\partial y_i \partial y_j} [b_{ij}(\mathbf{y},t) p(s,\mathbf{x};t,\mathbf{y})] = 0 \end{aligned} \tag{1.151}$$

All what we said about the initial and boundary conditions holds for vector diffusion processes (in the appropriate vectorial formulation). Similarly as in the one-dimensional case, the Fokker-Planck-Kolmogorov equation for the one-dimensional density function $f(\mathbf{x}, t)$ can be represented as the continuity equation for the "probabilistic mass".

$$\frac{\partial f(\mathbf{x}, t)}{\partial t} + \sum_{i=1}^{n} \frac{\partial \Phi_{x_i}}{\partial x_i} = 0 \tag{1.152}$$

where Φ_{x_i} denotes the component of the flux vector $\mathbf{\Phi}$ in x_i direction, $(i = 1, 2, \ldots, n)$ that is

$$\Phi_{x_i} = a_i(\mathbf{x}, t) f(\mathbf{x}, t) - \frac{1}{2} \sum_{j=1}^{n} \frac{\partial}{\partial x_j} \left[b_{ij}(\mathbf{x}, t) f(\mathbf{x}, t) \right] \tag{1.153}$$

The boundary conditions can take diffrent forms depending on the physical nature of the problem. If the vector process $\mathbf{X}(t)$ is allowed to take its values on entire R_n space, then usually the following conditions are assumed

$$f(\mathbf{x}, t) \Big|_{x_i = \pm\infty} = 0 \tag{1.154}$$

If the state space of the process is bounded the situation is more complex. Above all, it is important whether the diffusion matrix $\mathbf{B} = \{b_{ij}\}$ is degenerated or not. If $\mathbf{B}$ does not degenerate everywhere in the considered domain the equation (1.151) is of parabolic type. In the opposite case it belongs to the elliptic-parabolic type.

In general, the problem of solving the Fokker-Planck-Kolmogorov equation for multi-dimensional process $\mathbf{X}(t)$ is involved. Of course, the methods presented above for the one-dimensional case can be instructive in dealing with multi-dimensional problems, but their effectiveness is restricted to the simpler cases. In paper [25] by Morita and Hara the authors provide the solution of the F-P-K equation (1.151) in many dimensions in the form of the path integral.

As far as stationary solutions are concerned, all what we have said in the previous subsection about the conditions for existence of stationary diffusion process holds also in multi-dimensional case; analytical conditions (1.116) are valid only in scalar case.

For an exhaustive discussion of the methods associated with multi-dimensional F-P-K equations the reader is referred to books [15], [29].

7. PROCESSES WITH INDEPENDENT INCREMENTS; WIENER PROCESS AND POISSON PROCESS

7.1. Definition and general properties

Mathematical models of numerous real phenomena can be described by stochastic processes whose increments are independent random variables. The earliest and basic results associated with such processes are due to Levy; his analysis of "integrals of independent elements" (or "additive stochastic processes"), that is—in the present terminology—his analysis of independent—increment processes (initiated in 1934) provided the foundations and indicated the directions of the existing theory.

Definiton 1.13. *A stochastic process $X(t), t \in T = [0,\infty)$ is said to be an independent-increment process , if for arbitrary subset $\{t_o < t_1 < \cdots < t_n\}$ of the index set the random variables*

$$X(t_o), X(t_1) - X(t_o), \ldots, X(t_n) - X(t_{n-1}) \tag{1.155}$$

are independent, that is for arbitrary Borel sets $B_o, B_1, \ldots, B_n$ on R_1

$$P\{X(t_o) \in B_o, X(t_1) - X(t_o \in B_1, \ldots, X(t_n) - X(t_{n-1}) \in B_n\}$$

$$= P\{X(t_o) \in B_o\} \prod_{i=1}^{n} P\{X(t_i) - X(t_{i-1}) \in B_i\} \tag{1.156}$$

The definition above implies that all finite-dimensional distributions of the process $X(t)$ are completely determined by the distribution of the random variable $X(t)$ for each t and the distributions of the increments $X(t_2) - X(t_1)$ for all possible values of t_1, t_2 such that $t_1 < t_2$.

Indeed, if the process $X(t)$ has independent increments and $\varphi(\lambda) = E\{exp^{i\lambda X(t)}\}, \varphi_{t_1,t_2}(\lambda) = \{exp^{i\lambda[X(t_2)-X(t_1)]}\}$ then by making use of the independence of the random variables (1.155) one obtains the following representation of the finite-dimensional characteristic functions of the process

$$\varphi_{t_1,t_2,\ldots,t_n}(\lambda_1, \lambda_2, \ldots, \lambda_n) = E\left\{exp\left\{i\sum_{j=1}^{n} \lambda_j X(t_j)\right\}\right\}$$

$$= E\left(exp\left\{i(\lambda_1 + \lambda_2 + \cdots + \lambda_n)X(t_1) + i(\lambda_2 + \cdots + \lambda_n)\right.\right.$$

$$\left.\left.\times [X(t_2) - X(t_1)] + \cdots + i\lambda_n [X(t_n) - X(t_{n-1})]\right\}\right)$$

$$= \varphi_{t_1}(\lambda_1 + \cdots + \lambda_n)\varphi_{t_1,t_2}(\lambda_2 + \cdots + \lambda_n) \cdot \ \ldots \ \cdot \varphi_{t_{n-1},t_n}(\lambda_n) \tag{1.157}$$

Let us take two arbitrary time instants $t_1 < t_2$. It is clear that $X(t_2) = X(t_1) + [X(t_2) - X(t_1)]$, i.e. $X(t_2)$ is a sum of two independent random variables . Hence, the following relation holds for characteristic functions:

$$\varphi_{t_2}(\lambda) = \varphi_{t_1,t_2}(\lambda) \cdot \varphi_{t_1}(\lambda)$$

This leads to the conclusion:
if $X(t)$ is the independent increment process then the characteristic function of its

increment is determined by the one-dimensional characteristic function of the process (i.e. by the one-dimensional distribution); for each $t_1 < t_2$:

$$\varphi_{t_1,t_2}(\lambda) = \frac{\varphi_{t_2}(\lambda)}{\varphi_{t_1}(\lambda)} \tag{1.158}$$

Making use of this formula in (1.157) we come to the following assertion:
all finite-dimensional distributions of the independent-increment process are defined by the one-dimensional distribution . However, it should be noticed that one-dimensional distribution of the independent-increment process can not be given arbitrarily; the one-dimensional characteristic function has to be such that $\varphi_{t_1,t_2}(\lambda)$ given by (1.158) is also a characteristic function .

Let $X(t), t \geq 0$ be an independent-increment process. Assume that $P\{X(0) = 0\} = 1$. For each t_n, $X(t_n)$ can be respresented in the form

$$X(t_n) = \sum_{i=1}^{n} [X(t_i) - X(t_{i-1})] \tag{1.159}$$

that is, in the form of sum of mutually independent random variables. It indicates that an arbitrary continuous — parameter independent-increment process is Markovian.

If $X(t)$ is an independent-increment process and f is a function defined on the index set T of the process , then the process $X(t) - f(t)$ is also an independent-increment process. Often it turns out to be convenient to replace the process $X(t)$ by $X(t) - f(t)$ where $f(t)$ should be chosen in such a way that the realizations of a new process have simple continuity properties. If T has the smallest element a, then usually $X(t)$ is replaced by $Y(t) = X(t) - X(a)$; i.e. by a new process $Y(t)$ which is also an independent-increment process having the same increments as $X(t)$ and with the property that $P\{Y(a) = 0\} = 1$. Therefore, it is not restrictive if we add that $P\{X(a) = 0\} = 1$

The most important examples of an independent-increment process are the Wiener process and the Poisson process .

7.2. Wiener process

Definition 1.14. *A stochastic process* $W(t)$, $t \in [0, \infty)$ *is called* a **Wiener process** *or* Brownian motion process *if:*

a) $P\{W(0) = 0\} = 1$,

b) *for arbitrary* $0 < t_0 < t_1 < \cdots < t_n$ *the increments* $W(t_1) - W(t_0), W(t_2) - W(t_1), \ldots, W(t_n) - W(t_{n-1})$ *are independent,*

c) for arbitrary t and $h > 0$ the increment $W(t+h) - W(t)$ has a Gaussian distribution with:

$$E[W(t+h) - W(t)] = \mu h \qquad \mu - \text{real constant,}$$

$$E\left[[W(t+h) - W(t)]^2\right] = \sigma^2 h \qquad \sigma^2 - \text{positive constant,} \tag{1.160}$$

where μ is called the drift and σ^2 is called the variance.

Without restriction one can assume that $\mu = 0, \quad \sigma^2 = 1$. In what follows we consider this case (standard Wiener process).

The correlation function of a Wiener process is

$$\begin{aligned} K_W(t_1, t_2) &= E[W(t_1)W(t_2)] \\ &= E(W(t_1)\{[W(t_2) - W(t_1)] + W(t_1)\}) \\ &= E[W(t_1)]^2 + E\{W(t_1)[W(t_2) - W(t_1)]\} \\ &= E[W(t_1) - W(0)]^2 \end{aligned}$$

If $t_2 > t_1$, then according to (1.160) $K_W(t_1, t_2) = t_1$. In general,

$$K_W(t_1, t_2) = \min(t_1, t_2) \tag{1.161}$$

A Wiener process is — as any independent-increment process — also Markovian (with continuous index set and continuous state space). It is a homogeneous Markov process with the following transition probability ($s < t$)

$$P(s, x; t, B) = P(x, t-s, B) =$$

$$= \int_B \frac{1}{\sqrt{2\pi(t-s)}} exp\left[-\frac{(y-x)^2}{2(t-s)}\right] dy \tag{1.162}$$

Let us note that for an arbitrary subset $\{t_\circ < t_1 < \cdots < t_n\}$

$$W(t_n) = W(t_\circ) + [W(t_1) - W(t_\circ)] + \cdots + [W(t_n) - W(t_{n-1}]$$

where the particular components are Gaussian and independent. Therefore, the one-dimensional distributions of $W(t)$ are Gaussian.

Since the vector $[W(t_1), W(t_2), \ldots, W(t_n)]$ is a non-degenerate linear transformation of the Gaussian vector $\mathbf{\Delta W} = [W(t_1) - W(t_\circ), W(t_2) - W(t_1), \ldots, W(t_n) - W(t_{i-1})]$ one can easily find the expression for the finite-dimensional probability density of the Wiener process. The appropriate formulae are:

$$f_{\Delta W}(x_1, x_2, \ldots, x_n) =$$

$$= \prod_{i=1}^{n} \frac{1}{\sqrt{2\pi(t_i - t_{i-1})}} exp\left\{-\frac{x_i^2}{2(t_i - t_{i-1})}\right\} \tag{1.163}$$

$$f_{W_{t_1},\ldots,W_{t_n}}(x_1, x_2, \ldots, x_n) =$$

$$= \prod_{i=1}^{n} \frac{1}{\sqrt{2\pi(t_i - t_{i-1})}} exp\left\{-\sum_{i=1}^{n} \frac{(x_i - x_{i-1})^2}{2(t_i - t_{i-1})}\right\} \tag{1.164}$$

Hence, all finite-dimensional distributions of a Wiener process are also Gaussian.

It can be calculated that for any arbitrary n

$$E\left\{[W(t+h) - W(t)]^{2n}\right\} =$$
$$= \frac{1}{\sqrt{2\pi h}} \int_{-\infty}^{+\infty} z^{2n} e^{-\frac{z^2}{2h}} dz =$$
$$= (2n-1)!!h^n = 1 \cdot 3 \cdot \ \ldots \ \cdot (2n-1)h^n \tag{1.165}$$

In particular

$$E\left\{[W(t+h) - W(t)]^2\right\} = h$$
$$E\left\{[W(t+h) - W(t)]^4\right\} = 3h^2 \tag{1.166}$$

The following theorem characterizes an interesting property of a Wiener process (cf. Example 8 — for the proof).

Theorem 1.7. *Let $a = t_0 < t_1 < \cdots < t_n = b$ be an arbitrary partition of interval $[a, b]$ and $\Delta = \max(t_{i+1} - t_i)$. Then*

$$\underset{\Delta \to 0}{\text{l.i.m.}} \sum_{i=0}^{n-1} [W(t_{i+1}) - W(t_i)]^2 = b - a \tag{1.167}$$

The fundamental properties of the realizations of a Wiener process are summarized in the following theorem [16].

Theorem 1.8. *Almost all realizations (sample functions) of the Wiener process are continuous, nowhere differentiable and have unbounded variation in every finite interval.*

The continuity follows from the Kolmogorov criterion which will be formulated in chapter 2. The non-differentiability for fixed t can be made clear as follows: the distribution of the difference quotient

$$\frac{1}{h}[W(t+h) - W(t)]$$

is the Gaussian distribution $N(0, \frac{1}{h})$ which diverges as $h \to 0$; therefore this quotient can not converge with positive probability to a finite random variable in any probabilistic sense. The last assertion of the theorem follows from the theorem 1.7. and some additional considerations connected with almost sure convergence.

Definition 1.15. *A stochastic vector process* $\mathbf{W}(t) = [W_1, \ldots, W_m(t)]$ *is called an m-dimensional* Wiener process *(or Brownian motion process) if each component* $W_i(t), \quad i = 1, 2, \ldots, m$ *is a scalar Wiener process and the processes* $W_i(t)$ *are mutually independent.*

The properties of a Wiener process stated in theorem 1.8. have very significant implications both in theory and applications. For instance, it should be noticed that the non-differentiability of a Wiener process means that the Brownian particle has not a well defined velocity at any instant. In order to avoid this disadvantage L.S. Ornstein and E.G. Uhlenbeck have constructed a model of a Brownian motion in which the particle has a velocity. The basic feature of this model is that stochastic process modelling the Brownian phenomenon describes not a spatial coordinate of the particle but its velocity; the position of the particle is found by integration of an Ornstein-Uhlenbeck process . It is worthy to characterize briefly this idea.

Definition 1.16. *Let* $W(t), \quad t \in [0, \infty)$ *be a normalized Wiener process . The process*

$$U(t) = \sqrt{\alpha}\, e^{-\beta t} W(e^{2\beta t}) \tag{1.168}$$

is called an Ornstein-Uhlenbeck *process .*

It is clear from the definition that $U(t)$ is a Gaussian process. Therefore, it can be completely characterized by its average and correlation function

$$E\left[U(t)\right] = \sqrt{\alpha}\, e^{-\beta t} E\left[W(e^{2\beta t})\right] = 0 \tag{1.169}$$

$$\begin{aligned} &K_u(t_1, t_2) \\ &= \alpha e^{-\beta(t_1+t_2)} E\left[W(e^{2\beta t_1}) W(e^{2\beta t_2})\right] \\ &= \alpha e^{-\beta(t_1+t_2)} e^{2\beta \min(t_1, t_2)} = \alpha e^{-\beta|t_2 - t_1|} \end{aligned} \tag{1.170}$$

The last formulae and Gaussianity imply that the process $U(t)$ is strictly stationary. More, the process $U(t)$ is also Markovian since it has been obtained from a Markov process $W(t)$ by a one-to-one transformation.

If $U(t)$ characterizes the particle velocity, then its position at time t is expressed by the stochastic integral

$$X(t) = \int_0^t U(s)ds \tag{1.171}$$

where we have assumed that the position at $t = 0$ is zero. The process $X(t)$ is evidently Gaussian, but it is not Markovian. However, it can be shown that the two-dimensional process $[X(t), U(t)]$ is Markovian.

7.3. Poisson process

The second important example of an independent-increment process is the Poisson process. A general definition is the following.

Definition 1.17. *A stochastic process $N(t)$, $t \in [0,\infty)$ with independent increments is called a* **non-homogeneous Poisson process** *if for arbitrary t and s the difference $N(t) - N(s)$ have a Poisson distribution , that is there exists such a function $a(s,t)$ that for each integer number $k > 0$*

$$P\{N(t) - N(s) = k\} = \frac{[a(s,t)]^k}{k!} e^{-a(s,t)} \tag{1.172}$$

Without any restriction one can assume that $P\{N(0) = 0\} = 1$.Then, the values of the process $N(t)$ are also random variables with the Poisson distribution , that is

$$P\{N(t) = k\} = \frac{[\mu(t)]^k}{k!} e^{-\mu(t)} \tag{1.173}$$

It is clear that

$$\begin{aligned} \mu(t) &= E[N(t)] \\ a(s,t) &= E[N(t) - N(s)] = \\ &= \mu(t) - \mu(s) \end{aligned} \tag{1.174}$$

Therefore formula (1.172) can be rewritten in the form

$$P\{N(t) - N(s) = k\} = \frac{[\mu(t) - \mu(s)]^k}{k!} e^{-[\mu(t)-\mu(s)]} \tag{1.175}$$

If $\mu(t) = \nu t, \nu > 0$ then the process $N(t)$ is called a *homogeneous Poisson process*; in this case

$$P\{N(t) - N(s) = k\} = \frac{\nu^k (t-s)^k}{k!} e^{-\nu(t-s)} \tag{1.176}$$

Parameter ν is called the *intensity* of the homogeneous Poisson process .

For the homogeneous Poisson process the following formulae hold

$$\begin{aligned} P\{N(t) = k\} &= \frac{(\nu t)^k}{k!} e^{-\nu t} \\ E[N(t)] &= \nu t \\ \mathrm{var} N(t) &= \nu t \\ K_N(t_1,t_2) &= E\{[N(t_1) - \nu t_1][N(t_2 - \nu t_2]\} \\ &= \nu \min(t_1,t_2) \end{aligned} \tag{1.177}$$

A homogeneous Poisson process is an example of a Markov process with continuous parameter and with an integer-valued state space. Its transition probability is

$$\begin{aligned} P(i, t-s, B) &= \\ &= \sum_{\substack{j \in B \\ j \geq i}} \frac{\nu^{j-1}(t-s)^{j-i}}{(j-i)!} e^{-\nu(t-s)} \end{aligned} \tag{1.178}$$

where j and i denote non-negative integer numbers.

The following assertions characterize the basic properties of the Poisson process cf. [33]).

Theorem 1.9. *For an homogeneous Poisson process with intensity ν, the interarrival times $t_1, t_2, \ldots, t_n$ are independent and identically distributed random variables with the common distribution being exponential with parameter ν.*

It is, however, worthy adding that interarrivals times for an inhomogeneous Poisson process are not independent.

Theorem 1.10. *Almost all sample functions of the Poisson process are non-decreasing with integer increments; they increase only by jumps equal to one and the number of jumps in every finite interval is finite.*

7.4. Processes related to Poisson process

A. Compound Poisson process

Definition 1.18. *A stochastic process $X(t), t \geq 0$ is said to be a* compound Poisson process *if it can be represented, for $t \geq 0$, by*

$$X(t) = \sum_{n=1}^{N(t)} Y_n \tag{1.179}$$

where $N(t), \geq 0$ is a Poisson process , and $\{Y_n, n = 1, 2, \ldots, n\}$ is a family of independent random variables indentically distributed as a random variable Y; the process $N(t), t \geq 0$ and the sequence Y_n are assumed to be independent.

Theorem 1.11. *(cf. [33]). A compound Poisson process $X(t), t > 0$ has stationary independent increments, and characteristic function*

$$\varphi_{X(t)}(\lambda) = e^{\nu t[\varphi_Y(\lambda) - 1]}, \qquad t \geq 0 \tag{1.180}$$

where $\varphi_Y(\lambda)$ is the common characteristic function of the independent identically distributed random variables Y_n and ν is the mean rate of the occurence of the events. If $E\left[Y^2\right] < \infty$, then

$$\begin{aligned} E\left[X(t)\right] &= \nu t E\left[Y\right] \\ Var\left[X(t)\right] &= \nu t E\left[Y^2\right] \\ R_X(t_1, t_2) &= \nu E\left[Y^2\right] \min(t_1, t_2) \end{aligned} \tag{1.181}$$

B. Filtered Poisson process

Definiton 1.19. *A stochastic process $X(t), t > 0$ is called* a filtered Poisson process *if it is represented, for $t > 0$ by*

$$X(t) = \sum_{k=1}^{N(t)} w(t, \tau_k, Y_k) \tag{1.182}$$

where $N(t), t \geq 0$, is a Poisson process with intensity ν, $\{Y_k\}$ is a sequence of independent random variables , identically distributed as a random variable Y and independent of $N(t)$; $w(t, \tau, y)$ is a function of three real variables called a **response** *function .*

For most practical situations, it is sufficient to consider the subclass

$$X(t) = \sum_{k=1}^{N(t)} Y_k w(t, \tau_k) \tag{1.183}$$

An interpretation of (1.183) is the following: if τ_k represents the time at which an event (cf. arriving of random pulse) took place, Y_k represents the amplitude of signal associated with the event and $w(t, \tau_k)$ characterizes the values at time t of a signal originating at time τ_k then $X(t)$ describes the value at time t of the sum of signals arising from the events or pulses occuring in the interval $(0, t]$.

The following theorem cf. [33] gives the basic characteristics of a filtered Poisson process .

Theorem 1.12. *For any positive number t and real number λ, the characteristic function $\varphi_{X(t)}(\lambda)$ of the process (1.182) is*

$$\varphi_{X(t)}(\lambda) = exp\left\{\nu \int_0^t E\left[e^{i\lambda w(t,\tau,Y)} - 1\right] d\tau\right\} \tag{1.184}$$

For any $t_2 > t_1 > 0$ the mean, variance and covaraince function (if they exist) are given by

$$m_X(t) = \nu \int_0^t E\left[w(t, \tau, Y)\right] d\tau$$

$$Var\left[X(t)\right] = \nu \int_0^t E\left[w^2(t, \tau, Y)\right] d\tau$$

$$R_X(t_1, t_2) = \nu \int_0^{\min(t_1,t_2)} E\left[w(t_1, \tau, Y) w(t_2, \tau, Y)\right] d\tau \tag{1.185}$$

Often, the process $X(t)$ is represented by (1.183) where $w(t,\tau_k) = w(t-\tau_k)$. In this case

$$m_X(t) = \nu m_Y \int_{-\infty}^{+\infty} w(u)du$$

$$Var\left[X(t)\right] = \nu E\left[Y^2\right] \int_{-\infty}^{+\infty} w^2(u)du$$

$$R_X(t_1,t_2) = \nu E\left[Y^2\right] \int_{-\infty}^{+\infty} w(u)w(t_2-t_1+u)du \tag{1.186}$$

The formulae given in the above theorem are usually referred to as *Cambell's theorem* on the superposition of random impulses.

If $E[Y]=0$ and $w(t,-\tau)=\delta(t-\tau)$ one obtains a process which is usually called a *shot noise.*

8. POINT STOCHASTIC PROCESSES

So far we considered a Poisson process as an example of an independent increment process with integer values. The characteristic feature of a Poisson process $N(t), t \geq 0$ is the fact that for each t it characterizes the number of events that occur within the interval $(0,t]$. It turns out that a Poisson process is the simplest representant of a wide class of processes that are called *point* or *counting processes.*

A point process is best pictured as a collection of randomly distributed points in space. In general, there exists state space, say $\mathcal{X}$, and a set of points $\{x_n\}$ in $\mathcal{X}$ representing the locations of different elements or members of the population. If $\mathcal{X}$ is the real line R_1, the x_n are usually the instants t_n at which the "event" occur.

Analogously, as in the case of a Poisson process , a general point process can be characterized by its *counting properties* (the properties which relate to the numbers of events occuring within specified time intervals) or by its *interval properties* (properties relating to the relative spacings between events, i.e. the interarrival times).

A possible definition of a stochastic point process is as follows. Let, as always $(\Gamma,\mathcal{F},P)$ be a basic probability space.

Definition 1.20. *A* point process *is a stochastic process $N(t), t \geq 0$ defined on $(\Gamma,\mathcal{F},P)$ so that for almost all $\gamma \in \Gamma$, the mapping $t \to N(t,\gamma)$ is non-decreasing, right-continuous, non-negative, finite, integer valued and $N(0,\gamma)=0$.*

Such a process increases by jumps only, and the times of jumps are called *points* of process $N(t)$. Let $(\mathcal{X},\mathcal{A})$ be an arbitrary measurable space. By a *random counting measure* we mean a stochastic process $N(A), A \in \mathcal{A}$ taking non-negative integer values so that $N(A \cup B) = N(A) + N(B)$ for arbitrary disjoint $A, B \in \mathcal{A}$.

Any point process can be regarded as a random measure in R_1 by identifying $N(t) - N(s)$ with $N([s,t]), s < t$. If $N(t), t \geq 0$ is a point process , we write $N(A)$ for the number of points of $N(t)$ in the set A; if $N(A), A \in R_1$ is a random counting measure, we write $N(t)$ for $N((0,t])$. It is, therefore, seen that the analysis of general point processes is essentially concerned with random measures.

Definition 1.21. *A random counting measure* N *on* $(\mathcal{X}, \mathcal{A})$ *is called* a Poisson random measure *with mean measure* ν *if* $N(A_1), \ldots, N(A_m)$ *are independent when* $A_1, \ldots, A_m \in \mathcal{A}$ *are disjoint and*

$$P\{N(A) = k\} = \frac{[\nu(A)]^k}{k!} e^{-\nu(A)} \tag{1.187}$$

A Poisson process with intensity $\mu(t)$ is a Poisson measure on $(R_1^+, \mathcal{A}^+)$ with mean measure element $\mu(dx)$. For a more detailed discussion we refer to [24], [33].

Let us give now an idea on how the probabilistic stucture of a point process can be characterized in practice. Let $N(t)$ represent the numbers of events in $(0,t]$ so that $dN(t)$ represents the number of events in the interval $(t, t+dt)$. We assume, moreover, that the probability of occurance of one evnt in dt is proportional to dt, while the probability that there occur $n(n > 1)$ events in dt is of negligibly smaller order of magnitude then dt. Thus, if $P(n)$ is the probability that n events occur in dt, where n is a non-negative integer, then

$$\begin{aligned} &P(1) = f_1(t)dt \quad , \quad P(n) = o(dt) \quad , \quad n > 1 \\ &P(0) = 1 - \sum_{n \geq 1} P(n) = 1 - f_1(t)dt + o(dt) \end{aligned} \tag{1.188}$$

The average values of the powers of the variable $dN(t)$ is as follows

$$\begin{aligned} E[dN(t)] &= \sum_{k=0} kP(k) = P(1) + o(dt) = f_1(t) + o(dt) \\ E[dN(t)]^n &= \sum_{k=0} k^n P(k) = f_1(t) + o(dt) = E[dN(t)] + o(dt) \end{aligned} \tag{1.189}$$

Therefore, all moments of random variable $dN(t)$ are equal to the probability that it takes the value unity.

The function $f_1(t)$ is termed as the *product density of degree one* of the process $N(t)$. The *product* ***density of degree two*** **is defined by**

$$f_1(t_1, t_2)dt_1dt_2 = E[dN(t_1)dN(t_2)] \tag{1.190}$$

if the intervals do not overlap. But, when they overlap we have

$$E[dN(t_1)dN(t_2)]_{dt_1=dt_2} = E\left\{[dN(t)]^2\right\} = E[dN(t_1)] = f_1(t_1)dt_1 \tag{1.191}$$

In general, the product density of degree n is defined as

$$f_n(t_1, t_2 \ldots, t_n) dt_1 dt_2 \ldots dt_n = E\left[dN(t_1) dN(t_2) \ldots dN(t_n)\right] \tag{1.192}$$

if the intervals do not overlap; if they do, the product density of degree n becomes one of degree $n-1$. The function (1.192) represents the joint probability that one event occurs in dt_1, one in $dt_2, \ldots$, one in dt_n when $dt_1, dt_2, \ldots, dt_n$ are all separate non-overlapping intervals.

Let us notice, that if $N(t)$ is a homogeneous Poisson process with intensity ν, the product densities are

$$\begin{aligned} &f_1(t)dt = E\{dN(t)\} = d\left[EN(t)\right] = d(\nu t) = \nu dt, \\ &f_1(t) = \nu, \\ &f_n(t_1, t_2, \ldots, t_n) dt_1 dt_2 dt_n = E\{dN(t_1) \ldots dN(t_n)\} \\ &= E\left[dN(t_1)\right] \ldots E\left[dN(t_n)\right] = \nu^n dt_1 \ldots dt_n \end{aligned} \tag{1.193}$$

9. MARTINGALES

An important class of processes which includes, for instance,the processes with independent increments and Markov processes, is formed by martingales.

Let $(\Gamma, \mathcal{F}, P)$ denote a probability space, and let $X(t), t \in T \subseteq R_1$ denote, in general n-dimensional, stochastic process defined on $(\Gamma, \mathcal{F}, P)$. Let $\{\mathcal{F}_t\}_{t \in T}$ denote an increasing family of σ-algebras of $\mathcal{F}$ (that is, for every $s < t, \mathcal{F}_s \subset \mathcal{F}_t \subset \mathcal{F}$) such that for each t, $X(t)$ is $\mathcal{F}_t$-measurable; $\mathcal{F}_t$ characterizes a collection of events representing the known information at time t.

Definition 1.22. *A real-valued process $X(t)$ measurable with respect to $\mathcal{F}$ (adapted to $\mathcal{F}_t$) is called* a martingale *if*

$$E\{X(t) | \mathcal{F}_s\} = X(s) \qquad \textit{almost surely} \tag{1.194}$$

for all $s, t \in T$, and $s \leq t$.

A real-valued process $X(t)$ is called a *super-martingale* or a *submartingale* (with respect to σ-algebra $\mathcal{F}_t$) if

$$E\{X(t) | \mathcal{F}_s\} \leq X(s) \qquad \text{or} \qquad E\{X(t) | \mathcal{F}_s\} \geq X(s) \tag{1.195}$$

respectively, for all $s \leq t$, $\quad s, t \in T$.

A martingale is both super and sub-martingale. But, the inverse assertion is also valid, namely: if $X(t)$ is a supermartingale and submartingale (with respect to the same σ-algebra $\mathcal{F}_t$) then $X(t)$ is a martingale. Often, the term "martingale" is referred to processes for which relation (1.194) holds for the family of σ-algebras generated by the process $X(t)$ itself, i.e., if

$$\mathcal{F}_t = \mathcal{F}([0,t]) = \mathcal{F}(X_s : 0 \leq s \leq t)$$

what means that the history of the process $X(t)$ prior to the instant t is chosen as a condition.

Supermartingales and submartingales form a class of processes which are called semi-martingales (the definition — cf. [27], [38]).

It can be shown, that if $X(t), t \in T$ is a process with independent increments such that $E[X(t)]$ exists and is constant, then $X(t)$ is a martingale.

A Wiener process $W(t)$ defined on $(\Gamma, \mathcal{F}, P)$ and for $t \geq 0$ (i.e. $W(0) = 0$) is a martingale with respect to the family of σ-algebras generated by random events $\mathcal{F}[0,t]$. Indeed, for $0 < s < t$ we have

$$\begin{aligned} E\{W(t)|\mathcal{F}_{\leq t}\} &= E\left\{[W(s) + (W(t) - W(s))]\Big|\mathcal{F}_{\leq t}\right\} = \\ &= W(s) + E\left\{[W(t) - W(s)]\Big|\mathcal{F}_{\leq t}\right\} = \\ &= W(s) + E\{W(t) - W(s)\} = W(s) \end{aligned}$$

since random variable $W(t) - W(s)$ is independent of all events of σ-algebra $\mathcal{F}[(0,t)]$.

In a similar way we come to the assertion that the Poisson process $N(t)$, $t \geq 0$, $N(0) = 0$, after subtracting its mean νt is a martingale.

Some basic properties of martingales are as follows:

a) Sample functions of a separable martingale have no discontinuities of the second kind; so, they have, at worst, jumps.

b) If $X(t)$ is a martingale, then $E[X(t)]$, $t \in T$ is constant; if $X(t)$ — supermartingale, then $E[X(t)]$ does not increase; if $X(t)$ — is submartingale, then $E[X(t)]$ does not decrease.

c) If $X_1(t)$ and $X_2(t)$ are two martingales with respect to the same family $\mathcal{F}_t$, then $a_1X_1(t) + a_2X_2(t)$ (where a_1 and a_2 are fixed constants) is a martingale, and in particular, $X(t) - X(t_o)$ is a martingale.

d) For every martingale $X(t)$, the process $|X(t)|^p$, $p \geq 1$ is a submartingale whenever $X(t)$ has moment of order p.

e) The convergence theorem: let $X(t)$ be a supermartingale (with respect to $\mathcal{F}_t$) continuous from the right. If

$$\lim_{t \to 0} E[X(t)] < \infty$$

then there exists with probability one $\lim_{t \to 0} X(t) = X_o$ and

$$\lim_{t \to 0} E|X(t) - X_o| = 0.$$

f) If $X(t)$ is a separable martingale defined on $[t_o, T]$ then for arbitrary $C > 0$

$$P\left\{\sup_{t_o \leq t \leq T} |X(t)| > C\right\} \leq \frac{1}{C} E|X(T)|$$

g) If for some $p > 1$, $E|X(t)|^p < \infty$ for all $t \in [t_o, T]$ then

$$E\left\{\sup_{t_o \leq t \leq T} |X(t)|^p\right\} \leq \left(\frac{p}{p-1}\right)^p E|X(T)|^p$$

Martingales can be considered as abstract models of "fair games". Indeed, if $X(t)$ characterizes a capital of a player at moment t_{n+1}, then in accordance with the definition of a **martingale**, the mean value of a capital at t_{n+1} provided the capital at moment t_n was $X(t_n)$ is equal to $X(t_n)$ independently of the state of capital in the previous moments of time.

It is also worth saying that during long time main interst in martingales centred around purely mathematical properties (like those presented above). It was not much interest in applications because the role played by martingales was primarily to facilitate proofs and not to facilitate calculations. The new interest in martingales among practically oriented people is due to a large body of recent results which have made martingale theory an important analytical tool in studying of random processes including those described by differential equations.

10. GENERALIZED STOCHASTIC PROCESSES; WHITE NOISE

In applications one often has to do with processes $X(t)$ whose values defined by $X(t)$ at given instants t_1 and t_2 are independent if $|t_1 - t_2| > \epsilon$, where ϵ is a very small number (much smaller than all time intervals of physical interest). For example, this is the case where the process describes the force acting on the particle in a Brownian motion. A natural mathematical idealization of such process are processes with independent values or purely random processes. An important example here is a Gaussian "white noise", which is commonly understood as a process whose correlation function is expressed by a Dirac delta. It turns out that in order to obtain a mathematically satisfactorily theory of the processes just described one has to introduce a concept of generalized stochastic process.

The concept of a generalzed stochastic process was defined independently by Itô and Gelfand and it constitutes an extension of the notion of a generalized function introduced by Schwartz.

Let $D(T)$ be the space of all infinitely differentiable functions $\varphi(t)$ on T to R_1 vanishing identically outside a finite closed interval; it is a space of test functions.

The topology (the kind of convergence) in $D(T)$ is introduced in the same way as in the case of ordinary Schwartz distributions . The space $D(T)$ is a topological vector space. Let Z be a Hilbert space of all P-equivalent random variables defined on $(\Gamma, \mathcal{F}, P)$ having finite second moment.

Definition 1.23. *A* generalized stochastic process *on T is a continuous linear mapping Φ from $D(T)$ to Z; the value of a generalized stochastic process Φ at φ will be denoted by $\{\varphi, \Phi\}$ or by $\Phi(\varphi)$.*

An example of the generalized random process is as follows. Let $X(t)$ be a stochastic process with finite second moment (a mapping from T to Z). As we shall see in the next chapter, for a very wide class of processes and for $\varphi(t) \in D(T)$ there exists the following stochastic integral

$$\{\varphi, \Phi_X\} = Y_\varphi(\gamma) = \int_S \varphi(t)X(t,\gamma)dt \qquad \gamma \in \Gamma \tag{1.196}$$

where S is an arbitrary compact set in T ($X(t)$ is a locally integrable stochastic process on T). It is easy to show that the mapping (1.196) is linear and continuous. Therefore, every locally integrable stochastic process on T determines a generalized stochastic process in the form (1.196).

One of the important advantages of a generalized stochastic process is the fact that its derivative always exists and is itself a generalized stochastic process .

Definition 1.24. *The derivative Φ' with respect to t of a generalized stochastic process Φ in $D(T)$ is defined by means of the formula*

$$\{\varphi, \Phi'\} = -\left\{\frac{d\varphi}{dt}, \Phi\right\} \qquad \text{for all} \quad \varphi \in D(T) \tag{1.197}$$

It can be verified that the mapping $\varphi \to \left\{\frac{d\varphi}{dt}, \Phi\right\}$ is a continuous linear mapping from $D(T)$ to Z. The derivatives $\Phi^{(\alpha)}$ for any order $\alpha \geq 1$ are given by

$$\left\{\varphi, \Phi^{(\alpha)}\right\} = (-1)^\alpha \left\{\frac{d^\alpha \varphi}{dt^\alpha}, \Phi\right\} \tag{1.198}$$

It is clear from (1.196) that if a generalized process is defined by (1.196), or in other words — if it has point values — then the appropriate definition of $\{\varphi, \Phi'_X\}$ is

$$\{\varphi, \Phi'_X\} = -\int_S \varphi'(t)X(t,\gamma)dt \quad , \qquad \gamma \in \Gamma \tag{1.199}$$

which justifies (1.197). From the above it follows that an ordinary process $X(t)$ may also be regarded as always differentiable, provided its derivative be a generalized process instead of an ordinary process .

From the standpoint of correlation theory, a generalized stochastic process Φ or $\Phi(\varphi)$ is characterized by its mean

$$m_\Phi(\varphi) = E[\Phi(\varphi)] \tag{1.200}$$

and correlation operator

$$K(\varphi,\psi) = E[\Phi(\varphi)\Phi(\psi)] \quad , \quad \varphi,\psi \in D(T) \tag{1.201}$$

which is assumed to be linear and continuous with respect to φ and ψ.

Definition 1.25. *A generalized stochastic process $\Phi(\varphi)$ is* weakly stationary *if it has a mean value $m_\Phi(\varphi)$ and a correlation operator $K(\varphi_1,\varphi_2)$ such that for every τ*

$$m_\Phi(\Lambda_\tau\varphi) = m_\Phi(\varphi) \tag{1.202}$$

$$K(\Lambda_\tau\varphi_1,\Lambda_\tau\varphi_2) = K(\varphi_1,\varphi_2) \tag{1.203}$$

where φ_1 and φ_2 are arbitrary elements of $D(T)$ and Λ_τ is the shift or transition operator defined by

$$\Lambda_\tau\varphi(t) = \varphi(t+\tau) \tag{1.204}$$

A typical example of a stationary generalized stochastic process is Gaussian white noise $\xi(t), t \in T$.

Let $W(t)$, $t \in [0,\infty)$ be a Wiener process according to its definition in sec. 7.2; we assume $W(t) \equiv 0$ for $t < 0$. The generalized derivative $W'(t)$ is well defined by means of formula (1.199).

Definition 1.26. A Gaussian white noise ξ *or* $\xi(\varphi)$ *is a generalized stochastic process such that there exists a Wiener process $W(t)$, $t \in T$ satisfying $\xi(t) = W'(t)$, i.e.*

$$\{\varphi,\xi\} = \{\varphi,W'\} = -\left\{\frac{d\varphi}{dt},W\right\} \qquad \textit{for all} \quad \varphi \in D(T) \tag{1.205}$$

Note that for every $\varphi \in D(T)$, $\{\varphi,\xi\}$ is a Gaussian random variable , and that for every finite number of functions $\varphi_1,\varphi_2,\ldots,\varphi_n$ in $D(T)$, the random variables $\{\varphi_i,\xi\}\,;1 \le i \le n$ are jointly Gaussian. This is the reason why ξ is called a white Gaussian process .

Theorem 1.13. *If ξ is a white Gaussian noise, then*

$$E[\xi(\varphi)] = 0 \qquad \textit{for all} \quad \varphi \in D(T) \tag{1.206}$$

$$K_\xi(\varphi_1,\varphi_2) = \int_T \varphi_1(t)\varphi_2(t)dt, \qquad \varphi_1,\varphi_2 \in D(T) \tag{1.207}$$

Proof:

$$\begin{aligned} E\left[\xi(\varphi)\right] &= E\left[\{\varphi,\xi\}\right] \\ &= E\left[-\int_T \frac{d\varphi}{dt} W(t)dt\right] \\ &= -\int_T \frac{d\varphi}{dt} E\left[W(t)\right] = 0 \end{aligned}$$

since $E\left[W(t)\right] = 0$ for all $t \in T$ and φ is arbitrary. To show the second equality, let $\varphi_1, \varphi_2 \in D(T)$. Then

$$\begin{aligned} K_\xi(\varphi_1,\varphi_2) &= E\left[\{\varphi_1,\xi\},\{\varphi_2,\xi\}\right] \\ &= E\left[\int_T \frac{d\varphi_1}{dt} W(t_1)dt_1 \int_T \frac{d\varphi_2}{dt} W(t_2)dt_2\right] \\ &= \int_T \int_T \frac{d\varphi_1}{dt_1} E\left[W(t_1)W(t_2)\right] \frac{d\varphi_2}{dt_2} dt_1 dt_2 \\ &= \int_T \int_T \min(t_1,t_2) \frac{d\varphi_1}{dt_1} \frac{d\varphi_2}{dt_2} dt_1 dt_2 \end{aligned}$$

After integrating by parts we obtain the desired equality.

It is worth noting that formula (1.207) can be put in the form

$$K_\xi(\varphi_1,\varphi_2) = \int_T \int_T \delta(t-s)\varphi_1(t)\varphi_2(t)dsdt \tag{1.208}$$

what implies that the correlation function of the Gaussian white noise is the generalized function $\delta(t-s)$.

Remark 1. The n-dimensional Gaussian white noise can be defined analogously, i.e. as the generalized derivative of the n-dimensional Wiener process . It is simply a collection of n independent one-dimensional Gaussian white noises.

Remark 2. In applications a Gaussian white noise is often understood as a stochastic process characterizing the limiting properties ($\alpha \to \infty$) of Gaussian and stationary processes with the correlation function

$$K(\tau) = e^{-\alpha|\tau|}.$$

More often, when spectral description of a stationary process is used, a (Gaussian) white noise is visualized as a stationary process with constant spectral density on the entire real axis, i.e. $g(\omega) = g_0, \quad \omega \in (-\infty,+\infty)$. The term "white" is introduced to reflect the analogy with the "white light" in optics which has the property that its spectral composition is uniform over the visible portion of the electromagnetic spectrum.

11. PROCESSES WITH VALUES IN HILBERT SPACE

In this section we wish to describe shortly another generalization of the concept of a stochastic process which has turned out to be very suitable in building a consistent theory of a wide class of stochastic differential equations. The generalization which we have in mind is a stochastic process which takes its values (not, as usually, in R_1 or R_n) in some abstract space, for instance in a Banach space or — in particular — in a Hilbert space.

To be systematic we shall first introduce the concept of a generalized random variable .

Let $(\Gamma, \mathcal{F}, P)$ be the basic probability space and let $\mathcal{X}$ be a measurable metric space with the σ-algebra $\mathcal{F}_1$ of measurable subsets of $\mathcal{X}$.

Definition 1.27. *A measurable mapping $X(\gamma) : \Gamma \to \mathcal{X}$, that is a mapping such that the inverse image (under this mapping) of every measurable set $B \in \mathcal{F}_1$ belongs to $\mathcal{F}$, is called a $\mathcal{X}$-*valued random variable (random element *or* generalized random variable).

The above definition implies for every $B \in \mathcal{F}_1$

$$X^{-1}(B) = \{\gamma : X(\gamma) \in B\} \in \mathcal{F}$$

It is clear that in the case when $\mathcal{X} = R_1$ or $\mathcal{X} = R_n$ a generalized random variable is simply a one-dimensional and an n-dimensional random variable , respectively.

As for real valued random variables , the generalized random variable $X(\gamma) : \Gamma \to \mathcal{X}$ induces a probability measure P_X on $\mathcal{F}_1$ in such a way that:

$$P_X(A) = P\{\gamma : X(\gamma) \in A\} \qquad \text{for} \quad A \in \mathcal{F}_1$$

and $(\mathcal{X}, \mathcal{F}_1, P_X)$ is a probability space (generated by $X(\gamma)$).

Theorem 1.14. *A mapping $X(\gamma) : \Gamma \to \mathcal{X}$ is a generalized random variable if and only if for every functional g defined on $(\mathcal{X}, \mathcal{F}_1)$the mapping $g(X(\gamma))$ is an ordinary random variable .*

Definition 1.28. *An $\mathcal{X}$-valued random variable $X(\gamma)$ is said to be* Gaussian *if for each functional g defined on $\mathcal{X}$, $g(X(\gamma))$ is a real Gaussian random variable.*

Let $\mathcal{X}$ be a Banach space.

Definition 1.29. *A* mean *or* average *of $X(\gamma)$ is the following Bochner integral (if it exists)*

$$m_X = E[X(\gamma)] = \int_\Gamma X(\gamma) dP(\gamma) \tag{1.209}$$

If the above integral exists we say that $X(\gamma)$ belongs to $L_1(\Gamma, \mathcal{F}, P; \mathcal{X})$.

Moments of higher order of a generalized random variable are defined analogously to those in the classical theory; the difference is that the Lebesgue integral is replaced by the Bochner integral (cf. [20]) and the meaning of multiplication of elements has to be correctly defined.

Most of abstract spaces needed in applied problems are Hilbert spaces. In this case we can define the covariance operator for $\mathcal{X} = H$-valued random variables (cf. [7]).

Definition 1.30. *The covariance operator of Hilbert-valued random variables X, $X \in L^2(\Gamma, \mathcal{F}, P; H)$ is defined as*

$$Cov\{X\} = E\{[X - m_X] \circ [X - m_X]\} \tag{1.210}$$

where $u \circ v$, a bounded linear operator on H, is defined for all $u, v, \in H$ by

$$[u \circ v]\,\lambda = u(v, \lambda) \qquad \lambda \in H$$

and $(,)$ denotes an inner product of H.

The covariance operator $(\cdot)$ is symmetric and positive.

The following identities hold

$$\begin{aligned} E\{(X, X)\} &= \text{trace} \quad \text{Cov}\{X\} \\ E\{(AX, X)\} &= \text{trace} \quad \{A\ \text{Cov}\{X\}\} \end{aligned} \tag{1.211}$$

where A is a bounded linear operator on Hilbert space H.

Hilbert space random variables are uniquely specified by their characteristic functionals.

Definition 1.31. *If $X(\gamma)$ is a H-valued random variable (with induced probability measure $P_X(\nu)$ then its characteristic functional $\Phi_X : H \rightarrow R_1$ is defined for all $\lambda \in H$, by*

$$\begin{aligned} \Phi_X(\lambda) &= E\{exp[i(X, \lambda)]\} = \\ &= \int_H exp[i(x, \lambda)]\, dP_X(x) \end{aligned} \tag{1.212}$$

For H-valued Gaussian random variables (or, in other words, for the Gaussian measures on H) the characteristic functional takes an explicit form given by the following theorem

Theorem 1.15. *An H-valued random variable $X(\gamma)$ (or, a measure ν on H) is Gaussian if and only if its characteristic functional $\Phi_X(\lambda)$ has the form*

$$\Phi_X(\lambda) = exp\left\{i(\alpha, \lambda) - \frac{1}{2}(Q\lambda, \lambda)\right\} \tag{1.213}$$

where $\alpha \in H$ and Q is non-negative, self adjoint nuclear operator on H. The random variable $X(\gamma)$ has mean α and covariance operator Q.

Another useful property of Gaussian random variables is the following.

Let $X(\gamma)$ be a Gaussian H-valued random variable with zero mean and covariance Q. Then

$$E\left\{\|X\|^{2n}\right\} \leq (2n-1)!!(\text{trace } Q)^n \tag{1.214}$$

for any integer n, where equality holds for $n = 1$ and, as usual, $(2n-1)!! = 1 \cdot 3 \cdot \ldots \cdot (2n-1)$.

There are many possible types of convergence of $\mathcal{X}$-valued random variables . One of them is as follows.

Definition 1.32. *A sequence $\{X_n(\gamma)\}$ of $\mathcal{X}$-valued random variables converges to $\mathcal{X}$-valued random variable $X(\gamma)$ in the mean square sense if*

$$E\left[\|X_n(\gamma) - X(\gamma)\|^2\right] \longrightarrow 0 \quad \text{as} \quad n \to \infty \tag{1.215}$$

where $\|\cdot\|$ denotes a norm in $\mathcal{X}$.

Definition 1.33 *An $\mathcal{X}$-valued stochastic process $X(t,\gamma)$ on T is a mapping: $T \times \Gamma \to \mathcal{X}$, which is measurable with respect to the product measure $\mu \times P$, where μ is the Lebesgue measure on T.*

A Hilbert-valued Wiener process can be defined in various ways. A straightforward definition may be obtained by extending the classical difinition to a Hilbert space. Here we provide a definition which appears to be more useful in applications.

Definition 1.34. A H-valued Wiener process *$W(t)$ is defined as*

$$W(t,\gamma) = \sum_{i=0}^{\infty} \beta_i(t,\gamma) e_i \tag{1.216}$$

for almost all (t,γ), where $\beta_i(t,\gamma)$ are real independent Wiener processes with covariance $(t-s)\lambda_i$, $\{e_i\}$ is a orthonormal basis in H and $\sum_{i=1}^{\infty} \lambda_i < \infty$.

It can be shown that $W(t,\gamma)$ is Gaussian and

$$E\left[\|W(t,\gamma) - W(s,\gamma)\|^2\right] = \sum_{i=0}^{\infty} \lambda_i(t-s) \tag{1.217}$$

Other concepts associated with stochastic processes such as stationarity, Markov and martingale properties can be extended to H-valued stochastic processes (cf. [7]).

Remark: The defintion 1.33 implies that for each fixed t, $X(t,\gamma)$ is a generalized random variable. So, the states of the process are elements of $\mathcal{X}$; for instance — they can be functions of a spatial variable , as in the case where a process is governed by a partial differential equation .

12. STOCHASTIC OPERATORS

An interesting and effective approach to interpretation and analysis of a wide class of stochastic equations is connected with the theory of generalized random variables and stochastic (or random) operators. This theory was initiated in the fifties by E. Mourier, O. Hans, A. Spacek, and now it is known as a probabilistic functional analysis (cf. [2], [18]). Here we give the very basic concepts of the theory.

Let $\mathcal{Y}$ be a Banach space and $\mathcal{C}$ — the σ-algebra of its subsets. Let A be an operator on $\mathcal{X}$ into $\mathcal{Y}$ with domain $\mathcal{D}(A)$.

Definition 1.35. *A mapping $A : \Gamma \times \mathcal{D}(A) \to \mathcal{Y}$ is said to be a* stochastic (or random) operator *if $A(\gamma)x = y(\gamma)$ is a $\mathcal{Y}$-valued random variable for every $x \in \mathcal{D}(A)$, that is, for every $x \in \mathcal{X}$ and every $B \in \mathcal{C}$*

$$\{\gamma : A(\gamma)x \in B\} \in \mathcal{F}$$

A stochastic operator $A(\gamma)$ is said to be *linear* if $A(\gamma)[\alpha x_1 + \beta x_2] = \alpha A(\gamma)x_1 + \beta A(\gamma)x_2$ for every $\gamma \in \Gamma$, $x_1, x_2 \in \mathcal{X}$ and $\alpha, \beta \in R_1$.

A stochastic operator $A(\gamma)$ is *bounded* if there exists a non-negative real-valued random variable $C(\gamma)$ such that for almost all $\gamma \in \Gamma$ and all $x \in \mathcal{D}(A)$

$$\|A(\gamma)\| \le C(\gamma)\|x\| \tag{1.218}$$

A mapping $S(\gamma)$ is called the *inverse* of $A(\gamma)$ if

$$P\{\gamma : A(\gamma)[S(\gamma)x] = x \qquad \text{for every} \quad x \in \mathcal{X}\} = 1$$

Definition 1.36. *A stochastic operator $A(\gamma) : \Gamma \times \mathcal{X} \to \mathcal{X}$ is said to be a* stochastic contraction operator *if there exists a real-valued non-negative random variable $C(\gamma)$ such that for almost all $\gamma \in \Gamma$, $C(\gamma) < 1$ and*

$$\|A(\gamma)x_1 - A(\gamma)x_2\| \le C(\gamma)\|x_1 - x_2\| \tag{1.219}$$

for every $x_1, x_2 \in \mathcal{X}$ and for almost all $\gamma \in \Gamma$.

The examples below illustrate the concept of a stochastic operator.

1) Let $A = \{a_{ij}\}$ be an $n \times n$ matrix such that $a_{ij} = a_{ij}(\gamma)$, $\gamma \in \Gamma$ are random variables . Therefore, $A = A(\gamma)$ is a stochastic operator which maps $\Gamma \times R_n$ in R_n; in other words, $A(\gamma)$ is a stochastic matrix.
2) Let $Au = A_0 \frac{\partial^2 u}{\partial x^2}$ where u belongs to the set C_2 of continuous functions with continuous second order derivatives on $(-\infty, +\infty)$ and $A_0 = A(\gamma)$ is a random variable . $A = A(\gamma)$ is a stochastic operator which maps $\Gamma \times C_2$ in C. Analogously, the more general differential operator with random coefficients

$$A(\gamma)u(t) = \sum_{i=0}^{n} A_i(\gamma)\frac{d^i u(t)}{dt^i}$$

where $A_i(\gamma)$ are random variables can also be understood as a stochastic operator which maps, $\Gamma \times C_n$ in C where C_n is a space of continuous functions with continuous derivatives of n-th order.

3) Let $Ax = \int_a^b K(t,s)x(s)ds$ be an integral operator in which $K(t,s) = K(t,s;\gamma)$, $\gamma \in \Gamma$, is a stochastic process . Then $A(\gamma)$ is a stochastic integral operator which maps $\Gamma \times C[a,b]$ in $C[a,b]$.

EXAMPLES

In order to illustrate the concepts presented we shall consider some specific problems.

1. Let us consider two stochastic processes defined on $T = [0,1]$ in the following way

$$\begin{aligned} X(t,\gamma) &\equiv 0 \\ Y(t,\gamma) &= 0 \quad \text{if} \quad t \neq \tau(\gamma) \\ &= 1 \quad \text{if} \quad t = \tau(\gamma) \end{aligned} \tag{E.1}$$

where $\tau(\gamma)$ is a random variable with uniform distribution on $[0,1]$.

These two stochastic processes are equivalent since $P\{X(t,\gamma) \neq Y(t,\gamma)\} = P\{\tau(\gamma) = t\} = 0$; however, their sample functions are different: sample functions of $X(t,\gamma)$ are everywhere equal to zero, but sample functions of $Y(t,\gamma)$ are discontinuous with jumps at $t = \tau$. Finite dimensional distributions of $X(t,\gamma)$ and $Y(t,\gamma)$ are the same, namely

$$F_{t_1,\ldots,t_n}(x_1,\ldots,x_n) = \begin{cases} 1 & \text{for all} \quad x_j \geq 0 \\ 0 & \text{otherwise} \end{cases}$$

However,

$$\begin{aligned} &P\{X(t,\gamma) < 1 \quad \text{for all} \quad t\} = 1, \\ &P\{Y(t,\gamma) < 1 \quad \text{for all} \quad t\} = 0. \end{aligned} \tag{E.2}$$

2. Let $X(t)$ be a real Gaussian stationary process with mean zero and covariance function $K_X(\tau)$. Let

$$X(t) = A(t)\cos\Phi(t) \tag{E.3}$$

that is, let us treat $X(t)$ for each t as a projection on axis x of vector A forming the angle Φ with axis x. To specify uniquely process $X(t)$ let us introduce the second process

$$Y(t) = A(t)\sin\Phi(t) \tag{E.4}$$

such that

$$K_Y(\tau) = K_X(\tau) \quad , \qquad R_{XY}(0) = 0$$

So, $X(t)$ and $Y(t)$ can be interpreted as coordinates of a random point of Euclidean plane. It is clear that

$$P\{a \le A \le a + da\} =$$
$$= \int\int_{a \le \sqrt{x^2+y^2} \le a+da} \tilde{f}(x,y)dxdy$$

where

$$\tilde{f}(x,y) = \frac{1}{2\pi\sigma^2} e^{-(x^2+y^2)/2\pi\sigma^2}$$

Introducing the polar coordinates (a,φ) we have

$$f(a)da = \int_a^{a+da} \int_0^{2\pi} f(a,\varphi)d\varphi da$$

where

$$f(a,\varphi) = \frac{a}{2\pi\sigma^2} e^{-\frac{a^2}{2\sigma^2}}$$

The above formula implies that $A(t)$ and $\Phi(t)$ are for each fixed t independent and they have the following distributions

$$f(a) = \frac{a}{\sigma^2} e^{-\frac{a^2}{2\sigma^2}} \quad , \qquad a \ge 0$$

$$f(\varphi) = \frac{1}{2\pi} \quad , \qquad 0 \le \varphi \le 2\pi \tag{E.5}$$

It is clear that

$$|X(t)| \le A(t) \quad , \quad A(t) = [X(t) + Y(t)]^{\frac{1}{2}} .$$

Stochastic process $A(t)$ is often called the *envelope* of $X(t)$. Sample functions of $A(t)$ play the role of envelopes of sample functions $X(t)$.

For more detailed study of envelope process , the two-dimensional distribution density $f(a_1, a_2)$ of the process $A(t)$ is often necessary (i.e. the distribution of $[A(t), A(t+\tau)]$). After some transformations we get the formula (cf. [6])

$$f(a_1, a_2) = \frac{a_1 a_2}{\sigma^4 p^2} e^{-\frac{a_1^2 + a_2^2}{2\sigma^2 p}} \; I_0\left(\frac{a_1 a_2 \sqrt{1-p^2}}{\sigma^2 p^2}\right) \tag{E.6}$$

where $a_1 > 0, \quad a_2 > 0$, and

$$\sigma^2 = K_X(0) = K_Y(0) \quad , \quad p^2 = 1 - k^2(\tau) - r^2(\tau)$$
$$k(\tau) = \frac{1}{\sigma^2} K_X(\tau) = \frac{1}{\sigma^2} K_Y(\tau) \quad , \quad r(\tau) = \frac{1}{\sigma^2} R_{XY}(\tau)$$

and $I_0(z)$ is the Bessel function of zero order of imaginary argument.

3. Let $X(t)$ be a real, Gaussian and stationary process with mean equal to zero. Process of the form ("a periodic signal in noise")

$$\begin{aligned} Y(t) &= R\cos(\omega t + \varphi) + X(t) = \\ &= A\cos\omega t + B\sin\omega t + X(t) \end{aligned} \tag{E.7}$$

has different probabilistic meaning depending on the interpretation of R, ω and φ.

a) Let A, B be independent Gaussian random variables with zero means and equal variances and ω-constant. This means that R and φ are independent random variables , R having a Raleigh distribution and φ being uniformly distributed over $(0, 2\pi)$. In this case $Y(t)$ is a stationary and Gaussian.

b) Let R, ω and φ be deterministic constants. The resulting process $Y(t)$ is Gaussian but nonstationary .

c) Let R and ω be constants, but the phase φ is a random variable uniformly distributed over $(0, 2\pi)$ independent of $X(t)$. The process $Y(t)$ is stationary but no longer Gaussian.

4. Let us consider a process $X(t)$ defined by

$$X(t) = X_0(-1)^{N(t)} \tag{E.8}$$

where X_o is a random variable assuming two values: 1 and -1 and $P(X_o = 1) = P(X_o = -1) = \frac{1}{2}$; $N(t)$ is a Poisson process (independent of X_o) with intensity μ describing the times in which the process changes the values $\{1, -1\}$. Process $X(t)$ is commonly termed as a *random telegraph signal*.

It is clear that

$$E[X(t)] = E(X_o)\, E\left[(-1)^{N(t)}\right]$$
$$E[X(t_1)X(t_2)] = E(X_o^2)E\left[(-1)^{N(t_1)}(-1)^{N(t_2)}\right]$$

Let us denote $Z(t) = (-1)^{N(t)}$. To calculate the above moments we have to find the probabilities $P\{Z(t) = 1\}$ and $P\{Z(t) = -1\}$. Of course,

$$P\{Z(t) = 1\} = P\{\text{even number of changes in } [0,t]\} =$$
$$= e^{-\mu t} \sum_{k=0}^{\infty} \frac{(\mu t)^{2k}}{(2k)!} = e^{-\mu t} \cos h\mu t;$$

$$P\{Z(t) = -1\} = P\{\text{odd number of changes in } [0,t]\} =$$
$$= e^{-\mu t} \sum_{k=0}^{\infty} \frac{(\mu t)^{2k+1}}{(2k+1)} = e^{-\mu t} \sin h\mu t;$$

Therefore,

$$E[Z(t)] = P\{Z(t) = 1\} + (-1)P\{Z(t) = -1\} =$$
$$= e^{-\mu t}[\cos h\mu t - \sin h\mu t] = e^{-2\mu t}$$
$$m_X(t) = E[X(t)] = E(X_o)e^{-2\mu t} = 0;$$

$$E[Z(t_1)Z(t_2)] = P\{Z(t_1) = 1 = Z(t_2) \text{or} Z(t_1) = -1 = Z(t_2)\}$$
$$+ (-1)\{Z(t_1), Z(t_2) \text{are of opposite signs}\}$$
$$= P\{Z(t_1) = 1 = Z(t_2)\} + P\{Z(t_1) = -1 = Z(t_2)\}$$
$$- P\{Z(t_1) = 1 = -Z(t_2)\} - P\{Z(t_1) = -1 = -Z(t_2)\}$$

Let us assume that $t_1 < t_2$ (opposite case is treated similarly). Then

$$P\{Z(t_1) = 1 = Z(t_2)\} = P\{Z(t_1) = 1\}\, P\{Z(t_2) = 1 | Z(t_1) = 1\} =$$
$$= e^{-\mu t_1} \cos \mu t_1 P\{\text{even number of changes in} \quad (t_1, t_2)\} =$$
$$= e^{\mu t_1} \cos h\mu t_1 e^{-\mu(t_2 - t_1)} \cos h\mu(t_2 - t_1).$$

Similarly

$$P\{Z(t_1) = -1 = Z(t_2)\} = e^{-\mu t_1} \sin ht_1 e^{-\mu(t_2-t_1)} \cos h\mu(t_2 - t_1)$$
$$P\{Z(t_1) = -1 = -Z(t_2)\} = e^{-\mu t_1} \sin ht_1 e^{-\mu(t_2-t_1)} \sin h\mu(t_2 - t_1)$$
$$P\{Z(t_1) = 1 = -Z(t_2)\} = e^{-\mu t_1} \cos ht_1 e^{-\mu(t_2-t_1)} \sin h\mu(t_2 - t_1)$$

Now

$$E[Z(t_1)Z(t_2)] = e^{-2\mu(t_2-t_1)}, \qquad t_1 < t_2$$

and finally (since $E(X_\circ^2) = 1$)

$$K_X(t_1,t_2) = E[X(t_1)X(t_2)] = e^{-2\mu|t_2-t_1|} = e^{-2\mu|\tau|} \tag{E.9}$$

where $\tau = t_2 - t_1$. Therefore, process $X(t)$ is stationary in the wide sense.

5. Let $X(t)$ be a stochastic process defined as (random harmonic oscillation)

$$X(t) = Ae^{i\Omega t} \tag{E.10}$$

where A and Ω are independent random variables such that $E(A) = 0$, $\sigma_A^2 = D$ and Ω has a Cauchy distribution , that is

$$f(\omega) = \frac{\alpha}{\pi}\frac{1}{\alpha^2 + \omega^2}$$

Of course, $m_X(t) = 0$ and

$$K_X(t_1,t_2) = E[X(t_1)X(t_2)] = E(A^2)E\left[e^{i\Omega(t_1-t_2)}\right]$$

$$= \frac{D\alpha}{\pi}\int_{-\infty}^{+\infty}\frac{e^{i\omega(t_1-t_2)}}{\alpha^2+\omega^2}d\omega = De^{-\alpha|t_1-t_2|} = De^{-\alpha|\tau|} \tag{E.11}$$

The examples 4 and 5 indicate clearly that the same function can serve as a correlation function of essentially different stochastic processes. Realizations of the process in Example 4 are stepwise functions with constant values between jump instants, whereas realizations of the process in Example 5 are continuous (random oscillation with random amplitude, frequency and phase). It is also worth noting that a rate of decrease of the exponential function in Ex. 4 is determined by the intensity of Poissonian impulses, whereas in Ex. 5 this decrease is governed by the parameter α of the Cauchy distribution of random frequency.

6. As we have said in Sec. 10 a white noise is usually understood in applications as a stationary process $\xi(t), -\infty < t < +\infty$ with mean $E(\xi) = 0$ and with constant

spectral density $g_\xi(\omega) = g_o$ on the entire real axis. If $E[\xi(t)\xi(t+\tau)] = K_\xi(\tau)$ is the correlation function of $\xi(t)$, then

$$g_\xi(\omega) = \frac{1}{2\pi}\int_{-\infty}^{+\infty} K_\xi(\tau)e^{-i\omega\tau}d\tau = g_o = \nu\frac{1}{2\pi} \quad , \quad -\infty < \omega < +\infty \tag{E.12}$$

Such a process does not exist in the classical sense since (E.12) is compatible only with the following correlation function

$$K_\xi(\tau) = \nu\delta(\tau)$$

A constant ν is called the intensity of a stationary white noise. From (E.12) we also have

$$K_\xi(0) = \sigma_\xi^2 = \int_{-\infty}^{+\infty} g_\xi(\omega)d\omega = \infty$$

what is very unpleasent. In such a situation it is tempting to introduce other processes which have essentially constant spectral densities but finite variance. There are many ways in which this could be done. One alternative is the so called *band limited white noise* characterized by the following spectral density

$$g(\omega) = \begin{cases} c & , \quad |\omega| < \omega^* \\ 0 & , \quad |\omega| \geq \omega^* \end{cases} \tag{E.13}$$

Another probability is concerned with processes like these considered in Ex. 3 and Ex. 4. We mean the ordinary stationary processes with correlation function

$$K(\tau) = ae^{-b|\tau|}$$

Such a process has spectral density

$$g(\omega) = \frac{1}{\pi}\frac{ab}{b^2+\omega^2}$$

If we now let a and b tend to ∞ in such a way that $\frac{a}{b} \to \frac{1}{2}$ then we obtain

$$g(\omega) \longrightarrow \frac{1}{2\pi}$$

It is seen that a white noise can be approximated by essentially different physical processes. For the random telegraph signal considered in Ex. 4 a limit should be taken with respect to the intensity of Poissonian events ($\mu \to \infty$) whereas in Ex. 5 parameter α of the Cauchy distribution of random frequency should increase to infinity.

In spite of the difficulties which will arise due to the infinite variance, the concept of white noise is very important in theory and applications of stochastic processes.

The use of white noise in stochastic analysis resembles in many ways the use of the Dirac delta function in deterministic theory of linear systems.

7. In solving practical problems it is useful to replace (or approximate) a stochastic process by white noise. It is clear that this is justified only in situations where a correlation time of the process considered is very small. In such cases a stationary stochastic process $X(t)$ is often "approximated" by a white noise $\xi(t)$ with intensity

$$\nu = \int_{-\infty}^{+\infty} K_X(\tau)d\tau \tag{E.14}$$

Of course, the concept of a correlation time is not mathematically rigorous. It is simply a practical measure of the length of time within which the values of a stochastic process show significant correlation. Practically, a *correlation time* of a stationary process is estimated by the quantity

$$\tau_c = \frac{1}{2}\int_{-\infty}^{+\infty} \frac{K_X(\tau)}{\sigma_X^2} d\tau \tag{E.15}$$

For example, the correlation time of the process of Example 5 (i.e. $K_X(\tau) = De^{-\alpha|\tau|}$) is equal to $\frac{1}{\alpha}$, whereas the intensity ν of a corresponding white noise (for large α) is: $\nu = \frac{2D}{\alpha}$.

8. Let $X(t)$ be a Gaussian stationary stochastic process with mean m_X and correlation function $K_X(\tau)$. Determine the covariance function and spectral density of the process

$$Y(t) = X^2(t)$$

Let us consider first the case where $m_X = 0$. Since the above transformation is nonlinear, the process $Y(t)$ is not Gaussian; however $Y(t)$ like $X(t)$ is weakly stationary. Let

$$X(t) = X_1 \quad , \quad X(t+\tau) = X_2$$

One obtains

$$\begin{aligned} K_Y(\tau) &= E\left\{\left[X_1^2 - E(X_1^2)\right]\left[X_2^2 - E(X_2^2)\right]\right\} = \\ &= E\left[X_1^2 X_2^2\right] - \left[K_X(0)\right]^2 \end{aligned}$$

Since X_1 and X_2 are Gaussian random variables , their joint characteristic function is

$$\varphi(\lambda_1, \lambda_2) = exp\left\{-\frac{1}{2}(k_{11}\lambda_1^2 + k_{22}\lambda_2^2 + 2k_{12}\lambda_1\lambda_2)\right\}$$

where

$$k_{11} = k_{22} = K_X(0) \quad , \quad k_{12} = K_X(\tau)$$

Therefore

$$E\left[X_1^2 X_2^2\right] = \left.\frac{\partial^4}{\partial\lambda_1^2\partial\lambda_2^2}\varphi(\lambda_1, \lambda_2)\right|_{\lambda_1=\lambda_2=0}$$
$$= 2k_{12}^2 + k_{11}k_{12} = 2\left[K_X(\tau)\right]^2 + \left[K_X(0)\right]^2$$

Finally, we have the result

(E.16) $$K_Y(\tau) = 2\left[K_X(\tau\right]^2$$

Applying the Fourier transform to the last formula we get, after simple transformations, the following expression for the spectral density of $Y(t)$:

(E.17) $$g_Y(\omega) = 2\int_{-\infty}^{+\infty} g_X(\omega_1)g_X(\omega - \omega_1)d\omega_1$$

Let $m_X \neq 0$. Since $X^2(t)$ can be represented as

(*) $$X^2(t) = \left[X(t) - m_X\right]^2 + 2m_X\left[X(t) - m_X\right] + m_X^2$$

where process $X(t) - m_X$ has zero mean and hence $\left[X(t) - m_X\right]^2$ has the spectral density given by (E.17). Since for Gaussian processes $X(t) - m_X$ and $\left[X(t) - m_X\right]^2$ are uncorrelated the spectral density of $X^2(t)$ is equal to the sum of the spectral densities of the two first components in (*), that is

(E.18) $$g_{X^2}(\omega) = 2\int_{-\infty}^{+\infty} g_X(\omega_1)g_X(\omega - \omega_1)d\omega_1 + 4m_X^2 g_X(\omega)$$

9. Application of formulae (1.71), (1.71′) leads to the following mutual relationships between the most common correlation functions and the corresponding spectral densities

a) $K_X(\tau)$

$$= c\delta(\tau) \longleftrightarrow g_x(\omega) = \frac{c}{2\pi}$$

b) $K_X(\tau)$

$$= \sigma^2 \frac{\sin b\tau}{b\tau}, \quad b > 0 \longleftrightarrow g_x(\omega) = \begin{cases} \frac{\sigma^2}{2b} & , |\omega| < b \\ 0 & , |\omega| > b \end{cases}$$

c) $K_X(\tau)$

$$= \sigma^2 e^{-\alpha|\tau|} \quad , \alpha > 0 \longleftrightarrow g_x(\omega) = \frac{\alpha^2}{\pi} \frac{\alpha}{\omega^2 + \alpha^2}$$

d) $K_X(\tau)$

$$= \sigma^2 e^{-\alpha|\tau|} \cos \beta\tau \longleftrightarrow g_x(\omega) = \frac{\alpha\sigma^2}{\pi} \frac{\omega^2 + \alpha^2 + \beta^2}{(\omega^2 + \alpha^2 - \beta^2)^2 + 4\alpha^2\omega^2}$$

e) $K_X(\tau)$

$$= \sigma^2 e^{-\alpha|\tau|} (\cos \beta\tau + \frac{\alpha}{\beta} \sin \beta|\tau|) \longleftrightarrow g_x(\omega) = \frac{2\sigma^2\alpha}{\pi} \frac{\omega^2 + \alpha^2 + \beta^2}{(\omega^2 + \alpha^2 - \beta^2)^2 + 4\alpha^2\omega^2}$$

f) $K_X(\tau)$

$$= \sigma^2 e^{-\alpha|\tau|} (1 + \alpha|\tau|) \longleftrightarrow g_x(\omega) = \frac{1}{\pi} \frac{2\sigma^2\alpha^2}{(\omega^2 + \alpha^2)^2}$$

g) $K_X(\tau)$

$$= \sigma^2 e^{-\alpha^2\tau^2} \longleftrightarrow g_x(\omega) = \frac{\sigma^2}{2\alpha\sqrt{\pi}} e^{-\frac{\omega^2}{2\alpha^2}}$$

h) $K_X(\tau)$

$$= \sigma^2 e^{-\alpha^2\tau^2} \cos \beta\tau \longleftrightarrow g_x(\omega) = \frac{\sigma^2}{4\alpha\sqrt{\pi}} \left[e^{-\frac{(\omega+\beta)^2}{4\alpha^2}} + e^{-\frac{(\omega-\beta)^2}{4\alpha^2}} \right]$$

i) $K_X(\tau)$

$$= \sigma^2 J_o(a\tau), a > 0 (*) \longleftrightarrow g_x(\omega) = \frac{\sigma^2}{\pi\sqrt{a^2 - \omega^2}}, \quad |\omega| < a$$

$$= 0, \quad |\omega| > a$$

j) $K_X(\tau)$

$$= B^2 \cos b\tau \longleftrightarrow g_x(\omega) = \frac{B^2}{2} [\delta(\omega - b) + \delta(\omega + b)] .$$

(*) $J_o(\alpha\tau)$ is the Bessel function of zero order.

10. Let $W(t)$ be a standard Wiener process and let $a = t_o < t_1 < \cdots < t_n = b$ be an arbitrary partition of interval $[a, b]$ and $\Delta = \max(t_{i+1} - t_i)$. Then

$$\text{l.i.m.} \sum_{i=0}^{n-1} [W(t_{i+1}) - W(t_i)]^2 = b - a \tag{E.19}$$

To prove the above assertion we make use of the property of independence of increments of a Wiener process on non-overlapping intervals and formula (1.166). Explicit reasoning is as follows (var — denotes variance)

$$E\left\{\sum_{i=0}^{n-1} [W(t_{i+1}) - W(t_i)]^2\right\} = \sum_{i=0}^{n-1} \left\{E\,[W(t_{i+1}) - W(t_i)]^2\right\}$$

$$= \sum_{i=0}^{n-1} (t_{i+1} - t_i) = b - a;$$

$$\text{var}\left\{\sum_{i=0}^{n-1} [W(t_{i+1}) - W(t_i)]^2\right\} =$$

$$= E\left\{\sum_{i=0}^{n-1} [W(t_{i+1}) - W(t_i)]^2 - (a - b)\right\}^2$$

$$= \sum_{i=0}^{n-1} E\left\{[W(t_{i+1}) - W(t_i)]^2 - (t_{i+1} - t_i)\right\}^2$$

$$= \sum_{i=0}^{n-1} E\left\{[W(t_{i+1}) - W(t_i)]^4 - 2(t_{i+1} - t_i)\,[W(t_{i+1}) - W(t_i)]^2 + (t_{i+1} - t_i)^2\right\}$$

$$= \sum_{i=0}^{n-1} \{3(t_{i+1} - t_i)^2 - 2(t_{i+1} - t_i)^2 + (t_{i+1} - t_i)^2\}$$

$$= 2\sum_{i=0}^{n-1} (t_{i+1} - t_i)^2 \leq 2\max(t_{i+1} - t_i) \sum_{i=0}^{n-1} (t_{i+1} - t_i) = 2\Delta(b - a) \underset{\Delta \to 0}{\longrightarrow} 0.$$

11. Let us consider the Wiener process $W(t)$ with drift $a(x) = \mu$ and diffusion coefficient $b(x) = \sigma^2$ and suppose that there exists an absorbing barrier at $x = a$.

The transition density function of $W(t)$ is governed by the following initial-boundary value problem

$$\frac{\partial p}{\partial t} = \frac{1}{2}\sigma^2 \frac{\partial^2 p}{\partial x^2} - \mu \frac{\partial p}{\partial x}, \quad x < a$$
$$p(x,0) = \delta(x)$$
$$p(a,t) = 0, \quad t > 0 \tag{E.20}$$

What is the solution for $p(t_o = 0, x_o; x, t)$? Let us adopt a reasoning in [5].

As we know, the solution for the unrestricted process starting from x_o is given by formula (1.129). Let us look for the solution of (E.20) in the form of the linear combination

$$p(x,t) = p_{\text{ur}}(x,t) + Ap_{\text{ur}}(x,t) \tag{*}$$

where $p_{\text{ur}}(x,t)$ is the solution (1.129) for unrestricted process. The above combination satisfies the equation and initial condition. One should find now the appropriate values of A and x_o to make the combination (*) satisfy also the boundary condition. To do that the method of images is used. We imagine the barrier as a mirror and place an "image source" at $x = 2a$ (the image of the origin in the mirror). This leads us to the solution

$$p(x,t) = p_{\text{ur}}(0; x, t) + Ap_{\text{ur}}(2a; x, t) \tag{**}$$

Substitution for p_{ur} from (1.129) gives the expression which satisfies the boundary condition $p(a,t) = 0$ if A takes the value: $-exp(2\mu a/\sigma^2)$. Therefore, the solution of problem (E.20) is

$$p(x,t) = \frac{1}{\sigma\sqrt{2\pi t}}\left\{exp\left[-\frac{(x-\mu t)^2}{2\sigma^2 t}\right] - exp\left[\frac{2\mu a}{\sigma^2} - \frac{(x-2a-\mu t)}{2\sigma^2 t}\right]\right\} \tag{E.21}$$

12. Let us consider the diffusion process whose drift is equal to zero and the diffusion coefficient $b(x) = \alpha x(1-x)$, $\alpha > 0$. Let the state space of this process will be the interval $[0,1]$. Is the barrier $r_1 = 0$ accessible?

To answer the above question let us check the conditions (1.105), (1.106). In our case $l_1(x) = 1$ which is integrable on $[0,x], x' \in (0,1)$.

$$\int_I l_2(x)dx = \frac{1}{\alpha}\int_o^{x'} \frac{dx}{x(1-x)} = -\infty$$

This means that $l_2(x) \notin L(I)$ what implies that the barrier $r_l = 0$ is not regular. Let us check the integrability of $l_3(x)$. We have

$$l_3(x) = \frac{1}{\alpha}\int_{x'}^{x} \frac{dz}{z(1-z)} = \frac{1}{\alpha}\left(\log\frac{x}{1-x} - \log\frac{x'}{1-x'}\right)$$

and

$$\int_0^{x'} \log\frac{x}{x-1}dx = \int_0^{x'} \log x dx - \int_0^{x'} \log(x-1)dx$$

The integrals on the right are finite, so $l_3(x) \in L(0, x')$ what implies (according to (1.105)) that the barrier $r_1 = 0$ is an exit boundary.

13. Let us consider a spatially homogeneous one-dimensional diffusion process with infinitesimal coefficients

$$a(t,x) = a(t) \quad , \quad b(t,x) = b(t)$$

The forward and backward Kolmogorov equations are (cf. (1.94) and (1.93), respectively)

$$\frac{\partial p}{\partial t} = -a(t)\frac{\partial p}{\partial y} + \frac{1}{2}b(t)\frac{\partial^2 p}{\partial y^2}$$

$$\text{(E.22)} \qquad \frac{\partial p}{\partial s} = -a(s)\frac{\partial p}{\partial x} - \frac{1}{2}b(s)\frac{\partial^2 p}{\partial x^2}$$

where $s < t$ and $p = p(s,x;t,y) = p(s,t,y-x)$.

In order to obtain the solution of (E.22) we reduce the problem with time-varying coefficients to the case where $a(t) \equiv 0$ and $b(t) \equiv 1$; in this case (standard Wiener process) the solution has form (1.126) with t_o playing the role of s.

Let us introduce new variables

$$\xi = x - \int_0^s a(u)du \quad , \quad \eta = y - \int_0^t b(u)du$$

$$\text{(E.23)} \qquad s' = \int_0^s a(u)du \quad , \quad t' = \int_0^t b(u)du$$

The above transformation reduces system (E.22) to the equations in which drift and diffusion coefficients are equal to zero and one, respectively. Therefore, the final solution corresponding to system (E.22) is

$$\text{(E.24)} \qquad p(s,x;t,y) = \frac{1}{D\sqrt{2\pi}} exp\left\{-\frac{(y-x-m)^2}{2D^2}\right\}$$

where

$$m = m(s,t) = \int_s^t a(u)du \quad ,$$

$$D^2 = D^2(s,t) = \int_s^t b(u)du$$

14. Let $N(t)$, $t \geq 0$ be a homogeneous Poisson process with intensity λ. Let us assume that n Poisson events have occured in interval $[0,t]$ and $0 < s < t$. Find $P\{N(s) = k | N(t) = n\}$ for $0 < k < n$.

$$\begin{aligned}
&P\{N(s) = k|N(t) = n\} \\
&= [P\{N(t) = n\}]^{-1} P\{N(s) = k, N(t) = n\} \\
&= \left[\frac{e^{-\lambda t}(\lambda t)^n}{n!}\right]^{-1} P\{N(s) = k, N(t) - N(s) = n - k\} \\
&= n!(\lambda t)^{-n} e^{\lambda t} \frac{e^{-\lambda s}(\lambda s)^k}{k!} e^{-\lambda(t-s)} \frac{\lambda^{n-k}(t-s)^{n-k}}{(n-k)!} \\
&= \binom{n}{k} \left(\frac{s}{t}\right)^k \left(1 - \frac{s}{t}\right)^{n-k}
\end{aligned}$$

The result indicates that $P\{N(s) = k|N(t) = n\}$ is a binomial distribution with parameters n and $\frac{s}{t}$.

15. Let stochastic process $X(t)$ be defined as

$$X(t,\gamma) = \sum_{i=1}^{N(t)} Y_i(\gamma) \tag{E.25}$$

where $N(t)$ is the Poisson process and $Y_i(\gamma)$ are mutually independent and identically distributed random variables with given probability distribution.

Determine one-dimensional probability density of $X(t,\gamma)$ and the first-passage time distribution.

Let us make use of the moment generating function

$$M(s) = E\left[e^{-sX(t,\gamma)}\right] = \int_0^\infty e^{-sx} f_{X(t)}(x)dx$$

If $N(t)$ is homogeneous Poisson process with intensity λ_o we obtain

$$M(s) = \sum_{k=0}^{\infty} E\left\{e^{-sX(t,\gamma)}|N(t) = k\right\} P\{N(t) = k\}$$

$$= \sum_{k=0}^{\infty} \left[Ee^{-sY_i(\gamma)}\right]^k \frac{(\lambda_o t)^k}{k!} e^{-\lambda_o t} = e^{-\lambda_o t} \sum_{k=0}^{\infty} [G(s)]^k \frac{(\lambda_o t)^k}{k!} \tag{E.26}$$

where $G(s)$ is the moment generating function of random variables $Y_i(\gamma)$. In order to get $f_{X(t)}(x)$ we have to assume a specific analytical form of $G(s)$. In the case of exponential distribution we have

$$g(y) = ae^{-ay} \quad , \quad y > 0, a > 0;$$

$$G(s) = \frac{a}{s+a} \quad , \quad s > 0$$

The inverse of (E.26) gives the result

$$f_{X(t)}(x) = e^{-\lambda_o t - ax} \sum_{k=0}^{\infty} \frac{(a\lambda_o t)^{k+1} x^k}{k!(k+1)!} \quad , \quad x > 0$$

$$= \sqrt{\frac{a\lambda_o t}{x}} e^{-\lambda_o t - ax} I_1\left(2\sqrt{\lambda_o axt}\right) \tag{E.27}$$

where $I_1(\cdot)$ is a modified Bessel function of the first order.

Let T be a positive random variable characterizing a random time of reaching by process $X(t,\gamma)$ a fixed value ξ. Of course,

$$P(T > t) = P\{X(t,\gamma) < \xi\}$$

This means that the passage time distribution and the distribution of process $X(t,\gamma)$ are directly related to each other, namely

$$F_T(t) = 1 - F_{X(t)}(x)\Big|_{x=\xi}$$

It can be shown that

$$f_T(t) = \lambda_o e^{-\lambda_o t - a\xi} I_o\left(2\sqrt{\lambda_o ta\xi}\right) \tag{E.28}$$

where $I_o(\cdot)$ is a modified Bessel function of order zero.

Chapter II

STOCHASTIC CALCULUS: PRINCIPLES AND RESULTS

13. INTRODUCTORY REMARKS

As in the classical mathematical analysis in the theory of stochastic processes we come naturally to concepts such as: continuity, derivatives and integrals of a stochastic process (or, the Stieltjes-type integrals with respect to a stochastic process). Of course, these notions are of basic importance in the formulation and analysis of stochastic differential equations. It is, therefore, essential to define them precisely and to make clear the mutual relationships between various possible interpretations.

The part of stochastic process theory dealing with the problems analogous to those considered in classical calculus can be termed as the **stochastic calculus** (or random analysis (cf. [23])). Such an understanding of the term "stochastic calculus" differs from that which is often used in the literature (cf. [44], [58]) — where it is traditionally restricted to problems and constructions associated with the Itô integrals (with respect to a Wiener process), the Itô's differentials and their recent extensions (cf. [53]). However, it seems to be advantageous from the methodical point of view to include all problems associated with differentiation and integration of stochastic processes into the term "stochastic calculus", even if some results and constructions might be analogous to those in the classical calculus. In this chapter the results associated with analytical properties of sample functions (sample function calculus) are also included. For additional reading on the subject, and for the proofs of unproven theorems, the reader is referred to [6], [16], [23].

14. PROCESSES OF SECOND ORDER; MEAN SQUARE ANALYSIS

14.1. Preliminaries

Definition 2.1. *A stochastic process $X(t)$, $t \in T$ (in general complex) is called a* **second order** *or* Hilbert *stochastic process, if $E\{|X(t)|^2\} < \infty$.*

Without restriction we shall assume that the stochastic processes under consideration have zero mean values. Then the second-order moment $E[|X(t)|^2]$ is the variance and $K_X(t_1, t_2)$ is the correlation function of the process. The Schwarz inequality implies, that a second-order stochastic process always has a correlation function. Conversely, if there exists a finite $K_X(t_1, t_2)$ defined on $T \times T$, then $K_X(t,t) = E\left[|X(t)|^2\right] < \infty$.

The properties of a stochastic process expressed in terms of its second order moments are usually called *second-order properties* (cf. [6]). Random variables with finite second-order moments form a Hilbert space $L_2(\Gamma, \mathcal{F}, P)$. A second-order stochastic process can thus be considered as a vector function defined on T and taking its values in $L_2(\Gamma, \mathcal{F}, P)$. The inner product $(.,.)$ of two elements of $L_2(\Gamma, \mathcal{F}, P)$ is simply

$$(X_1, X_2) = E\left[X_1 \overline{X}_2\right] \tag{2.1}$$

where "–" denotes the complex conjugate. Therefore, the mean, correlation function and variance of a second-order stochastic process $X(t)$ are

$$m_X(t) = (X_t, 1) \tag{2.2}$$

$$K_X(t_1, t_2) = (X_{t_1}, X_{t_2}) \tag{2.3}$$

$$\sigma_X^2(t) = \|X_t\|^2 \tag{2.4}$$

where $\|\cdot\|$ denotes a norm in $L_2(\Gamma, \mathcal{F}, P)$ induced by the inner product (2.1).

It is clear from the above that the convergence of random variables in a mean square sense (being equivalent to convergence in $L_2(\Gamma, \mathcal{F}, P)$-norm) is a very natural concept of convergence in the theory of a second-order stochastic processes. This kind of stochastic convergence also turns out to be very handy, since basic properties of a process defined by use of this concept have their analogies in the calculus of ordinary (deterministic) functions.

In what follows we shall make use of the following criterion.

Theorem 2.1. *Let $X(t)$ be a second-order stochastic process. A process $X(t)$ converges in the mean square (m.s.) to a random variables $X(\gamma)$ as $t \to t_o$, if and only if, the correlation function $K_X(t_1, t_2)$ converges to a finite limit as $t_1, t_2 \to t_o$; then*

$$K_X(t_1, t_2) \to E|X(\gamma)|^2 \qquad \text{as} \quad t_1, t_2 \to t_o$$

Proof:

Let $X(t) \xrightarrow{\text{m.s.}} X, \quad \text{as} \quad t \to t_o$. Then

$$E\left\{X(t_1)\overline{X(t_2)} - X\overline{X}\right\} = E\left[X(t_1) - X\right]\left[\overline{X(t_2)} - \overline{X}\right] \\ + E\left\{\left[X(t_1) - X\right]\overline{X}\right\} + E\left\{\left[\overline{X(t_2)} - \overline{X}\right]X\right\}.$$

Using the Schwarz inequality one obtains*

$$K_X(t_1,t_2) = E[X(t_1),X(t_2)] \to E|X|^2 \quad \text{as} \quad t_1,t_2 \to t_o$$

Conversely, let $K_X(t_1,t_2) \to c$, as $t_1,t_2 \to t_o$, where $c < \infty$. Then

$$\begin{aligned} &E\,|X(t_1) - X(t_2)|^2 = \\ &= E\,[X(t_1) - X(t_2)]\left[\overline{X(t_1)} - \overline{X(t_2)}\,\right] = \\ &= E\,|X(t_1)|^2 - E\left[X(t_1)\overline{X(t_2)}\,\right] - E\left[X(t_2)\overline{X(t_1)}\,\right] + E\,|X(t_2|^2 = \\ &= K_X(t_1,t_1) - K_X(t_1,t_2) - (K_X(t_2,t_1) + K_X(t_2,t_2) \longrightarrow \\ &\longrightarrow c - c - c + c = 0 \quad \text{as} \quad t_1,t_2 \to t_o. \end{aligned}$$

This relationship and the Cauchy's (m.s.) convergence criterion imply that $X(t) \xrightarrow{\text{m.s.}} X$, as $t \to t_o$.

14.2. Mean-square continuity

Definition 2.2. *A second order stochastic process* $X(t)$, $t \in T$ *is* **continuous in a mean square sense** *(m.s. continuous) at* $t \in T$, *if*

$$E|X(t+h) - X(t)|^2 \to 0 \qquad \text{as} \qquad h \to 0 \tag{2.5}$$

A process $X(t)$ is m.s. continuous on T if it is m.s. continuous at every point $t \in T$.

Theorem 2.2. *(the* m.s. continuity criterion*) A second-order process* $X(t)$, $t \in T$ *is m.s. continuous at* $t \in T$ *if and only if the correlation function* $K_X(t_1,t_2)$ *is continuous at* $t_1 = t_2$.

Proof:

Let $K_X(t_1,t_2)$ be continuous at $t_1 = t_2$. Then

$$\begin{aligned} &E|X(t+h) - X(t)|^2 = \\ &= E\,[X(t+h) - X(t)]\left[\,\overline{X(t+h)} - \overline{X(t)}\,\right] = \\ &= K_X(t+h,t+h) - K_X(t+h,t) - K_X(t,t+h) + K_X(t,t) \xrightarrow{h \to 0} 0, \end{aligned}$$

therefore $X(t)$ is m.s. at t.

---

* The Schwarz inequality used in the Hilbert space theory is: $(x,y)^2 \leq (x,x)(y,y)$; here it means that: $E|XY| \leq \left[EX^2EY^2\right]^{1/2}$

Let us assume that $X(t)$ is m.s. continuous. Since

$$K_X(t+h,t+k) - K_X(t,t) = E\Big\{ [X(t+h) - X(t)]\,[\,\overline{X(t+k)} - \overline{X(t)}] + [X(t+h) - X(t)\,]\,\overline{X(t)} + \left[\overline{X(t+k)} - \overline{X(t)}\,\right] X(t) \Big\}$$

use of the Schwarz inequality implies that $K_X(t_1,t_2)$ is continuous at $t \in T$.

Theorem 2.3. *If $K_X(t_1,t_2)$ is continuous at every diagonal point $t_1 = t_2$, then it is continuous everywhere on $T \times T$.*

Proof:

$$\begin{aligned}
&|K_X(t_1+h,t_2+k) - K_X(t_1,t_2)| \\
&= \left| E\left[X(t_1+h)\overline{X(t_2+k)}\right] - E\left[X(t_1)\overline{X(t_2)}\right] \right| \\
&= \left| E\left\{[X(t_1+h) - X(t_1)]\,\overline{X(t_2+k)}\right\} + E\left\{X(t_1)\left[\overline{X(t_2+k)} - \overline{X(_2)}\right]\right\} \right|
\end{aligned}$$

Using the Schwarz inequality it is easily seen that the above expression tends to zero as h and k approach zero.

It should be emphasized that m.s. continuity of a process $X(t)$ implies that it is continuous with respect to probability. Indeed, the Tchebycheff inequality gives

$$P\{|X(t_\circ + h) - X(t_\circ)| > \varepsilon\} \le \frac{E\,|X(t_\circ+h) - X(t_\circ)|^2}{\varepsilon^2}$$

The m.s. continuity does not, however, imply continuity of the realizations of a process. As an example one can take the Poisson process. Its correlation function $K_X(t_1,t_2) = \nu \min(t_1,t_2)$ is continuous for all $t_1, t_2 > 0$, but almost all realizations of this process have discontinuities over a finite interval of time.

14.3. Mean-square differentiation

Definition 2.3. *A second-order stochastic process $X(t), t \in T$ has mean square derivative $X'(t)$ at $t \in T$ if*

$$\underset{h\to 0}{l.i.m.}\frac{1}{h}[X(t+h) - X(t)] = X'(t) \tag{2.6}$$

Higher order m.s. derivatives are defined analogously.

Theorem 2.4. *(the m.s. differentiation criterion): A second-order stochastic process $X(t), t \in T$ is m.s. differentiable at $t \in T$ if and only if there exists a finite second-order derivative of the correlation function $K_X(t_1, t_2)$ at point (t,t).*

Proof:
This theorem follows directly from Theorem 2.1., which asserts that condition (2.6) is satified if and only if

$$E\left[Y_h, \overline{Y}_k\right] \underset{\substack{h\to 0\\ k\to 0}}{\longrightarrow} \text{finite limit,}$$

where

$$Y_h(t) = \frac{1}{h}\left[X(t+h) - X(t)\right].$$

But

$$E\left[Y_h(t)\overline{Y}_k(t)\right] = E\left[\frac{X(t+h)-X(t)}{h} \cdot \frac{\overline{X(t+k)} - \overline{X(t)}}{k}\right]$$

$$= \frac{1}{hk}\left\{K_X(t+h, t+k) - K_X(t, t+k) - K_X(t+h, t) + K_X(t,t)\right\}.$$

Therefore, in order to ensure the existence of the m.s. derivative $X'(t)$ of process $X(t)$ it is necessary and sufficient that the above expression has a finite limit, i.e. that the second-order derivative defined as

$$\frac{\partial^2 K_X(t_1,t_2)}{\partial t_1 \partial t_2}$$

$$= \lim_{h,k\to 0} \frac{K_X(t+h, t+k) - K_X(t+h, t) - K_X(t, t+k) + K_X(t,t)}{hk}$$

exists and is finite at the point (t,t).

The basic properties of m.s. derivatives are analogous to those associated with ordinary derivatives in the classsical calculus. For example, it is easy to show (cf. [6]) that:

a) mean square differentiability of $X(t)$ at $t \in T$ implies m.s. continuity of $X(t)$ at t;
b) if at $t \in T, X_1(t)$ and $X_2(t)$ are two m.s. differentiable stochastic processes, then the m.s. derivative of $aX_1(t) + bX_2(t)$ exists at t and

$$\frac{d}{dt}\left[aX_1(t) + bX_(t)\right] = a\frac{dX_1(t)}{dt} + b\frac{dX_2(t)}{dt} \tag{2.7}$$

where a and b are constants.

If $X(t), t \in T$ is n-times m.s. differentiable on T, the means of these m.s. derivatives of $X(t)$ exist on T and they are given by

$$E\left[\frac{d^n X(t)}{dt^n}\right] = \frac{d^n}{dt^n} E\left[X(t)\right] \tag{2.8}$$

Theorem 2.5. *If $X(t)$ has m.s. derivative for all $t \in T$, then the partial derivatives* $\frac{\partial K_X(t_1,t_2)}{\partial t_1}$, $\frac{\partial K_X(t_1,t_2)}{\partial t_2}$, $\frac{\partial^2 K_X(t_1,t_2)}{\partial t_1 \partial t_2}$ exist and

$$K_{X'X}(t_1,t_2) = E\left[X'(t)\overline{X(t_2)}\right] = \frac{\partial}{\partial t_1}K_X(t_1,t_2) \tag{2.9}$$

$$K_{XX'}(t_1,t_2) = E\left[X(t_1)\overline{X'(t_2)}\right] = \frac{\partial}{\partial t_2}K_X(t_1,t_2) \tag{2.10}$$

$$\begin{aligned} K_{X'X'}(t_1,t_2) = K_{X'}(t_1,t_2) &= E\left[X'(t_1)\overline{X'(t_2)}\right] \\ &= \frac{\partial^2}{\partial t_1 \partial t_2}K_X(t_1,t_2) \end{aligned} \tag{2.11}$$

Proof:
If $X(t)$ has $X'(t)$ at $t \in T$, then $Y_h(t) \xrightarrow{\text{m.s.}} X'(t)$ as $h \to 0$. Therefore,

$$\begin{aligned} \lim_{h\to 0} \frac{K_X(t_1+h,t_2) - K_X(t_1,t_2)}{h} &= \lim_{h\to 0} E\left[\frac{X(t_1+h) - X(t_1)}{h}\overline{X(t_2)}\right] \\ = E\left[\lim_{h\to 0} \frac{X(t_1+h) - X(t_1)}{h}\overline{X(t_2)}\right] &= \\ = E\left[X'(t_1)\overline{X(t_2)}\right] \end{aligned}$$

The above implies that the derivative $\frac{\partial}{\partial t_1}K_X(t_1,t_2)$ exists and formula (2.9) holds.

Similarly, it can be shown that the derivative $\frac{\partial}{\partial t_2}K_X(t_1,t_2)$ exists and the formula (2.10) holds. Finally,

$$\begin{aligned} &\lim_{k\to 0} \frac{1}{k}\left\{\frac{\partial}{\partial t_1}K_X(t_1,t_2+k) - \frac{\partial}{\partial t_1}K_X(t_1,t_2)\right\} \\ &= \lim_{k\to 0} E\left[X'(t)\frac{\overline{X(t_2+k)} - \overline{X(t_2)}}{k}\right] = E\left[X'(t_1)\overline{X'(t_2)}\right] \end{aligned}$$

shows that formula (2.11) is valid.
Analogous results for the derivatives of higher order can be easily derived. The cross-correlation function of the derivatives of order k and l of a given process is expressed by the general formula

$$K_{X^{(k)}X^{(l)}}(t_1,t_2) = E\left[X^{(k)}(t_1)X^{(l)}(t_2)\right] = \frac{\partial^{k+l}}{\partial t_1^k \partial t_2^l}K_X(t_1,t_2) \tag{2.12}$$

if appropriate m.s. derivatives exist.

Let us assume that $K_X(t_1,t_2)$ has an infinite number of derivatives on $T \times T$. Let

$$X_n(t) = X(0) + \frac{t}{1}X(0) + \cdots + \frac{t^n}{n!}X^{(n)}(0) \tag{2.13}$$

Simple transformations of the expression

$$E\left|X(t) - X_n(t)\right|^2$$

lead to the following theorem (cf. [6]).

Theorem 2.6. *A second-order stochastic process $X(t)$, $t \in T$ is m.s. analytic if and only if its correlation function $K_X(t_1, t_2)$ is analytic at each diagonal point $(t,t) \in T \times T$.*

In the case where $X(t)$ is a *weakly stationary* stochastic process the criterion of m.s. differentiation is as follows.

Theorem 2.7. *A weakly-stationary stochastic process $X(t)$ is differentiable at $t \in T$ if and only if its correlation function $K_X(\tau)$, $\tau = t_2 - t_1$, has the derivative of a second order at $\tau = 0$.*

Formulae (2.9)-(2.12) reduce to the following ones

$$K_{X'X}(\tau) = E\left[X'(t_1)\overline{X(t_2)}\right] = \frac{\partial}{\partial t_1} K_X(t_2 - t_1) = -\frac{d}{d\tau} K_X(\tau) \tag{2.14}$$

$$K_{XX'}(\tau) = E\left[X(t_1)\overline{X'(t_2)}\right] = \frac{d}{d\tau} K_X(\tau) \tag{2.15}$$

$$K_{X'X'}(\tau) = E\left[X'(t_1)\overline{X'(t_2)}\right] = -\frac{d^2}{d\tau^2} K_X(\tau) \tag{2.16}$$

$$K_{X^{(k)}X^{(l)}}(\tau) = (-1)^k \frac{d^{k+l}}{d\tau^{k+l}} K_X(\tau) \tag{2.17}$$

If $X(t)$ is also real process, its correlation function $K_X(\tau)$ is an even function of τ. So, $\frac{dK_X(\tau)}{d\tau}$ is equal to zero at $\tau = 0$. Hence

$$K_{XX'}(0) = 0 \tag{2.18}$$

what implies, that in the case of a real m.s. differentiable weakly stationary process the values of a process and its derivative corresponding to the same instant are uncorrelated.

In the analysis of stationary processes an important role is played by the spectral density of a process. It can be shown (cf. Example 6) that the differentiability condition (in the mean square sense) expressed in terms of the spectral density is the following

$$\int_{-\infty}^{+\infty} \omega^2 g_X(\omega) d\omega < \infty \tag{2.19}$$

The spectral density of the derivative of a stationary process and the mutual spectral density of the derivatives of various orders are related to $g_X(\omega)$ by the formulae (cf. Example 6).

$$g_{X'}(\omega) = \omega^2 g_X(\omega) \tag{2.20}$$

$$g_{X^{(k)}X^{(l)}} = (i\omega)^k (i\omega)^l g_X(\omega) \tag{2.21}$$

In closing, let us note that m.s. differentiability does not imply sample function differentiability. If, however, a process $X(t)$ with probability one has a sample function derivative on T, then for every $t \in T$, the difference ratio $\frac{1}{h}[X(t+h) - X(t)]$ converges both in the mean square sense and with probability one as $h \to 0$; the m.s. derivative and the sample function derivative are equal with probability one.

14.4. Mean square stochastic integrals

A.

Let $[a,b] \subset T$ be a finite interval and let the points $t_0, t_1, \ldots, t_n$ define a partition of $[a,b]$ such that

$$a = t_0 < t_1 < \cdots < t_n = b.$$

Let $\xi \in [t_i, t_{i+1}]$. Let us form the approximating sum

$$I_R(n) = \sum_{i=0}^{n-1} X(\xi_i)(t_{i+1} - t_i) \tag{2.22}$$

Definition 2.4. *If there exists a m.s. limit of the sequence* $\{I_R(n)\}$ *as* $n \to \infty$ *and* $\max(t_{i+1} - t_i) \to 0$, *and it is independent on the sequence of partitions as well as the positions of* $\xi_i \in [t_i, t_{i+1}]$, *then this limit is called the* **mean square Riemann integral** *of a process* $X(t)$ *over the interval* $[a,b]$ *and it is denoted by*

$$I_R = \int_a^b X(t)dt \tag{2.23}$$

Theorem 2.8. (the m.s. integration criterion): A stochastic process $X(t)$ of second-order is m.s. Riemann integrable on $[a,b]$ if the ordinary double Riemann integral

$$Q_R = \int_a^b \int_a^b K_X(t_1, t_2)dt_1 dt_2 \tag{2.24}$$

exists; moreover

$$E\left|I_R\right|^2 = Q_R \tag{2.25}$$

Proof:

By virtue of the Theorem 2.1.$I_n \xrightarrow{\text{m.s.}} I$, if and only if $E\left[I_n \overline{I_m}\right] \to E[I\overline{I}]$ where I_m is the approximating sum corresponding to the partition $a = t'_0 < t'_1 < \cdots < t'_m = b$ of the interval $[a,b]$ and $t'_k \le \xi'_k \le t'_{k+1} (k = 0, 1, \ldots, m-1)$. Of course,

$$E\left[I_n \overline{I}_m\right] = \sum_{i=0}^{n-1} \sum_{k=0}^{m-1} K_X(\xi_i, \xi'_k)\left[t_{i+1} - t_i\right]\left[t'_{k+1} - t'_k\right]$$

If there exists a limit of the above sum, as $n \to \infty, m \to \infty$, $\max_{0 \le i \le n-1} |t_{i+1} - t_i| \to 0$, $\max_{0 \le k \le m-1} |t'_{k+1} - t'_k| \to 0$, then this limit is the double integral (2.24).

Analogously, one defines the integral

$$I_R^{\varphi} = \int_a^b \varphi(t) X(t) dt$$

$$= \underset{n\to}{\text{l.i.m.}} \sum_{i=0}^{n-1} \varphi(\xi_i)(X(\xi_i)(t_{i+1} - t_i) \tag{2.26}$$

where $\varphi(t)$ is an ordinary Riemann integrable function defined on $[a, b]$.

It can be easily shown that the integral (2.26) exists provided the double integral

$$Q_R^{\varphi} = \int_a^b \int_a^b \varphi(t_1)\overline{\varphi(t_1)} K_X(t_1, t_2) dt_1 dt_2 \tag{2.27}$$

exists; moreover $E\left|I_R^{\varphi}\right|^2 = Q_R^{\varphi}$.

Improper m.s. Riemann stochastic integrals on finite interval are defined in the usual manner. Also, the variability of the integration limits does not bring any essential change to the definition of m.s. Riemann integral except that in this case the value of the integral is not random variable but a stochastic process. Mean square integral with a variable limit of integration appear in a natural way when the process $Y(t)$ is generated by a linear system (filter) subjected to random excitation described by a second-order stochastic process $X(t)$, that is

$$Y(t) = \int_{t_0}^{t} p(t, t_1) X(t_1) dt_1 \tag{2.28}$$

where the function $p(t_1, t_2)$ is termed the *impulse response* of a system.

Using the approximating sums one can easily show that m.s. integrals possess the properties of ordinary integrals, such as, for example,

$$\int_a^b X(t)dt + \int_b^c X(t)dt = \int_a^c X(t)dt \tag{2.29}$$

$$\int_a^b [\alpha X_1(t) + \beta X_2(t)]\, dt = \alpha \int_a^b X_1(t) + \beta \int_a^b X_2(t) dt \tag{2.30}$$

Also, the formula for integration by parts is valid in its usual form.

Remark: In the analysis of stochastic processes there is also a need for m.s. Lebesque stochastic integral. Since a second-order stochastic process can be interpreted as a vector function with its values in a Hilbert space, the m.s. Lebesque integral is simply the *Bochner integral.*

B.

Let us consider briefly m.s. Riemann-Stieltjes stochastic integrals. Let $\varphi(t)$ be a non-random function defined on $[a, b]$.

Definition 2.5. *A m.s.* Riemann-Stieltjes integral *of a function $\varphi(t)$ with respect to a second-order stochastic process $X(t)$ over the interval $[a, b]$ is defined as m.s. limit (as $n \to \infty$ and $\max(t_{i+1} - t_i)) \to 0$) of the approximating sums*

$$I_{RS}(n) = \sum_{i=0}^{n-1} \varphi(\xi_i) [X(t_{i+1}) - X(t_i)] \tag{2.31}$$

and it is denoted by

$$I_{RS} = \int_a^b \varphi(t) dX(t) \tag{2.32}$$

It can be shown that the existence of the integral (2.32) is assured, if the ordinary double Riemann-Stieltjes integral

$$Q_{RS} = \int_a^b \int_a^b \varphi(t_1)\varphi(t_2) d_{t_1 t_2} K_X(t_1, t_2) \tag{2.33}$$

exists; moreover $E|I_{RS}|^2 = Q_{RS}$.

Riemann-Stieljes m.s. stochastic integrals (2.32) have the formal properties of ordinary Riemann-Stieltjes integrals. For example, the following formula of partial integration

$$\int_a^b \varphi(t) dX(t) = \varphi(b)X(b) - \varphi(a)X(a) - \int_a^b \varphi'(t)X(t)dt \tag{2.34}$$

holds, where $\varphi(t)$ has a bounded, Riemann integrable derivative $\varphi'(t)$ in $[a, b]$ and the correlation function $K_X(t_1, t_2)$ is continuous and of a bounded variation over $[a, b] \times [a, b]$.

It is obvious that the m.s. Riemann-Stieltjes integral needs not to be considered if the process $X(t)$ is m.s. differentiable, since in such case one can use m.s. Riemann integral of the process $\varphi(t)X'(t)$ instead of integral (2.32). Therefore, the m.s. Riemann-Stieltjes stochastic integral is of interest mainly in the case when the process $X(t)$ is non-differentiable. An important class of non-differentiable processes is formed by the processes with orthogonal increments. The Riemann-Stieltjes m.s. integrals with respect to a process with orthogonal increments are particularly important in the study of stationary stochastic processes.

Let us assume that $X(t)$, $t \in T$ has orthogonal increments. It is known (cf. [9]), that there exists a non-decreasing function $F(t)$, $t \in T$, such that the increment

$X(t) - X(s)$, $s < t$, of the process has variance equal to $F(t) - F(s)$, that is (assuming $m_X(t) = 0$)

$$E\,|X(t) - X(s)|^2 = F(t) - F(s) \tag{2.35}$$

In the case under consideration one can replace $X(t)$ in the approximating sum (2.31) by $X(t) - X(t_o)$, where t_o is an arbitrary but fixed point of T. The correlation function $K_X(t_1, t_2)$ of a process with orthogonal increments is of bounded variation over every finite domain, so that the m.s. Riemann-Stieltjes integral (2.32) exists for any $\varphi(t)$ satisfying the condition for existence of the integral (2.33).

Usually, the m.s. Stieltjes integral with respect to process $X(t)$ with orthogonal increments is defined gradually, starting from the definition of the integral for the case when $\varphi(t)$ is a step function.

a. Let $\varphi(t)$ be a step function, that is

$$\varphi(t) = \begin{cases} \varphi_i & t_i \le t \le t_{i+1}, \quad i = 0, 1, \ldots, n-1 \\ 0 & \text{otherwise} \end{cases} \tag{2.36}$$

For such a function the integral (2.32) is defined as

$$I_{RS}(\varphi) = \int_a^b \varphi(t)dX(t) = \sum_{i=0}^{n} \varphi_i \left[X(t_{i+1} - X(t_i)\right] \tag{2.37}$$

where $t_o = a$, and $t_{n+1} = b$. It is easily seen that

$$I_{RS}(c_1\varphi_1 + c_2\varphi_2) = c_1 I_{RS}(\varphi_1) + c_2 I_{RS}(\varphi_2) \tag{2.38}$$

and

$$E\,|I_{RS}|^2 = E\left|\int_a^b \varphi(t)dX(t)\right|^2$$

$$= \sum_{i=0}^{n} |\varphi_i|^2 \left[F(t_{i+1}) - F(t_i)\right] = \int_a^b |\varphi(t)|^2\, dF(t) \tag{2.39}$$

b. Now, the integral (2.32) can be defined for any arbitary, bounded and continuous function $\varphi(t)$. If $\varphi(t)$ has such a property, a sequence $\varphi_n(t)$ of step functions (of the form (2.36)) exists such that

$$\int_a^b |\varphi(t) - \varphi_n(t)|^2\, dF(t) \to 0, \qquad n \to \infty$$

The integral I_{RS} is then defined as a limit:

$$I_{RS}(\varphi) = \underset{n\to\infty}{\text{l.i.m.}} \int_a^b \varphi_n(t)dX(t) \tag{2.40}$$

This definition can be extended to arbitrary functions $\varphi(t)$ which are square-integrable with respect to $F(t)$, i.e. such that

$$\int_a^b |\varphi(t)|^2 \, dF(t) < \infty$$

The properties of the m.s. Riemann-Sieltjes intgeral (2.40) with respect to process $X(t)$ with orthogonal increments can be derived from the properties of the integral (2.37) through a limiting passage. It turns out that the relation (2.38) holds almost surely; the mean value of the integral (2.40) is equal to zero, and its variance is given by (2.39).

Remark: One can also define (cf. [23]) a m.s. Stieltjes integral of the

$$\int_T X(t) dY(t) \tag{2.42}$$

where $X(t)$ and $Y(t)$ are two second order stochastic processes. In such a case it is usually assumed that the process $X(t)$ and the increments of $Y(t)$ are independent.

14.5. Orthogonal expansions

In applications it is useful to represent a given stochastic process in the form of a sum of certain simpler processes. Among the possible representations an important role is played by the orthogonal expansions.

Theorem 2.9. *An arbitrary m.s. continuous second-order stochastic process $X(t)$, $t \in T = [a, b]$ can be represented in the form of a series*

$$X(t) = \sum_{k=1}^{\infty} Z_k(\gamma)\varphi_k(t) \tag{2.43}$$

which is m.s. convergent for each $t \in T$ and $Z_k(\gamma)$ are random variables such that:

$$E\left[Z_k \overline{Z}_l\right] = 0 \quad , \quad k \neq l \tag{2.44}$$

$$E\left|Z_k\right|^2 = \omega_k \tag{2.45}$$

where ω_k and $\varphi_k(t)$ are the eigenvalues and the eigenfunctions of the following integral equation

$$\omega\varphi(t_1) = \int_T K_X(t_1, t_2)\varphi(t_2) dt_2 \tag{2.46}$$

Proof:
Equation (2.46) is an integral equation with symmetric kernel and according to the theory of such equations (Merser's theorem) the kernel $K_X(t_1, t_2)$ being continuous can be represented in the form of a uniformly convergent on $T \times T$ series with respect to the eigenfunctions of equation (2.46), that is

$$K_X(t_1,t_2) = \sum_{n=1}^{\infty} \omega_n \varphi_n(t_1)\overline{\varphi_n(t_2)} \tag{2.47}$$

where

$$\omega_n \varphi_n(t_1) = \int_T K_X(t_1,t_2)\varphi_n(t_2)dt_2,$$
$$\int_T \varphi_n(t)\overline{\varphi_m(t)}dt = \delta_{nm}$$

and the eigenvalues ω_n are positive.

To show the validity of (2.43) let us consider the expression

$$E\left|X(t) - \sum_{k=1}^{n} Z_k\varphi_k(t)\right|^2$$
$$= K_X(t_1,t_2) - 2\sum_{k=1}^{n}\overline{\varphi_k(t)}E\left[X(t)\overline{Z}_k\right] + \sum_{k=1}^{n}\sum_{l=1}^{n}|\varphi_k(t)|^2 E\left[Z_k\overline{Z}_l\right] \tag{2.48}$$

Let us define

$$Z_n(\gamma) = \int_T X(t)\overline{\varphi_n(t)}dt$$

The average values occuring in (2.48) are

$$E\left[Z_k\overline{Z}_l\right] = \int_T\int_T K_X(t_1,t_2)\overline{\varphi_k(t_1)}\varphi_l(t_2)dt_2dt_1$$
$$= \int_T \overline{\varphi_k}(t_1)dt_1 \int_T K_X(t_1,t_2)\varphi_l(t_2)dt_2$$
$$= \omega_l \int_T \overline{\varphi_k(t_1)}\varphi_l(t_1)dt_1 = \omega_l\delta_{kl};$$
$$E\left[X(t)\overline{Z}_k\right] = E\left[X(t)\int_T \overline{X(s)}\varphi_k(s)ds\right]$$
$$= \int_T K_X(t,s)\varphi_k(s)ds = \omega_k\varphi_k(t)$$

Therefore, the mean square (2.48) is

$$E\left|X(t)-\sum_{k=1}^{n} Z_k\varphi_k(t)\right|^2 =$$

$$= K_X(t,t) - 2\sum_{k=1}^{n}\omega_k\,|\varphi_k(t)|^2 + \sum_{k=1}^{n}\omega_k\,|\varphi_k(t)|^2$$

and by virtue of (2.47) it tends to zero as $n \to \infty$.

The expansion (2.43) is called the *Karhunen-Loéve orthogonal expansion.*

It can easily be verified that the expansion (2.43) for the Wiener process on interval $[0,1]$ has the following form (cf. Example 11)

$$X(t) = \sqrt{2}\sum_{n=0}^{\infty} Z_n(\gamma)\frac{\sin(n+\frac{1}{2})\pi t}{(n+\frac{1}{2})\pi}. \tag{2.43'}$$

14.6. Transformations of second-order stochastic processes

Let A be an operator acting on a measurable second-order stochastic process $X(t)$ yielding the second-order process $Y(t)$, that is

$$Y(t) = A\,[X(t)] \qquad \text{or} \qquad Y(t) = A\,[X(s),t] \tag{2.49}$$

The basic problem which arises is the following: assuming that the probabilistic characteristics of $X(t)$ and the operator A are known one should determine appropriate characteristics of $Y(t)$.

The solvability of the above problem — its simplicity, accuracy, etc. — depends essentially upon the operator A and its properties, but most important is whether A is linear or not.

Let us assume that A is a *linear* operator; we shall visualize it as an integral or a differential-integral linear operator. Applying the averaging operator to both sides of (2.49) we have

$$E\,[Y(t)] = E\,[AX(t,\gamma)] \quad , \qquad \gamma \in \Gamma \tag{2.50}$$

The operator A acts on the variable $t \in T$, and the operator E — also linear — is an integral with respect to γ. Changing the order of these operators we get

$$E\,[Y(t)] = A\,\{E\,[X(t)]\} = Am_X(t) \tag{2.51}$$

Let us now determine the correlation function $K_Y(t_1, t_2)$. To do this let us multiply both sides of (2.49) by $X(t_1)$ using the homogeneity of the operator A (i.e. $A[cX(t)] = cA[X(t)]$). Making use of (2.51) one obtains

$$E[X(t_1)Y(t)] = A_t E[X(t_1)X(t)]$$

for arbitrary $t \in T$. Taking $t = t_2$ we have

$$K_{XY}(t_1, t_2) = A_{t_2} K_X(t_1, t_2) \tag{2.52}$$

Multiplying (2.49) by $Y(t_2)$ and averaging gives

$$E[Y(t)Y(t_2)] = A_t \{E[X(t)Y(t_2)]\}$$

For $t = t_1$ it yields

$$K_Y(t_1, t_2) = A_{t_1} [K_{XY}(t_1, t_2)] \tag{2.53}$$

By virtue of (2.52) and (2.53) one obtains the following relation between $K_X(t_1, t_2)$ and $K_Y(t_1, t_2)$

$$K_Y(t_1, t_2) = A_{t_1} A_{t_2} K_X(t_1, t_2) \tag{2.54}$$

Let us note, that the formulae obtained in the previous subsection — expressing the correlation functions of the derivative and the integral of a second-order stochastic process are the particular cases of (2.54). Formula (2.51) implies that the average of the derivative and integral of a given stochastic process is equal to the derivative and integral of the average, respectively.

If the operator A acting on $X(t)$ is non-linear, the determination of the correlation function of $Y(t)$ is not straightforward. The success in obtaining $K_Y(t_1, t_2)$ depends essentially on the particular form of A. The accurate expression for $K_Y(t_1, t_2)$ are obtainable only in particular cases; the basic simplifying assumption is the Gaussianity of the process $X(t)$ — cf. Example 6 in Chapter I.

14.7. Mean square ergodicity

In order that the theory of stochastic processes is useful in a description of physical phenomena one has to be able to determine from measurements (or observations) of a stochastic process $X(t)$ such probabilistic characteristics as the average value of $X(t)$, the correlation function or the distribution functions of a process. The question which arises is: if one has observed a single realization (sample function) of a stochastic process $X(t)$, under what conditions (if any) is it possible to take this realization as a base for estimation the ensamble averages such as $m_X(t)$ and

$K_X(t_1, t_2)$. The problem of determining conditions under which averages calculated from a single realization of a stochastic process can be identified with corresponding ensamble averages constitutes a subject of ergodic theorem.

In the theory of stationary processes the concept of ergodicity is most often related to strictly stationary stochastic processes; it takes origin in problems of statistical mechanics and in the famous Birkhoff-Neumann theorem. However, a stochastic process needs not be strictly stationary in order to formulate for it ergodic-type theorems. Here, we shall present the ergodic theorem for an arbitrary second order stochastic process; such theorem is commonly used in the practice of statistical inference.

Let $X(t), \quad t \in T = [a, b]$ be a stochastic process with $m_X(t) \neq 0$ and the correlation function $K_X(t_1, t_2)$. Let us denote $X^\circ(t) = X(t) - m_X(t)$.

Theorem 2.10. *Let $X(t)$ has continuous correlation function $K_X(t_1, t_2), \quad t_1, t_2 \geq 0$. Then, in the sense of mean square convergence*

(2.55)
$$\frac{1}{T} \int_0^T X^\circ(t)dt \longrightarrow 0 \quad \text{as} \quad T \to \infty$$

if and only if

(2.56)
$$\frac{1}{T^2} \int_0^T \int_0^T K_X(t_1, t_2)dt_1 dt_2 \longrightarrow 0 \quad \text{as} \quad T \to \infty$$

Proof:

Integral (2.56) exists, since $K_X(t_1, t_2)$ is continuous. The statement of the theorem follows directly from the relation

$$E\left|\frac{1}{T} \int_0^T X^\circ(t)\right|^2 = \frac{1}{T^2} \int_0^T \int_0^T K_X(t_1, t_2)dt_1 dt_2$$

If $m_X(t) = m_X = \text{const.}$, the above theorem implies that

(2.57)
$$\frac{1}{T} \int_0^T X(t)dt \xrightarrow{\text{m.s.}} m_X \quad \text{as} \quad T \to \infty$$

It is clear that the sufficient condition for (2.55) is

(2.58)
$$K_X(t_1, t_2) \to 0 \quad \text{as} \quad |t_1 - t_2| \to \infty$$

If somewhat stronger conditions are imposed on the correlation function, the ergodic relation (2.55) will also hold with probability one, that is for almost all sample functions of $X(t)$ (cf. [6]). The reader will easily note the analogy between ergodicity and law of large numbers; ergodic theorems constitute simply an extension of the law of large numbers to an uncountable family of random variables.

Let $X(t)$ be a Poisson process with parameter c. Let $X^\circ(t) = X(t) - m_X(t) = X(t) - ct$. In this case the ergodicity relation (2.55) is not satisfied. Indeed, since $K_X(t_1, t_2) = c\min(t_1, t_2)$,

$$\frac{1}{T^2}\int_0^T\int_0^T K_X(t_1,t_2)dt_1dt_2$$

$$= \frac{1}{T^2}\int_0^T dt_1\left\{\int_0^{t_1} K_X(t_1,t_2)dt_2 + \int_{t_1}^T K_X(t_1,t_2)dt_2\right\}$$

$$\frac{1}{T^2}\int_0^T dt_1\left\{\int_0^{t_1} ct_1dt_2 + \int_{t_1}^T ct_2dt_2\right\}$$

$$= \frac{1}{T^2}\cdot\frac{2}{3}cT^3 = \frac{2}{3}cT \to \infty$$

as $T \to \infty$.

Let us assume now that process $X(t)$ is weakly stationary. By virtue of general theorem 2.10 the **ergodicity** condition (2.56) takes the form

$$\lim_{T\to\infty}\frac{1}{T^2}\int_0^T\int_0^T K_X(t_2 - t_1)dt_2dt_1 = 0 \tag{2.56$'$}$$

After transformation (shown in Example 10) the above necessary and sufficient condition simplifies to the following one

$$\lim_{T\to\infty}\frac{1}{T}\int_0^T (1 - \frac{\tau}{T})K_X(\tau)d\tau = 0 \tag{2.56$''$}$$

It is clear that the above condition is satisfied if

$$K_X(\tau) \to 0 \quad , \quad \text{as} \quad |\tau| \to \infty$$

Weakly stationary processes for which conditions (2.56′) and (2.56″) hold are called *ergodic* (more exactly: *ergodic with respect to the mean*).

15. ANALYTICAL PROPERTIES OF SAMPLE FUNCTIONS

15.1. Sample function integration

Second-order properties of a stochastic process formulated by use of m.s. convergence in terms of correlation function do not provide, in general, sufficient information about the behaviour of the realizations of a process. This fact can be illustrated by a Poisson process which — as we have seen in the previous section — is m.s. continuous, but its sample functions are discontinuous. It is, therefore, of interest to throw some light on the analysis and the results associated with sample function properties.

In the theory and applications we often deal with ordinary Lebesque integral of realizations of a stochastic process. The mathematical base for sample function integration gives a theorem following directly from the known Fubini theorem.

Theorem 2.11. *Let $X(t,\gamma)$, $t \in T$, $\gamma \in \Gamma$ be a measurable stochastic process. Then almost all realizations of $X(t,\gamma)$ are mëasurable with respect to Lebesque measure μ on T. If $E[X(t,\gamma)]$ exists for $t \in T$, then this average defines a measurable (respect to μ) function of t. If A is a measurable set on T and*

$$\int_A E[|X(t,\gamma)|]\, dt < \infty \tag{2.59}$$

then almost all realizations of $X(t,\gamma)$ are Lebesque-integrable on A and

$$E\left[\int_A X(t,\gamma)dt\right] = \int_A E[X(t,\gamma)]\, dt \tag{2.60}$$

Indeed, according to the assumption, $X(t,\gamma)$ is a measurable process. So, the Fubini theorem implies that the "intersection" $X_\gamma(t)$ is, for almost all γ, a measurable function of t. The Fubini theorem also implies that if $E[X(t,\gamma)]$ exists, then it is a measurable function of t. Since the average of $X(t)$ is the integral

$$E[X(t,\gamma)] = \int_\Gamma X(t,\gamma)dP(\gamma)$$

the assumption (2.59) asserts that the double integral (first — with respect to γ, and then with respect to t) of $X(t,\gamma)$ is finite. This double integral taken in the opposite order is also finite; therefore, the integral

$$\int_A |X(t,\gamma)|\, dt$$

is finite for almost all $\gamma \in \Gamma$. This means that almost all realizations of $X(t)$ are Lebesque integrable on T. The Fubini theorem implies that the value of a double integral is independent of the order of the integration; so, we obtain (2.60).

Let us note that the basic assumption of the above theorem requires the process $X(t,\gamma)$ to be measurable. The following theorem (cf. [9]) gives a condition for an arbitrary stochastic process $X(t,\gamma)$ to be equivalent to measurable process $\tilde{X}(t,\gamma)$.

Theorem 2.12 *If for almost all t (with respect to measure μ) a stochastic process $X(t)$ is continuous in probability then a measurable and separable stochastic process $\tilde{X}(t,\gamma)$ exists which is stochastically equivalent to $X(t,\gamma)$.*

The condition of this theorem (continuity of $X(t)$ in probability) is not very restrictive. It is satisfied, for instance, by all processes whose correlation function is continuous.

If $\varphi(t)$ is a non-random Lebesque measurable function defined on $T = [a, b]$ and $X(t)$ is a measurable stochastic process, then the sample function integral

$$I_L = \int_a^b \varphi(t)X(t)dt \tag{2.61}$$

exists, and by virtue of Fubini theorem

$$\begin{aligned} &E\left[\int_a^b \varphi_1(s)X(s)ds \overline{\int_a^b \varphi_2(t)X(t)dt}\right] = \\ &= E\left[\int_a^b \int_a^b \varphi_1(s)\overline{\varphi_2(t)}X(s)\overline{X(t)}dsdt\right] = \\ &= \int_a^b \int_a^b \varphi_1(s)\varphi_2(t)K_X(s,t)dsdt \end{aligned} \tag{2.62}$$

where "—" denotes the complex conjugate.

15.2. Sample function continuity

What kind of conditions should one impose on a stochastic process so that its realizations are continuous? The answer to this question gives the following general theorem which we shall present here, without proof, in the formulation proposed by Loéve (cf. [6], [23]).

Theorem 2.13 *Let $X(t)$, $t \in T = [a, b]$ be a separable stochastic process. If for all $t, t + h$ in the interval $[a, b]$*

$$P\{|X(t+h) - X(t)| \geq g(h)\} \leq q(h) \tag{2.63}$$

where $g(h)$ and $q(h)$ are even functions of h, non-increasing as $h \to 0$ and such that

$$\sum_{n=1}^{\infty} g(2^{-n}) < \infty \quad , \sum_{n=1}^{\infty} 2^n q(2^{-n}) < \infty \tag{2.64}$$

then almost all realizations of a process $X(t)$ are continuous on T.

The above theorem and the Markov inequality:

$$P\{|X| \geq a\} \leq \frac{E|X|^p}{a^p} \quad ,p > 0$$

imply the following corollary which constitutes a useful sufficient condition for the validity of the theorem.

Corollary: *If $X(t)$, $t \in [a,b]$ is a separable stochastic process such that for certain number $p > 0$ and for all t, $t+h$ in $[a,b]$,*

$$E\,|X(t+h) - X(t)|^p \le \rho(h) \tag{2.65}$$

where

$$\text{a)}\quad \rho(h) = c|h|^{1+r} \qquad c > 0 \;,\; r > 0 \tag{2.66}$$

$$\text{b)}\quad \rho(h) = \frac{c|h|}{|\log|h||^{1+r}} \qquad r > p \tag{2.67}$$

then almost all realizations of a process $X(t)$ are continuous on $[a,b]$.

Indeed, after application of the Markov inequality, condition (2.63) will be satisfied if

$$q(h) = \frac{\rho(h)}{[g(h)]^p}.$$

Taking, in case a) $g(h) = h^\alpha$, $0 < \alpha < \frac{r}{p}$ one obtains $q(h) = c|h|^{1+r-\alpha p}$. Taking the case b) $g(h) = |\log|h||^{-\beta}$, $1 < \beta < \frac{r}{p}$ one gets $q(h) = \frac{h}{|\log|h||^{1+r-\beta p}}$. These expressions for $q(h)$ agree with the requirements of the theorem.

The above corollary with conditions (2.65)–(2.66) is known as the *Kolmogorov theorem.*

Simple calculations show that for the Wiener process

$$P\{W(t) - W(s) < x\} = \int_{-\infty}^{x} \frac{1}{\sqrt{2\pi|t-s|}} exp\left(-\frac{\xi^2}{2|t-s|}\right) d\xi$$

and

$$E\left[|W(t) - W(s)|^4\right] = \int_{-\infty}^{+\infty} \frac{\xi^4}{\sqrt{2\pi|t-s|}} exp\left(-\frac{1}{2}\frac{\xi^2}{|t-s|}\right) d\xi = $$
$$= 3(t-s)^2$$

which means that

$$E\left[|W(t+h) - W(t)|^4\right] = 3h^2$$

Therefore, Kolmogorov condition (2.65)–(2.66) is satisfied with constants $c = 3$, $p = 4$, $r = 1$. It proves that almost all realizations of the Wiener process are continuous on each finite interval.

If sample functions of a stochastic process are discontinuous it is of interest to know whether the discontinuities are of the first kind only (finite jumps). The following theorem given by Chentsov (cf. [6]) holds.

Theorem 2.14. *Let* $X(t), \quad t \in T = [a, b]$ *be a separable stochastic process. If for* $a \leq t_1 < t_2 < t_3 \leq b, t_3 - t_1 = h$

$$E\{|[X(t_3) - X(t_2)][X(t_2) - X(t_1)]|^p\} \leq c|h|^{1+r} \tag{2.68}$$

where p, r, c *are positive constants, then almost all realizations of* $X(t)$ *do not have discontinuities of the second kind (they have at most discontinuities of the first kind.).*

The conditions of this theorem are, in particular, satisfied for a process with independent increments and such that for any $t, \quad t + h \in [a, b]$

$$E[X(t+h) - X(t)]^2 < Ah$$

where A is a constant. It can be verified that a Poisson process satisfies the condition.

15.3. Sample function differentiation

We shall state here a general theorem (cf. [23]) giving a sufficient conditon in order that almost all realizations of a process are continuously differentiable. A more "effective" theorem expressing the differentiability in terms of the correlation function of a process is provided in the next subsection.

Theorem 2.15 *Let* $X(t), \quad t \in T = [a, b]$ *be a separable stochastic process. If for all* t *and* $t + T$ *the conditions of the theorem 2.13 are satisfied and for all* $t - h, t, t + h$ *in the interval* $[a, b]$

$$P\{|X(t+h) + X(t-h) - 2X(t)| \geq g_1(h)\} \leq q_1(h) \tag{2.69}$$

where g_1 *and* q_1 *are even functions of* h*, non-increasing as* $h \to 0$ *and such that*

$$\sum_{n=1}^{\infty} 2^n g_1(2^{-n}) < \infty, \quad \sum_{n=1}^{\infty} 2^n q_1(2^{-n}) < \infty \tag{2.70}$$

then all realizations of $X(t)$ *have continuous derivatives in* $[a, b]$.

15.4. Relation to second-order properties

Let us assume that $X(t)$ is a second-order stochastic process with mean $m_X(t)$ and correlation function $K_X(t_1, t_2)$. It is of interest to recognize the mutual connection between analytical properties of the sample functions of a process and its second-order properties.

A.

Let us consider first integration of a process $X(t)$. The realizations of m.s. integrable stochastic process can be non-integrable. But, if even the realizations are integrable nevertheless one should differentiate the integrals:

$$I = (\text{m.s.}) \int_a^b X(t,\gamma)dt,$$

$$I^* = \int_a^b X(t,\gamma)dt \tag{2.71}$$

where I is a m.s. Riemann integral of $X(t)$, that is it is a random variable defined as m.s. limit of appropriate approximating integral sums and I^* is a function making a correspondence between the elementary event γ and a Lebesque integral of the corresponding realization. In general, without assumption on measurability of a process, the integral I^* needs not to be a random variable. What is the relation between these two intgerals? The answer is given by the following theorem.

Theorem 2.16 *Let $X(t), t \in [a,b]$ be a measurable and m.s. continuous stochastic process. Then the integrals (2.71) exist and they are equal with probability one.*

Proof:

The existence of integrals I and I^* follows directly from the assumptions. To show the equality of I and I^* with probability one it is sufficient to show that

$$E\left[|I_n - I^*|^2\right] \xrightarrow{n\to\infty} 0$$

where I_n denotes an approximating sum of I. Of course,

$$E\left[|I_n - I^*|^2\right] =$$
$$= E\left[|I_n|^2\right] - E\left[I_n\overline{I}^*\right] - E\left[\overline{I}_n I^*\right] + E\left[|I^*|^2\right]$$

By virtue of (2.25)

$$E\left[|I_n|^2\right] \longrightarrow \int_a^b\int_a^b K_X(t_1,t_2)dt_1dt_2$$

Making use of (2.62) one has

$$E\left[|I^*|^2\right] = \int_a^b\int_a^b K_X(t_1,t_2)dt_1dt_2$$

It can be easily shown that $E\left[I_n\overline{I}^*\right]$ and $E\left[\overline{I}_n I^*\right]$ approach the same quantity equal to $E\,|I^*|^2$ as $n \to \infty$.

Remark: It can only be assumed that a process $X(t)$ is m.s. continuous. Hence, by virtue of theorem 2.12 process $X(t)$ has stochastically equivalent measurable representation.

B.

Let us consider now the questions associated with continuity of a process $X(t)$. The conditions for sample functions continuity stated in the previous subsection can be simplified for particular classes of second-order processes. Let $X(t)$ be real process. Then

$$E\left[X(t_2) - X(t_1)\right]^2 = \\ = K_X(t_2, t_2) - 2K_X(t_1, t_2) + K_X(t_1, t_1) \tag{2.72}$$

If the processes $X(t)$ is weakly stationary, then by virtue of (2.72) the Kolmogorov condition (2.65)–(2.66) takes the form ($p = 2$)

$$|K_X(0) - K_X(\tau)| \leq C\,|\tau|^{1+r} \tag{2.73}$$

Formulae (2.72) and (2.73) lead to the following assertion.

Theorem 2.17. *If a real, weakly stationary process $X(t)$ is m.s. differentiable, then its stochastically equivalent representation with continuous realizations exists.*

Let us assume that $X(t)$ is real Gaussian process. Its one-dimensional characteristic function is

$$\varphi(\lambda; t) = E\left[e^{i\lambda X(t)}\right] = e^{i\lambda m_X(t) - \frac{\lambda^2}{2}\sigma_X^2(t)}$$

Let us assume, that almost all realizations of $X(t)$ are continuous. From the Lebesque theorem on a limiting passage under an integral follows that $\varphi(\lambda; t)$ is a continuous function of t for arbitrary λ. This, however,implies continuity in t of the function $ln\varphi(\lambda; t) = i\lambda m_X(t) - \frac{\lambda^2}{2}\sigma_X^2(t)$ and therefore, continuity of the functions

$$m_X(t) = \frac{1}{i\lambda}\left[ln\varphi(\lambda, t) + \frac{\lambda^2}{2}\sigma_X^2(t)\right],$$

$$\sigma_X^2(t) = \frac{ln\varphi(\lambda, t) + ln\varphi(-\lambda, t)}{-\lambda^2}.$$

So, we have come to the following **corollary** : a necessary condition for almost all realizations of a real Gaussian process to be continuous is continuity of the average $m_x(t)$ and the variance $\sigma_X^2(t)$.

Let us assume now that $m_x(t) = 0$. By virtue of (2.72) and the Kolmogorov criterion (2.65)–(2.66) one ontains the following theorem.

Theorem 2.18. *$X(t), t \in [a,b]$ is a separable real Gaussian process and there exist constants $C > 0$ and $r > 0$ such that*

$$|\sigma_X^2(t_2) - \sigma_X^2(t_1) - 2K_X(t_1,t_2)| \leq C\,|t_2 - t_1|^{1+r} \tag{2.74}$$

then almost all realizations of $X(t)$ are continuous on $[a,b]$.

For Gaussian processes with independent increments the following necessary and sufficient condition can be easily derived.

Theorem 2.19 *In order that almost all realizations of a separable Gaussian process with independent increments $X(t), t \in [a,b]$ are continuous it is necessary and sufficient that the functions $m_X(t)$ and $\sigma_X^2(t)$ are continuous on $[a,b]$.*

C.

We shall close this section stating the theorem (cf. [23]) on the relationship between second-order properties and sample-function properties of a stochastic process; this theorem summarizes the most useful facts and includes also the relations associated with **differentiation**.

Let us denote:

$$\Delta_h X(t) = X(t+h) - X(t), \quad \Delta_h \Delta'_k K_X(t_1,t_2) = E\left[\Delta_h X(t_1) \Delta'_k \overline{X(t_2)}\right].$$

Theorem 2.20. *Let $X(t), t \in T = [a,b]$ be a separable, second-order stochastic process with continuous correlation function $K_X(t_1,t_2)$. Then*

1) m.s. *integral*

$$Y(t) = \int_a^t X(s)ds$$

exists; $X(t)$ is its m.s. derivative and if $Y_o(t)$ is a m.s. primitive of $X(t)$, then $Y_o(t) = Y(t) + Y_o(a)$; almost all realizations of $X(t)$ are integrable and m.s. integrable; the m.s. integral and the integral of sample functions are equal to each other with probability 1;

2) if for sufficiently small h, $\Delta_h \Delta'_k K_X(t,t) \leq ch^2$, then almost all realizations of $X(t)$ are continuous and $X(t)$ is a sample function derivative of its sample function primitive;

3) if the derivative $\frac{\partial^2}{\partial t_1 \partial t_2} K_X(t_1,t_2)$ exists, then $X(t)$ is m.s. differentiable, and if for sufficiently small h, $\Delta_h \Delta'_h \frac{\partial^2}{\partial t_1 \partial t_2} K_X(t_1,t_2) \leq c' h_2$, c' - constant, then almost all realizations of $X(t)$ are differentiable;

4) if there exists a finite derivative $\frac{\partial^{2n+2}}{\partial t_1^{n+1} \partial t_2^{n+1}} K_X(t_1,t_2)$ then almost all realizations ***of $X(t)$ are differentiable n times.***

Proof of the above group of assertions is not difficult. Some of the statements follow from the analysis presented in the previous sections.

16. ITÔ STOCHASTIC INTEGRAL

Among stochastic integrals (i.e. integrals of stochastic processes) a special and significant role is played by *Itô stochastic integral.* This integral is of main importance in interpretation and analysis of Itô stochastic differential and integral equations (cf. sec. 22). We shall consider here an integral of the form

$$I(\Phi) = \int_a^b \Phi(t,\gamma)dW(t,\gamma) \tag{2.75}$$

where $\Phi(t,\gamma)$ is a stochastic process satisfying appropriate conditions and $W(t,\gamma)$ is a Wiener process. Since the process $W(t)$ is non-differentiable and almost all its realizations have unbounded variation, the integral (2.75) can not be interpreted as the familiar Riemann-Stieltjes stochastic integral.

Let us note that in the case when Φ does not depend on γ, i.e. it is a deterministic function of t, the integral of the form (2.75) can be understood as m.s. Riemann-Stieltjes integral defined in section 14.4; in such a case only the orthogonality of the increments of the Wiener process need to be taken into account. If Φ is a continuously differentiable deterministic function, formula (2.34) for integration by parts can be used as a possible definition, i.e.

$$\int_a^b \Phi(t)dW(t,\gamma) =$$
$$= \Phi(b)W(b,\gamma) - \Phi(a)W(a,\gamma) - \int_a^b \Phi'(t)W(t,\gamma)dt$$

Integral (2.75) can be defined by the above formula also in case when Φ is a stochastic process $\Phi(t,\gamma)$ absolutely continuous with respect to t for each γ; each sample function is then differentiable almost everywhere with respect to t. However, if $\Phi(t,\gamma)$ is only continuous, or just integrable, this definition does not make sense. Let us add, that if $\Phi(t,\gamma)$ is a stochastic process independent of $W(t,\gamma)$, the integral (2.75) could be defined in a usual manner as m.s. Stieltjes integral (see — the remark in sec. 12.4).

In a general case, when Φ and W are not mutually independent and $\Phi(t,\gamma)$ is not absolutely continuous for each γ, integral (2.75) has to be defined in a special way. The proper meaning of integral (2.75) has first been proposed by K. Itô (cf. [46]). Here, we shall present briefly the idea of Itô integral, characterizing first the appropriate class of processes $\Phi(t,\gamma)$.

The essential feature of the construction of the Itô's integral is, that the dependence of $\Phi(t,\gamma)$ on $W(t,\gamma)$ should be non-anticipative; the random function $\Phi(t,\gamma)$ can depend, at most, on the present and past values of the Wiener process $W(\tau,\gamma)$, $\tau \le t$, but not on $W(\tau,\gamma)$ for $\tau > t$. To characterize precisely this kind of dependence, let us consider a probability space $(\Gamma, \mathcal{F}, P)$ on which the Wiener process $W(t,\gamma)$ is defined for $t \geq 0$ and a family $\{\mathcal{F}_t, t > 0\}$ of σ-algebras of subsets of Γ such that $\mathcal{F}_t \subset \mathcal{F}$ and:

a) for $t_1 < t_2$, $\quad \mathcal{F}_{t_1} \subset \mathcal{F}_{t_2}$;

b) for every $t \geq 0$, random variable $W(t,\gamma)$ is $\mathcal{F}_t$-measurable;

c) for $t_2 > t_1 \geq t$ the increments: $W(t_2,\gamma) - W(t_1,\gamma)$ are independent of $\mathcal{F}_t$.

Let us denote by $H_2[a,b]$, $\quad 0 \leq a \leq b$ the class of random processes (functions) $\Phi(t)$ defined on $[a,b]$ and satisfying the following conditions:

A) for each $t \in [a,b]$, $\quad \Phi(t,\gamma)$ is $\mathcal{F}_t$-measurable;

B) integral

$$\int_a^b \Phi^2(t,\gamma)dt$$

is finite with probability one.

The above conditions imply, that for each $t \in [a,b]$ the process $\Phi(t)$ is independent of the increments $W(t_{k+1}) - W(t_k)$ for all t_k, t_{k+1} such that $a \leq t \leq t_k \leq t_{k+1} \leq b$.

The Itô stochastic integral is defined for all $\Phi(t) \in H_2[a,b]$. As in the case of non-stochastic integrals, the Itô integral is introduced in two stages. Let $\Phi(t,\gamma)$ be a step function, that is $\Phi(t,\gamma) = \Phi(t_i,\gamma)$ for $t \in [t_i, t_{i+1}]$ where $a = t_1 < t_2 \cdots < t_{n+1} = b$ is a partition of $[a,b]$. In this case, the Itô stochastic integral is defined as follows:

$$\int_a^b \Phi(t,\gamma)dW(t) = \sum_{i=1}^{n} \Phi(t_i,\gamma)\,[W(t_{i+1}) - W(t_i)] \tag{2.76}$$

Let now $\Phi(t,\gamma)$ be an arbitrary process belonging to $H_2[a,b]$. It can be easily shown (cf. [16]), that there exists a sequence of step functions $\Phi_n(t,\gamma) \in H_2[a,b]$ for which

$$\lim_{n\to\infty} \int_a^b [\Phi(t,\gamma) - \Phi_n(t,\gamma)]^2\, dt = 0 \tag{2.77}$$

with probability one. The sequence

$$\int_a^b \Phi_n(t,\gamma)dW(t,\gamma)$$

is convergent in probability as $n \to \infty$ to a limit which does not depend on a particular choice of $\Phi_n(t,\gamma)$ such that (2.77) holds. This limit is called the **Itô stochastic integral** of the function $\Phi(t,\gamma) \in H_2[a,b]$ and is denoted as in (2.75) by $I(\Phi)$.

Making use of the definition (2.76) of the Itô integral of a step function and taking into account independence of random variables $\Phi(t_i,\gamma)$ of the increments $W(t_{i+1}) - W(t_i)$ and then performing a limiting passage (when arbitrary process $\Phi(t,\gamma) \in H_2[a,b]$ is considered) the following properties of the Itô integral can be derived:

1) if Φ_1, Φ_2 belong to $H_2[a,b]$ and α_1, α_2 are random variables such that $\alpha_1\Phi_1 + \alpha_2\Phi_2 \in H_2[a,b]$, then

$$\int_a^b [\alpha_1\Phi_1(t) + \alpha_2\Phi_2(t)]\, dW(t)$$

$$= \alpha_1 \int_a^b \Phi_1(t)dW(t) + \alpha_2 \int_a^b \Phi_2(t)dW(t) \tag{2.78}$$

2) if $\chi_{[a,b]}(t)$ is a characteristic function of the interval $[\alpha, \beta] \subset [a,b]$ then

$$\int_a^b \chi_{[a,b]}(t)dW(t) = W(\beta) - W(\alpha) \tag{2.79}$$

3) if $\Phi(t) \in H_2[a,b]$ and $\int_a^b E\left[\Phi^2(t)\right] dt < \infty$, then

$$E\left[\int_a^b \Phi(t)dW(t)\right] = 0,$$

$$E\left[\int_a^b \Phi(t)dW(t)\right]^2 = \int_a^b E\left[\Phi^2(t)\right] dt. \tag{2.80}$$

Let us consider the Itô stochastic integral as a function of the upper limit, that is

$$I(t) = \int_{t_o}^t \Phi(s,\gamma)dW(s,\gamma) \quad , \quad t \geq t_o \tag{2.81}$$

where $\Phi \in H_2[t_o, t]$; this integral is often called the Itô indefinite integral of Φ. For each t the quantity $I(t)$ is defined only with probability one, that is the process $I(t)$ is defined in the class of stochastic equivalence. Hence, it is natural to assume that a separable version of $I(t)$ is chosen. The following theorem holds (cf. [16]).

Theorem 2.21. *If $\Phi(t,\gamma) \in H_2[t_o, T]$, $t \in [t_o, T]$, and*

$$\int_{t_o}^t E\left[|\Phi(s,\gamma)|^2\right] ds < \infty, \qquad t \leq T$$

then integral (2.81) is a martingale and its sample functions are continuous with probability one; if for some natural number k,

$$\int_a^t E\left[|\Phi(s,\gamma)|^{2k}\right] ds < \infty, \qquad t_o \leq a \leq t \leq T,$$

then

$$E\left[|I(t)-I(a)|^{2k}\right] \leq$$

$$\leq [k(2k-1)]^{k-1}(t-a)^{k-1}\int_a^t E\left[|\Phi(s,\gamma)|^{2k}\right] ds \tag{2.82}$$

The concept of Itô integral can be naturally extended to random processes with values in R_n. Let $\mathbf{W}(t) = [W_1(t), \ldots, W_n(t)]$ be an n-dimensional Wiener process with mutually independent components $W_i(t)$, $i = 1, 2, \ldots, n$. Let $\mathbf{\Phi} = [\Phi_{ij}(t)]$ be a matrix $m \times n$ whose elements are processes belonging to $H_2[a, b]$. The Itô integral of $\mathbf{\Phi}$ with respect to $\mathbf{W}(t)$ is defined as

$$\int_a^b \mathbf{\Phi}(t)d\mathbf{W}(t) = \left\{\sum_{j=1}^n \int_a^b \Phi_{ij}(t)dW_j(t)\right\}_{i=1,2\ldots,m} \tag{2.83}$$

where each integral on the right is understood as the Itô integral of scalar process. The properties of a vector Itô integral are analogous to those presented above for a scalar processes; for example

$$E\left[\left|\int_a^b \mathbf{\Phi}(t)d\mathbf{W}(t)\right|^2\right] = \left[\int_a^b |\mathbf{\Phi}(t)|^2\, dt\right] \tag{2.84}$$

where

$$|\mathbf{\Phi}|^2 = \sum_{i=1}^n \sum_{j=1}^n \Phi_{ij}^2 \tag{2.85}$$

Let us take a look now at the specific **features of the Itô' integral**. Let us consider the integral

$$I(W) = \int_a^b W(s)dW(s) \tag{2.86}$$

where $W(t)$, $t \in T$ is the scalar Wiener process with unit variance. Let $a = t_1 < t_2 < \cdots < t_n < t_{n+1} = b$ be a partition of the interval $[a, b]$. One can construct two approximating sums

$$I_o(W) = \underset{n\to\infty}{\text{l.i.m.}} \sum_{i=1}^n W(t_i)\left[W(t_{i+1}) - W(t_i)\right] \tag{2.87}$$

$$I_1(W) = \underset{n\to\infty}{\text{l.i.m.}} \sum_{i=1}^n W(t_{i+1})\left[W(t_{i+1} - W(t_i)\right] \tag{2.88}$$

If the integral (2.86) could be defined as ordinary m.s. Stieltjes integral, the limits (2.87), (2.88) should be identical independently of the choice of intermediate points. The limits $I_o(W)$ and $I_1(W)$ are, however, not identical since by virtue of theorem 1.7.

$$I_1(W) - I_o(W) \tag{2.89}$$
$$= \underset{n\to\infty}{\mathrm{l.i.m.}} \sum_{i=1}^{n} [W(t_{i+1}) - W(t_i)]^2 = b - a$$

It is seen that the value of the m.s. limit of approximating sums depend on the choice of intermediate points τ_i. In order to obtain a unique definition of the integral one should select exactly one intermediate point from the interval $[t_i, t_{i+1}]$. According to (2.76) the Itô' integral is associated with the choice $\tau_i = t_i$. Other choices lead to other results. One can define a continuum of integrals using integrals defined by (2.87) and (2.88). Indeed,

$$\begin{aligned} I_\lambda(W) &= (1-\lambda)I_o + \lambda I_1 \\ &= \underset{n\to\infty}{\mathrm{l.i.m.}} \sum_{i=1}^{n} [\lambda W(t_{i+1}) + (1-\lambda)W(t_i)]\,[W(t_{i+1}) - W(t_i)] \\ &= \underset{n\to\infty}{\mathrm{l.i.m.}} \sum_{i=1}^{n} \left\{W(t_i)W(t_{i+1}) - W^2(t_i) + \lambda\,[W(t_{i+1}) - W(t_i)]^2\right\} \\ &= \underset{n\to\infty}{\mathrm{l.i.m.}} \sum_{i=1}^{n} \left\{\frac{1}{2}\left[W^2(t_{i+1}) - W^2(t_i)\right] + (\lambda - \frac{1}{2})\,[W(t_{i+1}) - W(t_i)]^2\right\} \\ &= \frac{1}{2}\left[W^2(b) - W^2(a)\right] + (\lambda - \frac{1}{2})(b-a). \end{aligned} \tag{2.90}$$

Taking $\lambda = 0$, i.e. $\tau_i = t_i$ we have the following value of the Itô integral (2.86)

$$I_o(W) = \frac{1}{2}\left[W^2(b) - W^2(a)\right] - \frac{b-a}{2} \tag{2.91}$$

However, if we evaluate the integral (2.86) according to the rules of ordinary calculus we obtain only the first component of (2.91); this value turns out to be $I_\lambda(W)$ when $\lambda = \frac{1}{2}$, i.e.

$$I_{\frac{1}{2}}(W) = \frac{1}{2}\left[W^2(b) - W^2(a)\right] \tag{2.92}$$

Let us verify now whether the usual formula for integration by parts holds in the case of stochastic integrals with respect to the Wiener process. This formula can be written as

$$\varphi(b)x(b) - \varphi(a)x(a) = \int_a^b \varphi(s)dx(s) + \int_a^b x(s)d\varphi(s)$$

The following identity holds

$$\Phi(t_{k+1})W(t_{k+1}) - \Phi(t_k)W(t_k) =$$
$$= \Phi(t_{k+1})\left[W(t_{k+1}) - W(t_k)\right] + W(t_k)\left[\Phi(t_{k+1}) - \Phi(t_k)\right] =$$

$$= W(t_{k+1})\left[\Phi(t_{k+1}) - \Phi(t_k)\right] + \Phi(t_k)\left[W(t_{k+1}) - W(t_k)\right]$$

Summing over k and taking m.s. limit as $\max_i |t_{i+1} - t_i| \to 0$ we obtain

$$\Phi(b)W(b) - \Phi(a)W(a) = I_1(\Phi) + I_0(W) = I_1(W) - I_0(\Phi) \tag{2.93}$$

The above formula means that in order to obtain a correspondence to the classical rule for integration by parts two stochastic integrals I_0 and I_1 are required. Let us notice, however, that for the symmetric integral

$$I_{\frac{1}{2}} = \frac{1}{2}(I_0 + I_1)$$

we obtain again the classical counterpart, i.e.

$$\Phi(b)W(b) - \Phi(a)W(a) = I_{\frac{1}{2}}(\Phi, dW) + I_{\frac{1}{2}}(W, d\Phi) \tag{2.94}$$

The symmetric integral $I_{\frac{1}{2}}$ defined in our case by

$$I_{\frac{1}{2}} = \underset{n\to\infty}{\text{l.i.m.}} \sum_{i=1}^{n} \frac{\Phi(t_i) + \Phi(t_{i+1})}{2}\left[W(t_{i+1}) - W(t_i)\right] \tag{2.95}$$

was first introduced by Stratonovich (cf. [56]). A general definition of the *Stratonovich integral* is the following.
Let $\mathbf{Y}(t)$ denote a m-dimensional diffusion process on the interval $[a, b]$ whose drift vector $\mathbf{A}(t, \mathbf{y})$ and diffusion matrix $\mathbf{B}(t, \mathbf{y})$ together with the derivatives $\frac{\partial \mathbf{B}(t,\mathbf{y})}{\partial y_j}$, for $j = 1, 2, \ldots, m$ are continuous in both arguments. Let $\mathbf{\Phi}(t, \mathbf{y})$ be a $l \times m$ matrix-valued function which is continuous in $\mathbf{y}$, and which satisfies, for $t \in [a, b]$, the conditions:

$$\text{there exist continuous partial derivatives} \frac{\partial \mathbf{\Phi}(t, \mathbf{y})}{\partial \mathbf{y}_j},$$

$$\int_a^b E\left[|\mathbf{\Phi}(s, \mathbf{Y}(s))\mathbf{A}(s, \mathbf{Y}(s))|\right] ds < \infty,$$

$$\int_a^b E\left[|\mathbf{\Phi}(s, \mathbf{Y}(s))\mathbf{B}(s, \mathbf{y}(s))\mathbf{\Phi}^*(s, \mathbf{Y}(s))|\right] ds < \infty;$$

the following limit

$$\text{l.i.m.}_{\Delta \to 0} \sum_{i=1}^{n} \mathbf{\Phi}\left(t_i, \frac{\mathbf{Y}(t_i) + \mathbf{Y}(t_{i+1})}{2}\right) [\mathbf{Y}(t_{i+1}) - \mathbf{Y}(t_i)] \tag{2.96}$$

where $\Delta = \max(t_{i+1} - t_i)$ is called the *Stratonovich stochastic integral* and it is denoted by

$$(S) \int_a^b \mathbf{\Phi}(s, \mathbf{Y}(s)) d\mathbf{Y}(s). \tag{2.97}$$

Stratonovich has shown that the relationship between this integral and the Itô integral defined in the case under consideration as

$$\int_a^b \mathbf{\Phi}(s, \mathbf{Y}(s)) d\mathbf{Y}(s) = = \text{l.i.m.}_{\Delta_n \to 0} \sum_{i=1}^{n} \mathbf{\Phi}(t_i, \mathbf{Y}(t_i)) [\mathbf{Y}(t_{i+1}) \mathbf{Y}(t_i)] \tag{2.98}$$

is given by the formula

$$(S) \int_a^b \mathbf{\Phi}(s, \mathbf{Y}(s)) s\mathbf{Y}(s) = \int_a^b (\mathbf{\Phi}(s, \mathbf{Y}(s)) d\mathbf{Y}(s) + \frac{1}{2} \sum_{j,k=1}^{m} \int_a^b [\Phi_{y_k}(s, \mathbf{Y}(s))]_{\cdot j}\, b_{jk}(s, \mathbf{Y}(s)) ds \tag{2.99}$$

where the l-vector $[\Phi_{y_k}]_{\cdot j}$ is the j-th column of the $l \times m$ matrix $\Phi_{y_k} = \frac{\partial \Phi_{ij}}{\partial y_k}$.
In the scalar case, when $l = m = 1$

$$(S) \int_a^b \Phi(s, Y(s)) dY(s) = = \int_a^b \Phi(s, Y(s)) dY(s) + \frac{1}{2} \int_a^b \frac{\partial \Phi(s, Y(s))}{\partial y} b(s, Y(s)) ds \tag{2.100}$$

In particular, the integrals of Itô and Stratonovich coincide when $\Phi(t, y) = \Phi(t)$.
It can be proven in general that the Stratonovich integral satisfies the formal rules of ordinary calculus (formulae (2.92) and (2.94) well illustrate it in a particular case) and therefore can be considered as more handy in practical applications in comparison with the Itô integral. Unfortunately, the disadvantage of the Stratonovich integral is that the formulae expressing the averaging of this integral are more complicated than these corresponding to Itô integral.

In closing it is worth noting that in many practical problems described by stochastic differential equations one is interested in the limit of a sequence of Riemann-Stieltjes integrals of the Itô form but with a sequence of smooth approximations $\{W_n(t,\gamma)\}$ replacing the Wiener process $W(t,\gamma)$. In the paper [59] it has been shown that if $W_n(t,\gamma)$ have piecewise continuous derivatives, then for example,

$$\int_a^b W_n(s,\gamma)W_n(s,\gamma) \longrightarrow \frac{1}{2}\left[W^2(b,\gamma) - W^2(a,\gamma)\right]$$

This means that a smoothing prior to evaluation of the stochastic integral leads, when we take the limit in the result, to the value which is compatible with the Stratonovich integration. Further discussion of this question is given in sec. 22.

17. STOCHASTIC DIFFERENTIALS. ITÔ FORMULA.

The concept of the Itô integral is closely related to the stochastic Itô differential. Let $X(t)$, $t \in [a,b]$ be a stochastic process.

Definition 2.6. *If functions $a(t)$ and $b(t)$ exist such that for arbitrary t_1, t_2 $(a \leq t_1 < t_2 \leq b)$*

$$X(t_2) - X(t_1) = \int_{t_1}^{t_2} a(t)dt + \int_{t_1}^{t_1} b(t)dW(t) \tag{2.101}$$

then we say that $X(t)$ has a **stochastic differential** *$dX(t)$ given by*

$$dX(t) = a(t)dt + b(t)dW(t) \tag{2.102}$$

To assure existence of the integrals occuring in (2.101) it is assumed that the process $b(t)$ belongs to $H_2[a,b]$, $a(t)$ is measurable with respect to $\mathcal{F}_t$ for each t and there exists with probability one a finite integral $\int_a^b |a(t)|\, dt$.

The operation of differentiation introduced by the above definition is linear. However, formulae for differentiation of a product and superposition differ from those in classical calculus. This is easily seen from the example given in the previous section. Indeed, formula (2.91) indicates that for arbitrary $t \in [a,b]$

$$\int_0^t W(t)dW(t) = \frac{1}{2}W^2(t) - \frac{t}{2}$$

and by virtue of definition 2.6 we have

$$W^2(t) = \int_0^t dt + 2\int_0^t W(t)dW(t)$$

or

$$d(W^2(t) = dt + 2W(t)dW(t)$$

The common rules lead, however, to the following expression

$$d(W^2(t)) = 2W(t)dW(t)$$

This specific feature of the stochastic differential is general.

Theorem 2.22. *Let $X(t)$ possess the stochastic differential*

$$dX(t) = a(t)dt + b(t)dW(t)$$

and let $f(t,x)$ be a continuous function in t and x together with its derivatives $\frac{\partial f}{\partial t}$, $\frac{\partial f}{\partial x}$, $\frac{\partial^2 f}{\partial x^2}$. Then the process $f(t,X(t))$ has a stochastic differential (with respect to the same Wiener process $W(t)$) given by

$$\begin{aligned} &df(t,X(t)) \\ &= \left[\frac{\partial f}{\partial t}(t,X(t)) + a(t)\frac{\partial f}{\partial x}(t,X(t)) + \frac{1}{2}b^2(t)\frac{\partial^2 f}{\partial x^2}(t,X(t))\right] dt \\ &+\frac{\partial f}{\partial x}(t,X(t))b(t)dW(t) \end{aligned} \tag{2.103}$$

This formula is called **Itô's formula.**

Let us notice that formula (2.103) differs from the standard calculus formula for total derivatives by the extra term

$$\frac{1}{2}b^2(t)\frac{\partial^2 f}{\partial x^2}(t,X(t))dt$$

This term gives often rise to errors in formal transformations of stochastic differential equations.

The outline of the proof of the Itô formula is as follows.

The increment of $f(X(t))$ on the interval $\Delta = [s,t]$ can be written as

$$\begin{aligned} &f(t,X(t)) - f(s,X(s)) = \\ &= \sum_{i=0}^{n-1} [f(t_{i+1}, X(t_{i+1})) - f(t_i, X(t_i))] \end{aligned}$$

where $s = t_o < t_1 < \cdots < t_n = t$ is a partition od Δ such that $\delta = \max_{0 \le i \le n-1}(t_{i+1} - t_i) \to 0$ as $n \to \infty$. By virtue of the Taylor formula (denoting $\Delta t_i = t_{i+1} - t_i$ and $\Delta X(t_i) = X(t_{i+1} - X(t_i)$) one has

$$\begin{aligned} &f(t_{i+1}, X(t_{i+1})) - f(t_i, X(t_{i+1})) \\ &= \frac{\partial f}{\partial t}(t_i, X(t_i))\Delta t_i + \frac{\partial f}{\partial x}(t_i, X(t_i))\Delta X(t_i) \\ &+\frac{1}{2}\frac{\partial^2 f}{\partial x^2}(t_i, X(t_i))[\Delta X(t_i)]^2 + R_i \end{aligned}$$

where R_i is the remainder term. It is natural to expect that the sums

$$\sum_{i=0}^{n-1} \frac{\partial f}{\partial t}(t_i, X(t_i))\,\Delta t_i, \quad \sum_{i=0}^{n-1} \frac{\partial f}{\partial x}(t_i, X(t_i))\,\Delta X(t_i)$$

are convergent to the integrals

$$\int_s^t \frac{\partial f}{\partial t}(\tau, X(\tau))\,d\tau,$$

$$\int_s^t \frac{\partial f}{\partial x}(\tau, X(\tau))\,dX(\tau)$$

$$= \int_s^t a(\tau)\frac{\partial f}{\partial x}(\tau, X(\tau))\,d\tau + \int_s^t b(\tau)\frac{\partial f}{\partial x}(\tau, X(\tau))\,dW(\tau)$$

respectively.

It can be shown that $\sum_{i=0}^{n-1} R_i$ tends to zero in probability. Let us consider in more details the sum

$$I_n = \frac{1}{2}\sum_{i=0}^{n-1} \frac{\partial^2 f}{\partial x^2}(t_i, X(t_i))\,[\Delta X(t_i)]^2$$

Since

$$[\Delta X(t_i)]^2 \approx$$
$$\approx a^2(t_i)\,[\Delta t_i]^2 + 2a(t_i)b(t_i)\Delta t_i W(t_i) + b^2(t_i)\,[\Delta W(t_i)]^2$$

where the two first terms are infinitely small of higher order than Δt, the asymptotic behaviour of I_n will be the same as

$$\tilde{I}_n = \frac{1}{2}\sum_{i=0}^{n-1} b^2(t_i)\frac{\partial^2 f}{\partial x^2}(t_i, X(t_i))\,[\Delta W(t_i)]^2$$

Let us take now a second partition $\tilde{\Delta} \subset \Delta$ of the interval $[s,t]$, $s = t_0 < t_1 < \cdots < t_N = t$ such that $\tilde{\delta}_N = \max_{0\le k\le N-1}(\tilde{t}_{k+1} - \tilde{t}_k) \to 0$ as $N \to \infty$ and each interval $[t_k, t_{k+1}]$ includes such a number n_k of points of the partition Δ, that $\min_k n_k \to \infty$ as $n \to \infty$. Then

$$\tilde{I}_n = \frac{1}{2}\sum_{k=0}^{N-1} \widetilde{\sum} b^2(t_i)\frac{\partial^2 f}{\partial x^2}(t_i, X(t_i))\,[\Delta W(t_i)]^2$$

where "$\sim$" in the second sum denotes summation over such i for which $t_i \in [\tilde{t}_k, \tilde{t}_{k+1}]$. Due to the continuity of $X(t)$ and $\frac{\partial^2 f}{\partial x^2}(t,x)$ the following asymptotic equation holds

$$\tilde{I}_n \approx \frac{1}{2} \sum_{k=0}^{N-1} b^2(\tilde{t}_k) \frac{\partial^2 f}{\partial x^2}(\tilde{t}_k, X(\tilde{t}_k)) \widetilde{\sum} [\Delta W(t_i)]^2$$

By virtue of the theorem 1.7 one obtains

$$\tilde{I}_n \approx \frac{1}{2} \sum_{k=0}^{N-1} b^2(\tilde{t}_k) \frac{\partial^2 f}{\partial x^2}(\tilde{t}_k, X(\tilde{t}_k)) (\tilde{t}_{k+1} - \tilde{t}_k)$$

Since $\tilde{I}_n$ constitutes an approximating sum of the integral

$$\frac{1}{2} \int_s^t b^2(\tau) \frac{\partial^2 f}{\partial x^2}(\tau, X(\tau))\, d\tau$$

we finally obtain

$$\begin{aligned} &f(t, X(t)) - f(s, X(s)) \\ &= \int_s^t \left[\frac{\partial f}{\partial t}(\tau,, X(\tau)) + a(\tau) \frac{\partial f}{\partial x}(\tau, X(\tau)) + \frac{1}{2} b^2(\tau, X(\tau)) \frac{\partial^2 f}{\partial x^2}(\tau, X(\tau)) \right] d\tau \\ &\quad + \int_s^t b(\tau) \frac{\partial f}{\partial x}(\tau, X(\tau))\, dW(\tau) \end{aligned}$$

which implies the Itô formula (2.103).

Analogously as in the above scalar case one can define a stochastic differential of a vector process.

Definition 2.7. *If* $\mathbf{X}(t)$ *is an* m*-dimensional stochastic process,* $t \in [a,b]$ *and for arbitrary* t_1, t_2, $(a \le t_1 < t < t_2 \le b$

$$\mathbf{X}(t_2) - \mathbf{X}(t_1) = \int_{t_1}^{t_2} \mathbf{A}(t) dt + \int_{t_1}^{t_2} \mathbf{B}(t) d\mathbf{W}(t) \tag{2.104}$$

where $\mathbf{A}(t) = [a_1(t), \ldots, a_m(t)]$, $\mathbf{B}(t) = [b_{ij}(t)]$ *is a* $(m \times n)$*-matrix and* $\mathbf{W}(t)$ *is an* n*-dimensional Wiener process then we say that* $\mathbf{X}(t)$ *has stochastic differential* $d\mathbf{X}(t)$ *given by*

$$d\mathbf{X}(t) = \mathbf{A}(t)dt + \mathbf{B}(t)d\mathbf{W}(t) \tag{2.105}$$

A multi-dimensional version of the theorem 2.22 is

Theorem 2.23. *Let an m-dimensional process* $\mathbf{X}(t)$ *possess stochastic differential (2.105) and let* $f(t,\mathbf{x})$ *be a continuous function on* $[a,b]\times R_m$ *with continuous partial derivatives* $\frac{\partial}{\partial t}f(t,\mathbf{x})$ $\frac{\partial}{\partial x_i}f(t,x)$, $\frac{\partial^2}{\partial x_i \partial x_j}f(t,\mathbf{x})$. *Then the process* $\mathbf{Y}(t)=f(t,\mathbf{X}(t))$ *has stochastic differential given by*

$$\begin{aligned} d\mathbf{Y}(t) = \Bigg[\frac{\partial f}{\partial t}(t,\mathbf{X}(t)) &+ \sum_{i=1}^{m}\frac{\partial}{\partial x_i}f(t,\mathbf{X}(t))\,a_i(t) \\ &+ \frac{1}{2}\sum_{l=1}^{n}\sum_{i,j=1}^{m}\frac{\partial^2}{\partial x_i\partial x_j}f(t,\mathbf{X}(t))\,b_{il}(t)b_{jl}(t)\Bigg]dt \\ &+ \sum_{l=1}^{n}\sum_{i=1}^{m}\frac{\partial f}{\partial x_i}(t,\mathbf{X}(t))\,b_{il}(t)dW_l(t) \end{aligned} \tag{2.106}$$

It is easily seen that in the case $m=n=1$ the above formula reduces to (2.103).

Let us consider now illustrative examples.

Let the function $f(t,\mathbf{x})$ in theorem 2.23 be as follows

$$f(t,\mathbf{x}) = f(\mathbf{x}) = f(x_1,x_2) = x_1\cdot x_2$$

By virtue of formula (2.106) one gets the formula for stochastic differential of a product. If

$$\begin{aligned} dX_1(t) &= a_1(t)dt + b_1(t)dW(t), \\ dX_2(t) &= a_2(t) + b_2(t)dW(t), \end{aligned}$$

then

$$\begin{aligned} & d\left(X_1(t)X_2(t)\right) \\ &= \left[X_1(t)a_2(t)+X_2(t)a_1(t)+b_1(t)b_2(t)\right]dt \\ &+ \left[X_1(t)b_2(t)+X_2(t)b_1(t)\right]dW(t) \\ &= X_1(t)dX_2(t)+X_2(t)dX_1(t)+b_1(t)b_2(t)dt \end{aligned} \tag{2.107}$$

This formula can be written down in the integral form

$$X_1(b)X_2(b) - X_1(a)X_2(a) =$$
$$\int_a^b X_1(t)dX_2(t) + \int_a^b X_2(t)dX_1(t) + \int_a^b b_1(t)b_2(t)dt$$

This is a rule for integration of Itô stochastic integrals by parts. In comparison with the corresponding formula for ordinary integrals (2.92) here we have the extra term $\int_a^b b_1(t)b_2(t)dt$.

Let us assume in formula (2.102) that $a(t) = 0, \quad b(t) = 1$ and $X(t) = W(t)$. Then the Itô formula (2.103) gives the expression for a stochastic differential of a smooth function of a Wiener process

$$df(t, W(t))$$
$$\left[\frac{\partial f}{\partial t}(t, W(t)) + \frac{1}{2}\frac{\partial^2 f}{\partial x^2}(t, W(t))\right] dt + \frac{\partial f}{\partial x}(t, W(t))\, dW(t)$$

In the particular case, when $f(t, x) = f(x)$ one obtains

$$df(W(t)) = \frac{1}{2}\frac{\partial^2 f}{\partial x^2}(W(t))\, dt + \frac{\partial f}{\partial x}(W(t))dW(t) \tag{2.108}$$

or in the integral form ($a = 0, \quad b = t$)

$$f(W(t)) = f(0) + \frac{1}{2}\int_0^t \frac{\partial^2 f}{\partial x^2}(W(s))\, ds + \int_0^t \frac{\partial f}{\partial x}(W(s))\, dW(s) \tag{2.109}$$

Let $f(x) = x^n, \quad n = 1, 2 \ldots; t \geq 0$. Formula (2.108) then yields

$$d(W^n(t)) = nW^{n-1}(t)dW(t) + \frac{n(n-1)}{2}W^{n-2}(t)dt \tag{2.110}$$

Let $f(x) = e^x$. In this case

$$Y(t) = f(W(t)) = e^{W(t)} \tag{2.111}$$

and formula (2.108) gives

$$dY(t) = e^{W(t)}dW(t) + \frac{1}{2}e^{W(t)}dt$$

Therefore, $Y(t)$ satisfies the following stochastic differential equation

$$dY(t) = \frac{1}{2}Y(t)dt + Y(t)dW(t), \qquad Y(0) = 1 \tag{2.112}$$

or, conversely, the above stochastic equation has the solution (2.111).

18. COUNTING STOCHASTIC INTEGRAL

In Section 8 devoted to point stochastic processes we introduced a concept of a counting random measure and its special case — a Poisson random measure. Here we assume that a random measure N with integer values is defined on the σ-algebra $\mathcal{B}$ of measurable sets of space $[0,\infty) \times R_n$. This measure has the property that its values are independent on non-overlapping sets belonging to $\mathcal{B}$ and for all sets of the form $[t_1, t_2] \times A$ (where A is a measurable set in R_n) the random variable $N([t_1, t_2] \times A)$ has the Poisson distribution with the intensity parameter $\nu(A)$.

It can be shown that (cf. [91]) for an arbitrary measurable set A from R_n process $\tilde{N}(A) = \tilde{N}([0,t] \times A) = N([0,t] \times A) - t\nu(A), \quad t \in [0,\infty)$ is a square-integrable martingale with respect to the family of σ-algebras $\mathcal{F}_t$. This statement implies that the integral (of the appropriate functions) with respect to Poisson random measure can be defined using the notion of a stochastic integral with respect to the martingale (cf. next Section). Here we indicate more elementary construction. The integral of interest is of general form

$$I_t(\Phi, A) = \int_{t_0}^{t} \int_A \Phi(s, \mathbf{u}, \gamma) N(ds, d\mathbf{u}) \tag{2.113}$$

where Φ is a suitable stochastic process depending on time variable s and space variable $\mathbf{u} \in R_n, \quad A \subseteq R_n$.

Like in definition of the Itô integral a delicate point associated with the definition of the above integral is the possible dependence of Φ on random measure N. To see the essence of the problem let us take a special case of (2.113) where Φ does not depend on $\mathbf{u}$, and N is a counting measure on time axis, that is $N([t_1, t_2])$ characterizes the number of events on $[t_1, t_2]$ regardless of their place in space. It is worth noting that the fact $\mathbf{u} \in R_n$ does not necessarily mean that $\mathbf{u}$ denotes a spatial variable; $\mathbf{u}$ denotes certain property of event which occurs at time (it can be, for instance, a place in which an event appears, but also a random strength of randomly occuring impuls etc.). The variable $\mathbf{u}$ is a mark which is used to identify random quantities associated with the event it accompanies.

Let us additionally assume (for illustration) that $\Phi(t, \gamma) = N(t, \gamma)$. So, we have the integral

$$I_t(N_t) = \int_{t_0}^{t} N(s) dN(s) \tag{2.114}$$

Evaluation of (2.114) according to the usual rules of integration would suggest that $I_t(N_t) = \frac{N_t^2}{2}$. However, more careful reasoning indicates that the integral (2.114) does not exist in the usual Riemann-Stieltjes sense because of simultaneous jumps in the integrand and the integrator. In addition, it turns out that (cf. [91]) the limit of approximating integral sums depends (like in the case of integral (2.86)) on the choice of intermediate points τ_i in elementary subinterval. In order to obtain a unique definition one should select exactly one intermediate point from the interval $[t_i, t_{i+1}]$ in the sum

$$S_n(\tau_i) = \sum_{i=1}^{n} N(\tau_i)\,[N(t_{i+1}) - N(t_i)]$$

The choice $\tau_i = t_i$ is consistent with integrals with respect to martingales; for this choice the term $N(\tau_i)$ does not anticipate the increment $N(t_{i+1})-N(t_i)$ since Poisson process has independent increments.

Let us come back to integral (2.113). The following conditions are assumed.

(i) $N = N(t, \mathbf{u}, \gamma)$ is a Poisson process defined on $[t_o, \infty) \times R_n$. The integral

$$I_N = \int_{t_1}^{t_2} \int_A N(dt, d\mathbf{u}) \tag{2.115}$$

characterizes a random number of events occuring in the interval $[t_1, t_2]$ with "marks" $\mathbf{u} \in A \subseteq R_n$.

(ii) For all intervals $[t_1, t_2]$ and $A \subseteq R_n$, there holds

$$E\left\{\int_{t_1}^{t_2} \int_A N(ds, d\mathbf{u})\right\} = P\{\mathbf{u} \in A\} \int_{t_1}^{t_2} \nu(s)ds < \infty$$

where $\nu(s)$ is the intensity function of the Poisson process.

(iii) The process $\Phi(t, \mathbf{u}, \gamma)$, $t \geq t_o$, $\mathbf{u} \in R_n$ does not anticipate the process $N(t, \mathbf{u}, \gamma)$. This means that $\Phi(t, \mathbf{u}, \gamma)$ is statistically independent of I_n for any choice of $\mathbf{u} \in R_n$ and $A \subseteq R_n$ and for all triples $t \leq t_1 \leq t_2$.

(iv) Almost all realizations of the process $\Phi(t, \mathbf{u}, \gamma)$ are continuous in $\mathbf{u}$, left continuous in t, bounded for $(t, \mathbf{u}) \in [t_1, t_2] \times A$, where $[t_1, t_2]$ is any interval in $[t_o, \infty)$ and A is any bounded set in R_n.

The above conditions ensure the existence of the counting integral defined as follows (cf. [33]).

Definition 2.8. *Let A be a set in R_n and consider the following sequence of partitions of $[t_o, t] \times A$. Let $A_{1_p}, A_{2_p}, \ldots$ for $p = 1, 2 \ldots$ be a sequence of pairwise disjoint partitions of A such that $A_{k_p} \cap A_{i_p} = \emptyset$ for $k \neq j$, $A = \bigcup_k A_{k_p}$, and $\max_k |A_{k_p}| \to 0$ as $p \to \infty$, where $|A_{k_p}|$ denotes the diameter of the set A_{k_p}.*

Similarly, let $t_o = t_{oq} < t_{1q} < \cdots < t_{qq} = t$, for $q = 1, 2, \ldots$ be a sequence of partitions of $[t_o, t]$ such that $\max_i |t_{jq} - t_{j-1,q}| \to 0$ as $q \to \infty$. Let

$$I_t^{(p,q)}(\Phi, A) =$$

$$= \sum_{j=1}^{q} \sum_{k=1}^{p} \Phi(t_{j-1,q}, \tilde{u}_{kp}, \gamma) \int_{t_{j-1,q}}^{t_{j,q}} \int_{A_{kp}} N(dt, du) \tag{2.116}$$

be an approximative sum, where $\tilde{u}_{kp}$ is an arbitrary point in $A_{p_{k_p}}$. If independently of the partition, $I_t^{(p,q)}(\Phi, A)$ converges with probability one to a unique finite limit as p and q tend to infinity, then we say that the counting integral of Φ with respect to N exists on $[t_o, t] \times A$ and the limit is denoted by (2.113).

Theorem 2.24. *Under the conditions(i)-(iv) given above, the counting integral $I_t(\Phi, A)$ exists and can be evaluated according to the formula*

$$I_t(\Phi, A) = \int_{t_o}^{t} \int_A \Phi(s, \mathbf{u}, \gamma) N(ds, d\mathbf{u}) =$$

$$= \begin{cases} 0 & , \quad I_N = 0 \\ \sum_{n=1}^{I_N} \Phi(\tau_n, \mathbf{u}_n, \gamma) & , \quad I_N \geq 1 \end{cases} \tag{2.117}$$

where I_N is given by (2.115), and τ_n and u_n are the occurence time and mark of the n-th event. Furthermore, if

$$\tilde{I}_t(\Phi, A) = I_t(\Phi, A) - \int_{t_o}^{t} \int_A \Phi(s, \mathbf{u}, \gamma) \nu(s) ds dP(\mathbf{u}) \tag{2.118}$$

then

$$E\left[\tilde{I}_t(\Phi, A)\right] =$$

$$= E\left[I_t(\Phi, A)\right] - \int_{t_o}^{t} \int_A E\left[\Phi(s, \mathbf{u}, \gamma)\right] \nu(s) ds dP(\mathbf{u}) = 0 \tag{2.119}$$

and

$$E\left[\left|\tilde{I}_t(\Phi, A)\right|^2\right] = \int_{t_o}^{t} \int_A E\left[|\Phi(s, \mathbf{u}, \gamma)|^2\right] \nu(s) ds dP(\mathbf{u}) \tag{2.120}$$

where $P(\mathbf{u})$ is the probability distribution function for the mark random variable $\mathbf{u}$ *and $|\cdot|$ denotes the norm of the vector.*

The continuity and boundedness assumption (iv) are not necessary for the definition of the counting integral. There exist other alternative constructions based on more relaxed assumptions.

Applications of the formula (2.117) to integral (2.114) gives the result

$$I_t(N_t) = \int_{t_o}^{t} N(s) dN(s) = \frac{1}{2} N_t^2 - \frac{1}{2} N_t \tag{2.121}$$

which indicates a difference with usual rules of integration. Also, a differentiation of random process defined by counting integrals does obey the rules of ordinary calculus. For example, the differential of the process Z_t, $t \geq t_o$ defined by

$$Z_t = \frac{1}{2} N_t + \int_{t_o}^{t} N(s) dN(s)$$

is not $N_t dN_t$ which is customary to the rules of ordinary calculus. Since by virtue of (2.121), $Z_t = \frac{1}{2}N_t^2$

$$dZ_t = Z_{t+\Delta t} - Z_t = \frac{1}{2}\left[N_{t+\Delta t}^2 - N_t^2\right]$$
$$= \frac{1}{2}\left[(dN_t + N_t)^2 - N_t^2\right] = \frac{1}{2}(dN_t^2) + N_t dN_t$$

In ordinary calculus the term $\frac{1}{2}(dN_t)^2$ is of second order compared to $N_t dN_t$ and it is neglected. Here, $(dN_t)^2 = dN_t$ since dN_t is equal to one or zero depending on whether or not an event occurs in $[t, t+\Delta t]$. Therefore,

$$dZ_t = \left(\frac{1}{2} + N_t\right) dN_t \tag{2.122}$$

This example indicates that differentiation of random processes defined with use of the counting integral is governed by its own rules. We shall now clarify this problem in more general formulation which is used in analysis of stochastic differential equations.

Let vector random functions $\mathbf{A}(t)$, $\mathbf{B}(t)$, $\mathbf{C}(t)$ satisfy all conditions stated in the previous and this section assuring existence of the integrals below.

If there exist such functions $\mathbf{A}(t)$, $\mathbf{B}(t)$, $\mathbf{C}(t)$ that for arbitrary $t \in [0, T]$

$$\mathbf{X}(t) =$$
$$= \mathbf{X}(0) + \int_0^t \mathbf{A}(\tau)d\tau + \int_0^t \mathbf{B}(\tau)d\mathbf{W}(t) + \int_0^t \int_{R_n} \mathbf{C}(\tau, \mathbf{u})N(d\tau, d\mathbf{u}) \tag{2.123}$$

then we say that $\mathbf{X}(t)$ has the *stochastic* **differential $d\mathbf{X}(t)$ given by**

$$d\mathbf{X}(t) =$$
$$= \mathbf{A}(t)dt + \mathbf{B}(t)d\mathbf{W}(t) + \int_{R_n} \mathbf{C}(t, \mathbf{u})N(d\tau, d\mathbf{u}) \tag{2.124}$$

The Itô formula (2.106) can be extended to the case considered here.

Theorem 2.25. *Let an m-dimensional random process $\mathbf{X}(t)$ has stochastic differential (1.124) where*

$$\mathbf{A}(t) = [a_1(t), \ldots, a_m(t)]$$
$$\mathbf{B}(t) = [b_{ij}(t)] \textit{is an } (m \times n)\textit{-matrix}$$
$$\mathbf{C}(t) = [c_1(t, \mathbf{u}), \ldots, c_m(t, \mathbf{u})]$$
$$\mathbf{W}(t) - \textit{an } n\textit{-dimensional Wiener process}$$

and $f(t,\mathbf{x}) = f(t, x_1, \ldots, x_m)$ is a twice continuously differentiable function with respect to x_i, $(i = 1, \ldots, m)$ and having $\frac{\partial}{\partial t} f(t,\mathbf{x})$. Then the process $\mathbf{Y}(t) = f[t, \mathbf{X}(t)]$ has the stochastic differential given by

$$
\begin{aligned}
d\mathbf{Y}(t) = \Bigg[& \frac{\partial f}{\partial t}(t, \mathbf{X}(t)) \\
& + \sum_{i=1}^{m} \frac{\partial}{\partial x_i} f(t, \mathbf{X}(t))\, a_i(t) \\
& + \frac{1}{2} \sum_{i,j=1}^{m} \frac{\partial^2}{\partial x_i \partial x_j} f\left(t, \mathbf{X}(t) \sum_{l=1}^{n} b_{il}(t) b_{jl}(t)\right) \Bigg] dt \\
& + \sum_{l=1}^{n} \sum_{i=1}^{m} \frac{\partial f}{\partial x_i}(t, \mathbf{X}(t))\, b_{il}(t) dW_l(t) \\
& + \int_{R_n} \left[f(t, \mathbf{X}(t) + \mathbf{C}(t, \mathbf{u})) - f(t, \mathbf{X}(t)) \right] N(dt, d\mathbf{x})
\end{aligned}
\tag{2.125}
$$

It is seen that the above formula reduces to the Itô formula (2.106) if the term associated with counting integral vanishes.

It is worth noting that when the mark space R_n reduces to a single point (i.e. $R_n = 1$) so that all the marks are unity, and the diffusion term vanishes ($B(t) = 0$) differential (2.124) is

$$
d\mathbf{X}(t) = \mathbf{A}(t)dt + \mathbf{C}(t)dN_t \tag{2.126}
$$

and

$$
\begin{aligned}
d\mathbf{Y}(t) = df\left[t, \mathbf{X}(t)\right] = \Bigg[& \frac{\partial}{\partial t} f(t, \mathbf{X}(t)) \\
& + \sum_{i=1} \frac{\partial}{\partial x_i} f\Big(t, \mathbf{X}(t)\Big) a_i(t) \Bigg] dt \\
& + \left[f\Big(t, \mathbf{X}(t) + \mathbf{C}(t)\Big) - f\Big(t, \mathbf{X}(t)\Big) \right] dN_t
\end{aligned}
\tag{2.127}
$$

where N_t is the number of points which occur during $[t_0, t]$.

We shall use the above extended Itô formula in the next chapter.

19. GENERALIZATIONS

The concepts and results presented in this chapter have been generalized in various important directions. As far as stochastic calculus for regular processes is concerned the extentions to higher dimensions (i.e. to multi-dimensional stochastic processes) are straightforward, although they can be connected with some difficulties when we would be interested, for example, in analytical properties of sample functions.

If $X(t,\gamma)$ is a Hilbert space valued process there are various types of continuity, differentiation and integration. In defintions the absolute value should be replaced by the norm induced by the inner product. For example, an H-valued process $X(t,\gamma)$ is said to be continuous in mean square if

$$E\left[\|X(t+\Delta t)-X(t)\|_H^2\right] \longrightarrow 0 \quad \text{as} \quad \Delta t \to 0 \tag{2.128}$$

Of special interest are various *generalizations of the Itô stochastic calculus.* Three directions are of particular importance.

1) The first extention deals with Itô type integral, in which the Wiener process $W(t)$ is replaced by a class of martingales (cf. Sec. 9). So, the following integral

$$I(\Phi) = \int_a^b \Phi(t,\gamma)dM(t,\gamma) \tag{2.129}$$

is defined, where $M(t,\gamma)$ is a sample — continuous second-order martingale. Like the Itô integral with respect to the Wiener process, integral (2.129) is defined in two steps:

a) if Φ is a step function, then

$$\tilde{I}(\Phi) = \sum_i \Phi_i(\gamma)\left[M(t_{i+1},\gamma) - M(t_i,\gamma)\right]$$

b) if

$$\int_a^b |\Phi_n(t,\gamma) - \Phi(t,\gamma)|\, dM(t,\gamma) \overset{n\to\infty}{\longrightarrow} 0 \quad \text{with pr-ty 1}$$

then

$$\tilde{I}(\Phi_n) \overset{n\to\infty}{\longrightarrow} I(\Phi) \qquad \text{in probability.}$$

Also the Itô formula (the Itô differentiation rule) has been extended to this case.

It is clear that (2.129) is more general than the integral (2.75). For example, the process $M(t,\gamma)$ can be taken in the form of integral (2.81) which, by virtue of theorem 2.21, is a martingale. This extension includes also a wide class of integrals with respect to jump processes and is related to the integrals with respect random measures (cf. [24]).

2) The second important generalization is the Itô type integral with respect to a Hilbert space valued Wiener process (and also, with respect to a Hilbert space valued martingale). The way of constructing such an integral is similar to that used in the scalar case presented in Sec. 16. Of course, there are differences connected with the fact that $W(t)$ is a Hilbert valued process (cf. [7]. [77]).

It is worth noting that the Itô integral with respect to a Hilbert valued Wiener process can also be defined using the definition 1.33 (formula 1.216) of an H-valued Wiener process, that is as

$$\int_a^b \Phi(t)dW(t) = \sum_{i=0}^{\infty} \int_a^b \Phi(t)e_i d\beta_i(t) \tag{2.130}$$

where $\beta_i(t)$ are real, independent Wiener processes, $\{e_i\}$ is an orthogonal basis in H and Φ satisfies appropriate conditions. The Itô' differential and the Itô' formula have also been extended to this abstract case.

The Stieltjes type integrals with respect to more general abstract valued stochastic processes have been constructed (cf. [49], [50], [57] and references therein). These concepts are of basic importance for formulation and analysis of stochastic partial differential equations via stochastic equations in abstract spaces.

3) The third generalization which should be mentioned is the Itô integral with respect to the Wiener random field (on plane or, on more-dimensional space) — cf. [41], [54]. In the case of the Wiener field $W(x, y)$ on R_2 ,the integral

$$\int_{I_x \times I_y} \Phi(x, y)dW(x, y) \tag{2.131}$$

is defined first for step functions, constant on rectangles, and then extended in the standard way to arbitrary functions $\Phi(x, y)$ belonging to the class $H_2(I_x \times I_y)$. Similar as in the construction presented in Section 16 the dependence of $\Phi(x, y)$ on $W(x, y)$ should be non-anticipative, what means here that Φ can depend, at most, on "present" and "past" values of the Wiener field $W(x, y, \gamma)$. The sense of the "past", "present" and "future" can be defined in different ways. A simple way is to define a partial ordering which in the case of the positive quadrant $\{z = (x, y), \quad x, y \geq 0\}$ means that: $z_1 < z_2 \Leftrightarrow x_1 \leq x_2$ and $y_1 \leq y_2$.
It is clear that integral (2.131) is necessary in formulation and analysis of partial stochastic differential equations of the Itô type (cf. [140]).

EXAMPLES

To illustrate the concepts and theorems presented in this chapter we shall consider some particular problems and examples.

1. **Given a stochastic process**

$$X(t,\gamma) = t + X_1(\gamma)\cos t + X_2(\gamma)\sin t$$

where the vector $[X_1(\gamma), X_2(\gamma)]$ has the mean value $\left[-\frac{1}{2}, 1\right]$ and the correlation matrix

$$\begin{bmatrix} 3 & -2 \\ -2 & 2,9 \end{bmatrix}$$

Find the mean, covariance function and variance of $Y(t,\gamma) = \frac{dX}{dt}$. Direct application of formula (2.8) and (2.11) gives

$$E\left[Y(t,\gamma)\right] = 1 + \frac{1}{2}\sin t + \cos t$$

$$K_Y(t_1,t_2) = 3\left(\sin t_1 + \frac{2}{3}\cos t_1\right)\left(\sin t_2 + \frac{2}{3}\cos t_2\right) + \frac{47}{30}\cos t_1 \cos t_2$$

$$\sigma_Y^2(t) = 3 - \frac{1}{10}\cos^2 t + 4\sin t\cos t$$

2. **Let $X_1(t)$ and $X_2(t)$ be two stochastic processes with the following correlation functions and zero means:**

$$K_{X_1}(\tau) = \sigma^2 e^{-\alpha|\tau|}, \quad K_{X_2} = \sigma^2 e^{-\alpha^2\tau^2}$$

Are these processes differentiable in the mean square sense?

Theorem 2.7 implies that process $X_1(t)$ is non-differentiable (in m.s.) since the derivative of $\sigma^2 e^{-\alpha|\tau|}$ is discontinuous at $\tau = 0$ and, therefore, its second derivative at $\tau = 0$ does not exist. It is easily seen that process $X_2(t)$ has m.s. derivative.

3. **Let us consider the Wiener process $W(t)$.** From its definition follows that $E\left[W(t+h) - W(t)\right]^2 = h$. Therefore,

$$E\left[\frac{W(t+h) - W(t)}{h}\right]^2 = \frac{1}{h}.$$

This quantity becomes infinitely large as $h \to 0$; so, the Wiener process is m.s. non-differentiable. This property follows also from theorem 2.4; the correlation function of the Wiener process is

$$K_W(t_1, t_2) = \min(t_1, t_2)$$

and

$$\frac{\partial K_W(t_1, t_2)}{\partial t_1} = \begin{cases} 1 & , \quad t_1 < t_2 \\ 0 & , \quad t_1 > t_2 \end{cases}$$

which implies that the mixed second-order derivative of $K_W(t_1, t_2)$ does not exist.

Let us regard the above function as a generalized function. Then

$$\frac{\partial^2 K_W(t_1, t_2)}{\partial t_1 \partial t_2} = -\delta(t_1, t_2)$$

where $\delta(t_1, t_2)$ is Dirac's delta. This formal calculation confirms our assertion in Section 10 that white noise is the generalized derivative of the Wiener process.

4. Let a stochastic process $X(t)$ be defined as

$$X(t, \gamma) = |\sin t| \sin(\omega t + \Psi(\gamma))$$

where ω is a positive constant and $\Psi(\gamma)$ is a random variable with uniform distribution on $[0, 2\pi]$.

Is this process m.s. differentiable?

Due to the uniformity of the distribution of $\Psi(\gamma)$

$$E[X(t, \gamma)] = \frac{1}{2\pi} |\sin t| \int_0^{2\pi} \sin(\omega t + \psi) d\psi \equiv 0$$

$$K_X(t_1, t_2) = \frac{|\sin t_1| \, |\sin t_2|}{2} \cos[\omega(t_2 - t_1)]$$

$$\frac{\partial K_X(t_1, t_2)}{\partial t_1} = \frac{|\sin t_2|}{2} \Big\{ \operatorname{sign}(\sin t_1) \cos t_1 \cos \omega(t_2 - t_1) + \\ + \omega |\sin t_1| \sin \omega(t_2 - t_1) \Big\}.$$

Of course,

$$\frac{\partial K_X(t_1, t_2)}{\partial t_1} = 0 \qquad \text{for } t_1 \text{ such that}$$

$$\operatorname{sign}(\sin t_1) = 0,$$

$$\sin t_1 = 0$$

For t_1 such that sign $(\sin t_1) = +1$

$$\frac{\partial K_X(t_1,t_2)}{\partial t_1} =$$

$$= \frac{|\sin t_2|}{2}\{\cos t_1 \cos\omega(t_2 - t_1) + \omega \sin t_1 \sin\omega(t_2 - t_1)\}$$

Therefore,

$$\lim_{t_2\to t_1} \frac{\partial K_X(t_1,t_2)}{\partial t_1} = \frac{\sin t_1 \cos t_1}{2} = \left.\frac{\partial K_X(t_1,t_2)}{\partial t_1}\right|_{t_2=t_1}$$

which means that $\frac{\partial K_X(t_1,t_2)}{\partial t_1}$ is continuous and differentiable on the axis $t_2 = t_1$; so, there exists $\frac{\partial^2 K_X}{\partial t_1 \partial t_2}$ on this axis. The same result is obtained for the case, when $\text{sign}(\sin t_1) = -1$.

The conclusion is (cf. theorem 2.4) that a given process is differentiable in the mean square sense for all t. However, it is easily seen that its sample functions have not derivative for $t = k\pi$.

5. Let $X(t,\gamma)$ be twice differentiable in m.s. and have covariance function $K_X(t_1,t_2)$.
   Let

$$Y(t,\gamma) = f(t)X(t,\gamma) + \varphi(t)\frac{d^2X(t,\gamma)}{dt^2}$$

where $f(t)$ and $\varphi(t)$ are non-random functions of t.

Find the mutual covariance function of $X(t,\gamma)$ and $Y(t,\gamma)$.

$$R_{XY}(t_1,t_2) = f(t_1)f(t_2)K_X(t_1,t_2) + f(t_1)\varphi(t_2)\frac{\partial^2 K_X(t_1,t_2)}{\partial t_2^2}$$
$$+ f(t_2)\varphi(t_1)\frac{\partial^2 K_X(t_1,t_2)}{\partial t_1^2} + \varphi(t_1)\varphi(t_2)\frac{\partial^4 K_X(t_1,t_2)}{\partial t_1^2 \partial t_2^2}.$$

6. Given stochastic process $X(t)$ having m.s. derivatives $X'(t)$. Find the spectral density of $X'(\tau)$

According to formula (2.16)

$$K_{X'}(\tau) = -\frac{d^2}{d\tau^2} K_X(\tau)$$

But the correlation function of $X(t)$ has the representation (1.71). Differentiation of (1.71) with respect to τ gives

$$K_{X'}(\tau) = \int_{-\infty}^{+\infty} g_X(\omega)\omega^2 e^{i\omega\tau} d\omega$$

On the other hand, due to stationarity of derivative we have

$$K_{X'}(\tau) = \int_{-\infty}^{+\infty} g_{X'}(\omega) e^{i\omega\tau} d\omega$$

Now, the uniqueness of the Fourier representation of a function implies that

$$g_{X'}(\omega) = \omega^2 g_X(\omega)$$

The above argument indicates also that for stationary process $X(t)$ to be m.s. differentiable the condition must be satisfied:

$$\int_{-\infty}^{+\infty} \omega^2 g_X(\omega) d\omega < \infty$$

7. Given two stochastic weakly stationary processes $X(t,\gamma)$ and $Y(t,\gamma)$ with the following spectral densities

$$g_X(\omega) = \frac{a}{\omega^2 + b^2}, \qquad a, b - \text{constants}$$

$$g_Y(\omega) = \frac{a(\omega^2 + b^2)}{(\omega^2 + c^2)\left[(a_1\omega + b_1)^2 + c_1^2\right]^2}, \qquad a_1, b_1, c_1 - \text{constants}$$

Are these processes differentiable in the mean square?

The process $X(t,\gamma)$ is not differentiable, since $\omega^2 g_X(\omega)$ is not integrable with respect to ω on the entire real axis. Process $Y(t,\gamma)$ is differentiable since $\omega^2 g_Y(\omega) \sim \frac{1}{\omega^2}$ which is an integrable function.

8. Let

$$Y(t) = AX(t) = a_1\dot{X}(t) + b_1\int_0^t e^{-t_1}X(t_1)dt_1 + C$$

where $X(t)$ is a given process with the correlation function $K_X(t_1, t_2)$ and a_1, b_1, C being constants. Find the correlation function of $Y(t)$.
By virtue of (2.54)

$$K_Y(t_1, t_2)$$

$$= a_1^2\frac{\partial^2 K_X}{\partial t_1 \partial t_2} + b_1^2\int_0^{t_2}\int_0^{t_1} e^{-\lambda(t'+t'')}K_X(t', t'')dt'dt''$$

$$+a_1 b_1\left[\int_0^{t_1} e^{-\lambda t'}\frac{\partial K_X(t_1', t_2)}{\partial t_2}dt' + \int_0^{t_2} e^{-\lambda t''}\frac{\partial K_X(t_1, t'')}{\partial t_1}dt''\right].$$

9. Show that (cf. formulae (2.56′)–(2.56″)

$$I = \frac{1}{T^2}\int_0^T\int_0^T K_X(t_2 - t_1)dt_2 dt_1$$

$$= \frac{2}{T}\int_0^T\left(1 - \frac{\tau}{T}\right)K_X(\tau)d\tau$$

where $\tau = t_2 - t_1$, and $X(t)$ is weakly stationary real process.

$$I = \frac{1}{T^2}\int_0^T\int_{-t_1}^{T-t_1} K_X(\tau)d\tau dt_1 = \frac{1}{T^2}\int_0^T\int_{-z}^{T-z} K_X(\tau)d\tau dz$$

$$= \frac{1}{T^2}\left[\int_0^T\int_0^{T-z} K_X(\tau)d\tau dz + \int_0^T\int_{-z}^0 K_X(\tau)d\tau dz\right]$$

Introducing in the first integral new variable $z' = T - z$, and in the second integral the variable $\tau' = -\tau$ we obtain

$$I = \frac{1}{T^2}\left[-\int_T^0\int_0^{z'} K_X(\tau)d\tau dz' - \int_0^T\int_z^0 K_X(-\tau)d\tau' dz\right]$$

$$= \frac{1}{T^2}\left[\int_0^T\int_0^z K_X(\tau)d\tau dz + \int_0^T\int_0^z K_X(-\tau)d\tau dz\right]$$

$$= \frac{1}{T^2}\int_0^T\int_0^z [K_X(\tau) + K_X(-\tau)]\,d\tau dz$$

$$= \frac{2}{T^2}\int_0^T\int_0^z K_X(\tau)d\tau dz$$

For arbitrary function $f(z)$ the integration by parts gives

$$\int_0^T f(z)dz = Tf(T) - \int_0^T zf'(z)dz$$

Taking

$$f(z) = \int_0^z K_X(\tau)d\tau$$

we have

$$\begin{aligned} I &= \frac{2}{T^2}\left\{T\int_0^T K_X(\tau)d\tau - \int_0^T zK_X(z)dz\right\} \\ &= \frac{2}{T^2}\left\{T\int_0^T K_X(\tau)d\tau - \int_0^T \tau K_X(\tau)d\tau\right\} \\ &= \frac{2}{T^2}\left\{\int_0^T (T-\tau)K_X(\tau)d\tau\right\} \\ &= \frac{2}{T}\int_0^T \left(1-\frac{\tau}{T}\right) K_X(\tau)d\tau. \end{aligned}$$

10. Let us consider the process $Y(t)$ defined by the following non-linear transformation of the Gaussian process $X(t)$ with zero mean and given correlation function $K_X(t_1, t_2)$:

$$Y(t) = \int_0^t e^{X(\alpha)}d\alpha$$

Find the mean and correlation function of process $Y(t)$.
According to the definition of the mean

$$m_Y(t) = \int_0^t \int_{-\infty}^{+\infty} \frac{e^x}{[2\pi K_X(\alpha,\alpha)]^{1/2}} exp\left[-\frac{x^2}{2K_X(\alpha,\alpha)}\right] dx d\alpha$$

To calculate the integral with respect to x we make use of the formula

$$(*) \qquad \int_{-\infty}^{+\infty} e^{-az^2-2bz-c}dz = \sqrt{\frac{\pi}{a}}e^{-\frac{ac-b^2}{a}}$$

We have

$$m_Y(t) = \int_0^t e^{\frac{1}{2}K_X(\alpha,\alpha)}d\alpha$$

In order to evaluate the correlation function of $Y(t)$ let us denote e^x by $F(x)$. We have

$$\tilde{K}_Y(t_1,t_2) = \int_0^{t_1}\int_0^{t_2} E\{F[X(\alpha_1)]F[X(\alpha_2)]\}\,d\alpha_1 d\alpha_2$$
$$= \int_0^{t_1}\int_0^{t_2}\int_{-\infty}^{+\infty}\int_{-\infty}^{+\infty} \frac{e^{x_1+x_2}}{2\pi\sqrt{\Delta}} e^{-Ax_1^2+2Bx_1x_2-Cx_2^2} dx_1 dx_2 d\alpha_1 d\alpha_2$$

where

$$\Delta = K_X(\alpha_1,\alpha_1)K_X(\alpha_2,\alpha_2) - K_X^2(\alpha_1,\alpha_2)$$

$$A = \frac{K_X(\alpha_1,\alpha_1)}{2\Delta}, \quad B = \frac{K_X(\alpha_1,\alpha_2)}{2\Delta}, \quad C = \frac{K_X(\alpha_2,\alpha_2)}{2\Delta}$$

To calculate the integrals with respect to x_1 and x_2 we use again formula (*). We have the result

$$\tilde{K}_Y(t_1,t_2) = \int_0^{t_1}\int_0^{t_2} exp\left\{\frac{1}{2}[K_X(\alpha_1,\alpha_1)+K_X(\alpha_2,\alpha_2)] + K_X(\alpha_1,\alpha_2)\right\} d\alpha_1 d\alpha_2$$

Since

$$m_Y(t_1)m_Y(t_2) =$$
$$= \int_0^{t_1}\int_0^{t_2} e^{\frac{1}{2}[K_X(\alpha_1,\alpha_1)+K_X(\alpha_2,\alpha_2)]} d\alpha_1 d\alpha_2$$

we finally get

$$K_Y(t_1,t_2) = \tilde{K}_Y(t_1,t_2) - m_Y(t_1)m_Y(t_2)$$
$$= \int_0^{t_1}\int_0^{t_2} e^{\frac{1}{2}[K_X(\alpha_1,\alpha_1)+K_X(\alpha_2,\alpha_2)]}\left\{e^{K_X(\alpha_1,\alpha_2)} - 1\right\} d\alpha_1 d\alpha_2$$

11. Let $W(t)$ be the standard Wiener process defined on the interval $[0,1]$. Construct the Karhunen-Loeve expansion for this process.

According to the theorem 2.9 one should find eigenvalues and eigenfunctions of the integral equation (2.46). In our case

$$W(0) = 0, \quad E[W(t)] = 0, \quad E[W(t)]^2 = t, \quad K_W(t_1,t_2) = \min(t_1,t_2)$$

The eigenvalues ω_k and eigenfunctions $\varphi_k(t)$ are defined by the integral equation

$$(*) \qquad \omega_n \varphi_n(t) = \int_0^1 \min(t,s)\varphi(s)ds = \int_0^t s\varphi_n(s)ds + \int_t^1 t\varphi_n(s)ds$$

Differentiation of both sides of the above equation gives

$$(**) \qquad \omega_n \varphi_n'(t) = \int_t^1 \varphi_n(s)ds$$

what implies that $\varphi_n'(1) = 0$. From equation (*) follows that $\varphi_n(0) = 0$. After differentiation of (**) with respect to t we have

$$\omega_n \varphi_n'' = -\varphi_n(t)$$

The solution of this equation which satisfies the boundary conditions: $\varphi_n(0) = 0, \quad \varphi_n'(1) = 0$ is

$$\varphi_n(t) = \sqrt{2}\sin\left(n + \frac{1}{2}\right)\pi t,$$

$$\omega_n^{-1} = \left(n + \frac{1}{2}\right)\pi^2, \qquad n = 0, 1, \ldots$$

Therefore, the Karhunen-Loeve expansion of the Wiener process has the form

$$W(t) = \sqrt{2}\sum_{n=0}^{\infty} Z_n(\gamma)\frac{\sin\left(n+\frac{1}{2}\right)\pi t}{\left(n+\frac{1}{2}\right)\pi}.$$

12. Let $W(t)$ be a standard Wiener process.
 Find the joint distribution of $W(t)$ and $Y(t) = \int_0^t W(s)ds, \quad t \geq 0$.
 The distribution $F_t(\omega, y)$ is Gaussian and

$$E[W(t)] = E[Y(t)] = 0$$

Covariance matrix is

$$\begin{bmatrix} K_W(t,t) & \int_0^{t_1} K_W(t,t_2)dt_2 \\ \int_0^{t_2} K_W(t_1,t)dt_1 & \int_0^t \int_0^t K_W(t_1,t_2)dt_1dt_2 \end{bmatrix}$$

$$= \begin{bmatrix} t & \frac{1}{2}t_1^2 \\ \frac{1}{2}t_2^2 & \frac{1}{3}t^3 \end{bmatrix}$$

since $K_W(t_1,t_2) = \min(t_1,t_2)$. The probability density is:

$$f_t(w,y) = f_t(x,y) =$$

$$= \left(\frac{2\pi t^2}{\sqrt{12}}\right)^{-1} exp\left\{-(6t^{-3}x^2 - 6t^{-2}xy + 2t^{-1}y^2)\right\}$$

13. Let us consider process $X(t)$ given by

$$X(t) = X(t_o) + \int_{t_o}^{t} a(t)dt + \int_{t_o}^{t} b(t)dW(t)$$

Calculate the Itô differential of $Y(t) = e^{X(t)}$.
The differential of $X(t)$, according to the definition 2.6 is

$$dX(t) = a(t)dt + b(t)dW(t)$$

For $f(x) = e^x$ the Itô's formula gives

$$dY_t = \left[a(t)e^{X_t} + \frac{1}{2}b^2(t)e^{X_t}\right] dt + e^{X_t}b(t)dW_t$$

$$= \left[a(t)Y_t + \frac{1}{2}b^2(t)Y_t\right] dt + b(t)Y_t dW_t$$

Let us assume that $2a(t) = -b^2(t)$. In this case

$$dY_t = b(t)Y_t dW_t,$$
$$Y_{t_o} = e^{X_{t_o}}$$

This is a stochastic equation for the process $Y(t)$. We easily conclude that its solution is:

$$Y_t = Y_{t_o} exp\left\{-\frac{1}{2}\int_{t_o}^{t} b^2(s)ds + \int_{t_o} b(s)dW(s)\right\}$$

For $b(t) \equiv 1$ and $t_o = 0$ we have the equation

$$(*) \qquad dY_t = Y_t dW_t \quad , \quad Y_o = 1$$

and the corresponding solution

$$Y_t = exp\left(W_t - \frac{t}{2}\right) \quad , \quad t \geq 0$$

It is worth noting that equation (*) treated according to the ordinary calculus would give the solution: $exp(W_t)$.

In the Itô calculus the role of the exponential function (in the analysis of linear equations) is played by the function $exp\left(W_t - \frac{t}{2}\right)$.

14. Let the process $X(t)$ be given by the equation

$$dX_t = a_o X_t dt + b_o X_t dW(t)$$

where a_o and b_o are constant. Let us assume additionally that the fourth moment $E\left[X_t^4\right]$ is bounded.

Calculate the mean square of X_t, i.e. $E\left[X_t^2\right]$. Let us take $Y_t = f(X_t) = X_t^2$. Then by virtue of the Itô formula (2.103) we get

$$\begin{aligned} d(X_t^2) &= \left(2a_o X_t^2 + b_o^2 X_t^2\right) dt + 2b_o X_t^2 dW(t) = \\ &= (2a_o + b_o^2) X_t^2 dt + 2b_o X_t^2 dW(t) \end{aligned}$$

This means that

$$X_t^2 = X_o^2 + (2a_o + b_o^2) \int_0^t X_s^2 ds + 2b_o \int_0^t X_s^2 dW(s)$$

Averaging gives

$$E\left[X_t^2\right] = E\left[X_o^2\right] + (2a_o + b_o^2) \int_0^t E\left[X_s^2\right] ds + 2b_o E\left[\int_0^t X_s^2 dW(s)\right]$$

Since $\left[X_t^4(s)\right]$ is bounded, so $X_t^2 \in H_2[0,t]$ and by virtue of (2.80)

$$E\left[\int_0^t X_s^2 dW(s)\right] = 0$$

The final result is

$$E\left[X_t^2\right] = E\left[X_o^2\right] + (2a_o + b_o^2) \int_0^t E\left[X_s^2\right] ds$$

Let us denote: $m_2(t) = E\left[X_t^2\right]$. Then the above equation can be written as

$$\begin{aligned} &\frac{dm_2(t)}{dt} = (2a_o + b_o^2) m_2(t), \\ &m_2(0) = E\left[X_o^2\right] \end{aligned}$$

Therefore,

$$m_2(t) = m_2(0)exp\left\{(2a_o + b_o^2)t\right\}$$

It is seen that when $2a_o + b_o^2 < 0$, then $m_2(t) - 0$ as $t \to \infty$ (i.e. the solution is stable in the m.s. sense).

15. Let process $X(t)$, $t \geq t_o$ has the differential

$$dX(t) = a(t)dt + \int_{R_1} c(t,u)N(dt,du)$$

Calculate the differential of $Y(t) = f[X(t)] = [X(t)]^k$.
Application of the Itô formula (1.125) gives

$$dY(t) = kX_t^{k-1}a(t)dt + \int_{R_1}\left[(X_t + c(t,u))^k - X_t^k\right]N(dt,du)$$

Alternatively, (using the definition of a stochastic counting differential),

$$X_t^k = \int_{t_o}^t kX_s^{k-1}a(s)ds + \int_{t_o}^t\int_{R_1}\left[(X_t + c(t,u))^k - X_t^k\right]N(ds,du)$$

In particular, for $k = 2$ and $a(t) = 0$ and $c(t,u) = 1$ we have $X_t = N_t$ and

$$dY_t = dN_t^2 = \left[(N_t+1)^2 - N_t^2\right]dN_t.$$

Chapter III

STOCHASTIC DIFFERENTIAL EQUATIONS: BASIC THEORY

20. INTRODUCTORY REMARKS

The basic theoretical problems concerned with stochastic differential equations are, generally speaking, the same as those in the case of deterministic differential equations, namely: existence and uniqueness of a solution, analytical properties of the solutions, dependence of the solutions on parameters and initial values etc. Yet, the introduction of random elements into appropriate differential equations leads to new probabilistic problems and specific difficulties. For instance, the sense itself of a stochastic differential equation should be clearly defined since it can be different depending on the understanding of a stochastic process and its derivatives. For the analysis of stochastic differential equations, however, a crucial point is regularity of random functions occuring in a given equation.

Let us consider the equation (in vector form)

$$\frac{\mathbf{Y}(t)}{dt} = \mathbf{F}\left[\mathbf{Y}(t), \mathbf{X}(t)\right], \quad \mathbf{y}(t_o) = \mathbf{Y}_o \tag{3.1}$$

or, its linear version

$$\frac{d\mathbf{Y}(t)}{dt} + \mathbf{A}(t)\mathbf{Y}(t) = \mathbf{X}(t), \quad \mathbf{Y}(t_o) = \mathbf{Y}_o \tag{3.2}$$

where $\mathbf{X}(t)$ and/or $\mathbf{A}(t)$, $\mathbf{Y}_o$ are random.

If the random processes occuring in a differential equation (for example, $\mathbf{X}(t)$, $\mathbf{A}(t)$ in (3.2)) are sufficiently regular then the majority of problems can be analyzed by use of the methods which are analogous to those in deterministic theory of differential equations; such equations we shall call the **regular stochastic differential equations**. Of course, these equations also provide some probabilistic problems which have not any counterpart in classical theory.

The situation is different if a very irregular random process occurs in the equations. Usually, these irregular elements are generalized random processes of the white noise type. Such equations (for example, Langevin equation for the displacement of a particle in Brownian motion) can serve as models of dynamical systems subjected to rapidly-varying random excitations. The equations of this type are written formally in the form

$$\begin{aligned} &\frac{d\mathbf{Y}(t)}{dt} = \mathbf{m}[t, \mathbf{Y}(t)] + \boldsymbol{\sigma}[t, \mathbf{Y}(t)]\boldsymbol{\xi}(t) \\ &\mathbf{Y}(t_\circ) = \mathbf{Y}_\circ \end{aligned} \tag{3.3}$$

where $\boldsymbol{\xi}(t)$ is Gaussian white noise, which — as we know — does not exist in the conventional sense. Strictly speaking, expression (3.3) as it stands should be regarded as a "pre-equation" which needs the appropriate interpretation. Although a white noise is a generalized stochastic process (cf. Sec. 10, Chapter I) the indefinite integral $\int_0^t \boldsymbol{\xi}(s)ds$ can nevertheless be identified as the Wiener process $\mathbf{W}(t)$, and "pre-equation" (3.3) is usually interpreted as follows:

$$\begin{aligned} &d\mathbf{Y}(t) = \mathbf{m}[t, \mathbf{Y}(t)]\,dt + \boldsymbol{\sigma}[t, \mathbf{Y}(t)]\,d\mathbf{W}(t) \\ &\mathbf{Y}(t_\circ) = \mathbf{Y}_\circ \end{aligned} \tag{3.4}$$

This is the **Itô stochastic differential equation** for the process $\mathbf{Y}(t)$. It is an abbreviated form of the Itô integral equation

$$\mathbf{Y}(t) = \mathbf{Y}_\circ + \int_{t_\circ}^{t} \mathbf{m}[s, \mathbf{Y}(s)]\,ds + \int_{t_\circ}^{t} \boldsymbol{\sigma}[s, \mathbf{Y}(s)]\,d\mathbf{W}(s) \tag{3.5}$$

where the second integral is the Itô stochastic integral defined in Sec. 16. Because of the special properties of the Wiener process and the Itô integral the theory of the Itô stochastic differential equations differs significantly from the theory of regular stochastic equations.

The theory of stochastic differential equations of the form (3.4) was initiated as a method of constructing a diffusion Markov process for given coefficients of drift and diffusion. This approach — originating from Bernstein (who, according to [90] used the term "stochastic differential equation" for the first time (1934) in his study of a certain difference scheme associated with Markov chains) and successfully developed by Itô, Gikhman and others — turned out to be extremely fruitful. Today, the theory of stochastic Itô equations and its recent extensions (e.g. the Itô stochastic equations in Hilbert space — cf. Sec. 23) constitute a wide and very important branch of the contemporary theory of stochastic processes. This theory also turned out to be very useful in variety of applications.

In spite of the above distinction between regular and Itô stochastic differential equations it is useful (especially for applications) to reflect in classification of stochastic differential equations the manner in which the random elements occur in the equation. Accordingly, equations of the form (3.1) can be classified into three classes: *equations with random initial conditions, equations with random inhomogeneous part* and *equations with random coefficients.* Of course, each of the above can be non-linear or linear.

Our objective in this chapter is to give a concise presentation of the basic results of the theory of stochastic differential equations with emphasis on various possible interpretations. We hope that this "basic theory" lightens the mathematical structure of the theory and at the same time, indicates the mathematical frame for various applications.

21. REGULAR STOCHASTIC DIFFERENTIAL EQUATIONS

21.1. Mean-square theory

Let us consider the equation for the n-dimensional stochastic process $\mathbf{Y}(t)$:

$$\frac{d\mathbf{Y}(t)}{dt} = \mathbf{F}\left[t, \mathbf{Y}(t); \gamma\right], \quad t \in T, \quad \mathbf{Y}(t_\circ) = \mathbf{Y}_\circ \tag{3.6}$$

where $\mathbf{f}(t, \mathbf{y})$ is a given function of $(t, \mathbf{y})$.
According to the definition, the stochastic process $\mathbf{Y}(t)$, $t \in T$ is a mapping of the interval T into a space S of random variables. Hence, equation (3.6) is a differential equation for a function with values in an abstract space S. Such equations have been extensively studied in the case where S is a Banach space. Assuming that process $\mathbf{Y}(t)$ is a second-order stochastic process, the space S is simply a Hilbert space $L_2(\Gamma, \mathcal{F}, P)$ of random variables with finite second-order moments. The derivative in equation (3.6) is then interpreted as the mean square derivative of $\mathbf{Y}(t)$ on T.

If the vector transformation $\mathbf{F}\left[t, \mathbf{Y}(t)\right]$ yields a second-order n-dimensional vector process for each $t \in T$, we shall denote this fact writing that $\mathbf{F} : T \times L_2(\Gamma, \mathcal{F}, P) \longrightarrow L_2(\Gamma, \mathcal{F}, P)$.

Definition 3.1. *The second-order stochastic process* $\mathbf{Y}(t)$, *that is a mapping:* $T \longrightarrow L_2(\Gamma, \mathcal{F}, P)$ *is a* mean square (m.s.) solution *of initial-value problem (3.6) on* T *where* $\mathbf{F}$ *is a mapping:* $T \times L_2(\Gamma, \mathcal{F}, P) \to L_2(\Gamma, \mathcal{F}, P)$, *and* $Y(t_\circ) = Y_\circ \in L_2(\Gamma, \mathcal{F}, P)$ *if:*

a) $\mathbf{Y}(t)$ *is m.s. continuous on* T,
b) $\mathbf{Y}(t_\circ) = \mathbf{Y}_\circ$,
c) $\mathbf{F}\left[t, \mathbf{Y}(t)\right]$ *is mean square derivative of* $\mathbf{Y}(t)$ *on* T.

It follows from the considerations concerned with the mean square calculus, that $\mathbf{Y}(t)$ is a m.s. solution of equation (3.6) if and only if, for all $t \in T$

$$\mathbf{Y}(t) = \mathbf{Y}_\circ + \int_{t_\circ}^{t} \mathbf{F}\left[s, \mathbf{Y}(s)\right] ds \tag{3.7}$$

where the integral is a m.s. stochastic integral.

In the above interpretation one can easily state the basic existence and uniqueness theorem, which constitutes a mean square version of the appropriate theorem for the Banach space valued functions (cf. [20]).

Theorem 3.1. *If* $\mathbf{F} : T \times L_2(\Gamma, \mathcal{F}, P) \to L_2(\Gamma, \mathcal{F}, P)$ *satisfies the m.s. Lipschitz condition*

$$\|\mathbf{F}(t, \mathbf{Y}_1) - \mathbf{F}(t, \mathbf{Y}_2)\| \leq k(t)\|\mathbf{Y}_1 - \mathbf{Y}_2\| \tag{3.8}$$

where

$$\int_T k(t)dt < \infty, \quad \|\mathbf{Y}\| = \{E|\mathbf{Y}|_n^2\}^{1/2}$$

then there exists a unique m.s. solution of (3.6) for any initial condition $\mathbf{Y}_\circ \in L_2(\Gamma, \mathcal{F}, P)$.

The above theorem is a direct extension of the classical Picard theorem. In the proof, the only significant difference compared with the Picard proof lies in the replacement of the ordinary norm in R_n by the mean square norm.

It is worth emphasizing, that this existence and uniqueness theorem in its general form has rather limited applicability. Its first disadvantage is the difficulty in showing that a function of a second-order stochastic process is a second-order process itself. Moreover, the m.s. Lipschitz condition turns out to be too restrictive. This can be illustrated by the following first order differential equation (cf. [133]) for scalar process $Y(t)$:

$$\frac{dY}{dt} = A(\gamma)Y(t), \quad Y(t_\circ) = Y_\circ(\gamma)$$

This equation has for almost all realizations a unique solution

$$Y(t, \gamma) = Y_\circ(\gamma)e^{tA(\gamma)} \tag{*}$$

As it will be stated later, each m.s. solution is equivalent to a sample function solution. So, the above process could be the only possible m.s. solution. However, process (*) belongs for each t, to $L_2(\Gamma, \mathcal{F}, P)$ if and only if $A(\gamma)$ is bounded almost surely. Thus, even Gaussian coefficients are not allowed. In the light of these restrictions it is more appropriate to consider the question of existence and uniqueness for particular classes of stochastic differential equations (cf. Strand [133]).

If we restrict our attention to the linear systems with constant random coefficients, the following result holds (cf. [133]).

Theorem 3.2. *Let*

$$\begin{aligned} \frac{d\mathbf{Y}}{dt} &= \mathbf{A}(\gamma)\mathbf{Y}(t) + \mathbf{X}(t, \gamma), \\ \mathbf{Y}(t_\circ) &= \mathbf{Y}_\circ(\gamma) \end{aligned} \tag{3.9}$$

where $\mathbf{A}(\gamma) = a_{ij}(\gamma)$ *is* $n \times n$ *matrix of random variables with finite moments and* $\mathbf{X}(t,\gamma)$ *is a second order m.s. integrable process, and* $\mathbf{Y}_o(\gamma) \in L_2(\Gamma, \mathcal{F}, P)$. *Suppose that* $\mathbf{Y}_o(\gamma)$ *and* $\mathbf{X}(t,\gamma)$ *are independent of* $\mathbf{A}(\gamma)$. *If moment generating functions*

$$L_{ij}(s) = E\left\{e^{s a_{ij}(\gamma)}\right\}$$

are analytic functions for $|s| < R$, *then there exists a unique m.s. solution of (3.9) on the interval* $|t - t_o| < \frac{R}{2n}$.

Let us consider, for illustration, the equation ($n = 1$)

$$\frac{dY}{dt} = A(\gamma)Y(t), \qquad Y(0) = Y_o(\gamma)$$

where $A(\gamma) \geq 0$ is a positive random variable independent of $Y_o(\gamma)$ and $Y_o \in L_2(\Gamma, \mathcal{F}, P)$. Process $Y_o(\gamma) exp\,[tA(\gamma)]$ is a second-order if and only if:

$$\frac{E\left\{A^{2n}(\gamma)\right\} t^n}{n!} \qquad \text{converges as} \quad n \to \infty.$$

This in turn is true if and only if $|t| < R$, where R is the radius of convergence of $E\left\{exp\,[sA(\gamma)]\right\}$.

If $A(\gamma)$ is exponentially distributed with density e^{-x}, then $R = 1$, and the m.s. solution exists on the interval $\left[0, \frac{1}{2}\right]$.

21.2. Sample function solutions

Another natural interpretation of regular stochastic differential equations can be based on the fact, that a stochastic process is a function $X(t,\gamma)$ of two arguments $t \in T, \gamma \in \Gamma$ which for a fixed γ gives a deterministic function of t. In other words, $Y(t)$ in equation (3.6) can be understood as a family of its sample functions. Hence, if the sample functions of $Y(t)$ are assumed to be regular enough, equation (3.6) can be interpreted as a family of equations for sample functions.

Defintion 3.2. *Let* $\mathbf{F}$ *be a mapping:* $T \times R_n \times \Gamma \longrightarrow R_n$ *and* $Y_o : \Gamma \longrightarrow R_n$. *The process* $\mathbf{Y}(t) : T \times \Gamma \longrightarrow R_n$ *is a* sample function solution *of the initial-value problem:*

$$\frac{d\mathbf{Y}(t,\gamma)}{dt} = \mathbf{F}\left[t, \mathbf{Y}(t,\gamma), \gamma\right]$$

$$\mathbf{Y}(t_o, \gamma) = \mathbf{Y}_o(\gamma) \tag{3.10}$$

if for almost all $\gamma \in \Gamma$ *the following conditions are satisfied:*

(a) $\mathbf{Y}(t,\gamma)$ *is absolutely continuous(in* t*) on* T,
(b) $\mathbf{Y}(t_0,\gamma) = \mathbf{Y}_0(\gamma)$,
(c) $\frac{\partial \mathbf{Y}(t,\gamma)}{\partial t} = \mathbf{F}[t, \mathbf{Y}(t,\gamma), \gamma]$ *for almost all* $t \in T$.

Below we give the theorem of the existence and uniqueness of a sample function solution (cf. Bunke [70]).

Theorem 3.3. *If the following conditions (a), (b), (c) are satisfied:*
(a) *function* $\mathbf{F} : T \times R_n \times \Gamma \longrightarrow R_n$ *is, for each* $(t, \mathbf{y}) \in T \times R_n$, $\mathcal{F}$*-measurable,*
(b) *function* $\mathbf{F}_\gamma : T \times R_n \longrightarrow R_n$ *is continuous for almost all* $\gamma \in \Gamma$,
(c) *for almost all* $\gamma \in \Gamma$ *there exists a continuous function* $K_\gamma(t) : T \longrightarrow R_1$ *such that for each* $t \in T$ *and arbitrary* $\mathbf{y}_1, \mathbf{y}_2 \in R_n$ *the condition is satisfied*

$$|\mathbf{F}(t,\mathbf{y}_1,\gamma) - \mathbf{F}(t,\mathbf{y}_2,\gamma)| \leq K_\gamma(t)\, |\mathbf{y}_1 - \mathbf{y}_2| \tag{3.11}$$

where $|\cdot|$ *means a norm in* R_n, *then the initial value problem has a unique sample function solution defined on* T.

It is clear that in order to proof the existence theorem in the case considered we should show that for all $\gamma \in \Gamma$ the corresponding deterministic equations have solutions defined on T, and that these solutions are sample functions of a certain stochastic process. It is important to notice in this context that solutions of deterministic equations (for fixed $\gamma \in \Gamma$) may exist though a common interval T (independent of γ) on which almost all solutions would be defined may not exist. To illustrate this assertion let us consider the following stochastic initial value problem

$$\frac{dY}{dt} = A(\gamma)Y^2(t) \quad , \quad Y(0) = 1$$

where $A(\gamma)$ is a random variable with exponential distribution. For each $\gamma \in \Gamma$ there exists a unique solution of the corresponding deterministic problem

$$y_\gamma(t) = \frac{1}{1 - A(\gamma)t}$$

defined for values of $t \in [0, A^{-1}(\gamma)]$.
However, it does not exist any interval $[0,t]$ on which the solutions would be defined for almost all $\gamma \in \Gamma$.

Another situation is such that deterministic solutions (for different $\gamma \in \Gamma$) though defined on a common interval can not be sample functions of any stochastic process. Indeed, let us take the example.

Let $B \subset \Gamma$ but $B \notin \mathcal{F}$. Let function $Y(t,\gamma)$ be defined for $t \in [0,\infty)$ and $\gamma \in \Gamma$ by the formula

$$Y(t,\gamma) = \begin{cases} 0 & \text{for} \quad \gamma \in B \\ \dfrac{t^2}{4} & \text{for} \quad \gamma \in \Gamma - B \end{cases}$$

This function is for each $\gamma \in \Gamma$ a solution of the initial value problem:

$$\frac{dY}{dt} = \sqrt{|Y(t)|} \quad , \quad Y(0) = 0.$$

However, function $Y(t,\gamma)$ defined above is not a stochastic process (for example, probability of the event that $Y(t,\gamma)$ assumes value zero is not defined, either the probability $P\{\Gamma - B\}$ is not determined). Therefore, $Y(t,\gamma)$ can not be a sample function solution of the above initial value problem.

The following theorem gives the relationship between m.s. solution and sample function solution of equation (3.6) — (cf. [133]).

Theorem 3.4. *If* $\mathbf{Y}(t)$ *is a m.s. solution of equation (3.6) then there exists an equivalent process* $\tilde{\mathbf{Y}}(t)$ *defined as a product measurable mapping on* $T \times \Gamma$ *into* R_n *which is a sample function solution of (3.6). Conversely, if* $\tilde{\mathbf{Y}}(t)$ *is a sample function solution of (3.6), the equivalent process* $\mathbf{Y}(t)$ *is a m.s. solution if and only if*

$$\int_T \|F(t, Y(t,\gamma), \gamma)\|_{L_2} \, dt < \infty. \tag{3.12}$$

This theorem directly follows from the interrelationship of m.s. integrals and sample function integrals.

The above theorem gives a base for understanding equation (3.6) as a family of equations for sample functions $\mathbf{y}_\gamma(t)$ of the process $\mathbf{Y}(t,\gamma)$, i.e.

$$\frac{d\mathbf{y}_\gamma(t)}{dt} = \mathbf{F}\left[t, \mathbf{y}_\gamma(t)\right]$$

for allmost all $\gamma \in \Gamma$. Such an interpretation is very often used in applications where, in the first step, the solution for the sample functions is looked for and then the moments of the solution are determined (exactly or approximately) by averaging, that is by integration with respect to γ.

21.3. Analysis via stochastic operators

A general and uniform approach to interpretation and analysis of regular stochastic differential equations is based on the concept of a stochastic operator (cf. Sec. 12) and uses the methods of the probabilistic functional analysis (cf. [2], [68]). In order to illustrate this approach let us consider the following three cases.

1. Let $\mathcal{X}$ be a Banach space and let $X(\gamma)$ be a generalized random variable with its values in $\mathcal{X}$. Let A be a deterministic operator acting in $\mathcal{X}$. The following equation

$$AY = X(\gamma) \tag{3.13}$$

is a stochastic equation for unknown element $Y \in \mathcal{X}$; as usual $\gamma \in \Gamma$ where $(\Gamma, \mathcal{F}, P)$ is a basic probability space.

2. Let A be a stochastic operator mapping $\Gamma \times \mathcal{X}$ in $\mathcal{X}$, and let X be a deterministic element of $\mathcal{X}$, then

$$A(\gamma)Y = X \tag{3.14}$$

is a stochastic operator equation for Y.

3. Let $A = A(\gamma)$ be a stochastic operator mapping $\Gamma \times \mathcal{X}$ in $\mathcal{X}$ and let $X = X(\gamma)$ be a generalized random variable with values in $\mathcal{X}$, then

$$A(\gamma)Y = X(\gamma) \tag{3.15}$$

is the most **general stochastic operator equation for element $Y \in \mathcal{X}$.**

In order to make equations (3.13) and (3.14) consistent we assume that in equation (3.13) the deterministic operator A is an operator valued random variable which assumes a given value with probability one. Similarly, in equation (3.14) we assume that X is a Banach space-valued random variable (or function) which assumes a given value with probability one. In this sense equations (3.13) and (3.14) are special cases of (3.15). When referring to equations of the form (3.14) and (3.15) we use the term *stochastic* or (*random*) *operator equations*.

Since γ is an element of a measurable space $(\Gamma, \mathcal{F})$ on which a complete probability measure P is defined, it is clear that a stochastic operator equation is a family of equations. This family of equations has only one member in the case where P is Dirac measure, that is $P\{\gamma_0\} = 1$ and $P\{\Gamma - \{\gamma_0\}\} = 0$. In this case, of course, we are dealing with a deterministic equation.

Let us define now a concept of solution of a stochastic operator equation.

Definition 3.3. *Any mapping $Y(\gamma) : \Gamma \to \mathcal{X}$ which satisfies the equality $A(\gamma)Y(\gamma) = X(\gamma)$ for every $\gamma \in \Gamma$, where $P\{\Gamma_0\} = 1$ is said to be a* wide sense solution *of equation (3.15).*

If, in addition, a wide sense solution is *measurable* with respect to P, then it is a stochastic solution.

Definition 3.4. *A generalized random variable $Y(\gamma)$ with values in $\mathcal{X}$ such that*

$$P\{\gamma : A(\gamma)Y(\gamma) = X(\gamma)\} = 1 \tag{3.16}$$

is called a stochastic solution *of equation (3.15).*

The examples presented in Sec. 21.2 provide wide-sense solutions which are not stochastic solutions. Therefore, an important question in analysis of stochastic operator equations is: when does existence of wide-sense solution imply the existence of a stochastic solution of the same equation?

The analysis of stochastic operator equations must not only, establish existence and uniqueness but it has also to assure the measurability of the solutions. The probabilistic versions of classical (associated with deterministic operator equations) theorems are often obtained by making use of the classical results together with appropriate measure theoretic analysis.

Within the stochastic operator formulation the theorems on existence and uni queness of a solution for a wide class of stochastic equations can be stated as appropriate probabilistic fixed-point theorems (cf. [2]).

The probabilistic version of the principle of contractive mappings is as follows.

Theorem 3.5. *Let $A(\gamma)$ be a stochastic contractive operator mapping $\Gamma \times \mathcal{X}$ into $\mathcal{X}$. Then there exists a generalized random variable $\xi(\gamma)$ with values in $\mathcal{X}$ such that*

$$P\{\gamma : A(\gamma)\xi(\gamma) = \xi(\gamma)\} = 1; \tag{3.17}$$

the random variable $\xi(\gamma)$ is defined uniquely in the sense that if $\eta(\gamma)$ is another random variable satisfying (3.17) then $P\{\gamma : \xi(\gamma) = \eta(\gamma)\} = 1$.

The random variable $\xi(\gamma)$ is called the *fixed point* of the stochastic operator $A(\gamma)$. This fixed point can be obtained by the method of successive approximations starting from an arbitrary generalized random variable $X_0(\gamma)$ with values in $\mathcal{X}$.

It is clear that the above theorem can serve as a base for proving the existence and uniqueness of a solution of a wide class of stochastic equations. One needs only prove that an operator in question is a contractive operator. This theorem constitutes also a base for constructing a solution by successive approximations (cf. Chapter V, Sec. 32.1.).

The theory of stochastic operator equations includes many interesting problems such as: measurability of a solution, existence of an inverse operator, properties of a solution and approximative schemes for a solution and their convergence. An interested reader is referred to the very extensive literature on this subject (cf. [2], [18], [68], [117]).

Let $\mathcal{X} = C[a,b]$ be a space of continuous functions on $[a,b]$ and let $C^{(n)}[a,b]$ denote the subspace of $C[a,b]$ consisting of all functions whose first n derivatives are continuous. If we define a stochastic differential operator $A(\gamma)$ as a mapping $\Gamma \times C^{(n)}[a,b] \to C[a,b]$, then the following equation

$$A(\gamma)Y(t) = \sum_{k=0}^{n} a_k(t,\gamma)\frac{d^kY}{dt^k} = X(\gamma), \qquad t \in [a,b] \tag{3.18}$$

where the coefficients $a_k(t,\gamma)$ are real-valued random processes constitutes an example of equation (3.15).

As in determinstic case, an initial value problem for stochastic differential equation (3.18) can be represented in the form of a stochastic Volterra equation. This fact is useful in practice, since the integral equation formulation has many advantages (cf. Sec. 26.3.4. in Chapter IV).

It is clear that stochastic integral equations of the Fredholm and Volterra type, i.e. the equations

$$Y(t,\gamma) = X(t,\gamma) + \int_a^b R(t,\tau,\gamma)Y(\tau,\gamma)d\tau \tag{3.19}$$

$$Y(t,\gamma) = X(t,\gamma) + \int_a^t R(t,\tau,\gamma)Y(t,\gamma)d\tau \tag{3.20}$$

can easily be put in the stochastic operator setting.

Let us consider a stochastic integral equation with random inhomogeneous term, i.e.

$$Y(t,\gamma) - AY(\tau,\gamma) = X(t,\gamma) \tag{3.21}$$

where A is deterministic Fredholm operator over a finite interval $[a,b]$

$$AY(t) = \int_a^b R(t,\tau)Y(\tau)d\tau \tag{3.22}$$

and $X(t,\gamma)$, $t \in [a,b]$ is a given second-order stochastic process which is assumed to be continuous in mean square. The following theorem holds [68].

Theorem 3.6. *If $R(t,\tau)$, $t,\tau \in [a,b]$ is a Fredholm kernel such that $|b-a| \max R(t,\tau) < 1$ and $X(t,\gamma)$ is a second-order m.s. continuous stochastic process, then the random process $Y(t,\gamma)$ defined as*

$$Y(t,\gamma) = X(t,\gamma) - \int_a^b \Lambda(t,\tau)X(t,\gamma)d\tau \tag{3.23}$$

satisfies the Fredholm equation (3.21) on $[a,b] \times \Gamma$; the resolvent kernel $\Lambda(t,s)$ is given by the Neumann series

$$\Lambda(t,\tau) = -\sum_{k=1}^{\infty} R^{(k)}(t,\tau) \tag{3.24}$$

where the iterated kernels $R^{(1)}(t,\tau)$, $R^{(2)}(t,\tau), \ldots$ are defined as follows

$$\begin{aligned} R^{(1)}(t,\tau) &= R(t,\tau), \\ R^{(2)}(t,\tau) &= \int_a^b R(t,\xi)R(\xi,\tau)d\xi \\ &\vdots \\ R^{(k)}(t,\tau) &= \int_a^b R^{(k-1)}(t,\xi)R(\xi,\tau)d\xi. \end{aligned} \tag{3.25}$$

The above result can be specialized to the case in which $R(t,\gamma)$ is a Volterra kernel, i.e. the equation has the form (3.20) in which $R(t,\tau,\gamma) = R(t,\tau)$.

An important class of problems is formed by the boundary-value problems for ordinary differential equations. For example, in the case of second-order equation we are often interested in a solution of the equation

$$Ay = \frac{d}{dt}\left[p(t)\frac{dy}{dt}\right] + q(t)y = 0 \tag{3.26}$$

which, at the end of a considered time interval $[a,b]$ satisfies the conditions

$$\begin{aligned} \alpha_1 y(a) + \beta_1 \dot{y}(a) &= 0, \\ \alpha_2 y(b) + \beta_2 \dot{y}(b) &= 0 \end{aligned} \tag{3.27}$$

If $p(t)$ and/or $q(t)$ in the above equation are random processes or $\alpha_i, \beta_i \quad (i = 1,2)$ in the boundary conditions are random variables we have a *stochastic boundary-value problem*. Such problems can also be formulated in the stochastic operator setting. By use of the associated Green function they can be reduced to the stochastic Fredholm integral equations.

A related problem associated with (3.26), (3.27) is the eigenvalue problem consisting in finding a non-trivial solution $y(t)$ of the equation

$$\frac{d}{dt}\left[p(t)\frac{dy}{dt}\right] + q(t)y + \lambda r(t)y = 0 \tag{3.26'}$$

which satisfies the boundary conditions (3.27); the function $r(t)$ is assumed to be continuous and positive on the interval $[a,b]$ and λ is a scalar which can be real or complex.

If equation (3.26′) or boundary conditions (3.27) contain random elements we have *stochastic eigenvalue problem*; in such a case, eigenvalues are random variables. The stochastic eigenvalue problems can also be reduced to stochastic integral equations (cf. [68] and [128]).

21.4. Asymptotic analysis

21.4.1. Introductory remarks

In the previous sections we focussed our attention on the possible interpretations of regular stochastic differential equations. Although these interpretations and the theorems associated indicate also the ways of solving such equations we wish to present here additionally an approach which has proved to be very useful in analysis of applied problems. We have in mind the analysis dealing with the properties of a solution when a certain parameter occuring in the equation tends to zero (and when $t \to \infty$). Such an asymptotic analysis is known in the theory of deterministic differential equations and leads to interesting results.

As an illustration, we shall remember the so-called *averaging principle* for deterministic equations formulated for the first time by N.N. Bogoliubov. The idea is as follows.

Given a system of ordinary differential equations with small parameter

$$\frac{d\mathbf{y}}{dt} = \varepsilon \mathbf{F}[\mathbf{Y}, t], \qquad \mathbf{Y}(0) = \mathbf{Y}_0 \tag{3.28}$$

Let us assume that there exists a mean value of $\mathbf{F}(\mathbf{y}, t)$ with respect to time

$$\lim_{T \to \infty} \frac{1}{T} \int_0^T \mathbf{F}(\mathbf{y}, t) dt = \mathbf{F}_0(\mathbf{y}) \tag{3.29}$$

The Bogoliubov avergaging principle states that on large time intervals of order $\frac{1}{\varepsilon}$ the solution of equation (3.28) can be approximated by the solution of the problem

$$\frac{d\mathbf{Y}_1}{dt} = \varepsilon \mathbf{F}_0(\mathbf{Y}_1), \qquad \mathbf{Y}_1(0) = \mathbf{Y}_0 \tag{3.30}$$

in the sense that for arbitrary constants $L > 0$ and $\delta > 0$ there exists such $\varepsilon_0 = \varepsilon_0(\delta, L)$ that

$$\max_{t \in [0, \frac{L}{\varepsilon}]} |\mathbf{Y}(t) - \mathbf{Y}_1(t)| < \delta$$

for all $\varepsilon < \varepsilon_0$.

It has been shown by Gikhman (cf. [88]) that the above averaging principle can be considered as a stronger form of the theorem on continuous dependence of a solution upon the parameter.

The averaging principle remains also valid in the case of the following system of equations (in R_n)

$$\frac{d\mathbf{Y}^\varepsilon}{dt} = \varepsilon \mathbf{F}[\mathbf{Y}^\varepsilon(t), \mathbf{X}(t)], \qquad \mathbf{Y}^\varepsilon(0) = \mathbf{Y}_0 \tag{3.31}$$

where $\mathbf{X}(t), t \geq 0$ is a ceratin function with values in R_n, ε is a small parameter and $\mathbf{F}(\mathbf{y}, \mathbf{x}) = [F_1(\mathbf{y}, \mathbf{x}), \ldots, F_n(\mathbf{y}, \mathbf{x})]$. If functions $F_i(\mathbf{y}, \mathbf{x})$ increase moderately then on every finite interval $[0, T]$ solution of (3.31) converges to $\mathbf{Y}^0(t) \equiv \mathbf{Y}$ as $\varepsilon \to 0$.

However, usually we are interested in the behaviour of a solution on large time intervals which are of order ε^{-1} or larger since the effects generated by $X(t)$ will manifest themselves after longer time.

In order to study a solution on intervals $[0, T\varepsilon^{-1}]$ it is is convenient to introduce new variables.

Let $\mathbf{Z}^\varepsilon(t) = \mathbf{Y}^\varepsilon(\frac{t}{\varepsilon})$. Then equation (3.31) will take the form

$$\begin{aligned} \frac{d\mathbf{Z}^\varepsilon(t)}{dt} &= \mathbf{F}\left[\mathbf{Z}^\varepsilon(t), \mathbf{X}\left(\frac{t}{\varepsilon}\right)\right], \\ \mathbf{Z}^\varepsilon(0) &= \mathbf{Y}_\circ \end{aligned} \tag{3.32}$$

Now, the time interval does not depend on ε and investigation of system (3.32) on finite time interval is equivalent to studying system (3.31) on time intervals of order ε^{-1}.

If function $\mathbf{F}(\mathbf{y}, \mathbf{x})$ is continuous with respect to $\mathbf{y}$ and $\mathbf{x}$, bounded an satisfies the Lipschitz condition with respect to $\mathbf{y}$ and there exists a function $\mathbf{F}_\circ(\mathbf{y})$ such that (uniformly in $\mathbf{y}$)

$$\lim_{T\to\infty} \frac{1}{T}\int_0^T \mathbf{F}(\mathbf{y}, \mathbf{X}(s))\, ds = \mathbf{F}_\circ(\mathbf{y}) \tag{3.33}$$

then the solution $\mathbf{Z}^\varepsilon(t)$ is convergent, as $\varepsilon \to 0$, to the solution of equation

$$\begin{aligned} \frac{d\mathbf{Z}_1}{dt} &= \mathbf{F}_\circ\left[\mathbf{Z}_1(t)\right], \\ \mathbf{Z}_1(0) &= \mathbf{Y}_\circ \end{aligned} \tag{3.34}$$

Often, process $\mathbf{X}^\varepsilon(t) = \mathbf{X}\left(\frac{t}{\varepsilon}\right)$ is interpreted as a fast motion, whereas $\mathbf{Z}(t)$ as a slow motion.

21.4.2. Avergaging principle for stochastic equations

Let us assume now that a fast variable $\mathbf{X}(t)$ in (3.32) is a stochastic process with values in R_l. If function $\mathbf{F}(\mathbf{y}, \mathbf{x})$ satisfies the conditions stated above and condition (3.33) is satisfied with probability one for arbitrary $\mathbf{y} \in R_n$, then the classical (for deterministic equations) averaging principle implies that trajectories of $\mathbf{Z}^\varepsilon(t)$ converge uniformly (on each finite interval) with probability one to the solution of equation (3.34).

More relaxed form of the averaging principle for stochastic equations can be formulated as follows (cf. Viencel [136]).

Theorem 3.7. *Let the following assumption be satisfied:*

(i) function $F(\mathbf{y}, \mathbf{x})$ satisfies the Lipschitz condition

$$|\mathbf{F}(\mathbf{y}_1, \mathbf{x}_1) - \mathbf{F}(\mathbf{y}_2, \mathbf{x}_2)| \le K\left(|\mathbf{y}_1 - \mathbf{y}_2| + |\mathbf{x}_1 - \mathbf{x}_2|\right)$$

(ii) there exists such function $\mathbf{F}_o(\mathbf{y})$ *with values in* R_n *that for arbitrary* $\delta > 0$ *and* $\mathbf{y} \in R_n$ *uniformly in* t

$$\lim_{T\to\infty} P\left\{\left|\frac{1}{T}\int_t^{t+T} \mathbf{F}(\mathbf{y}, \mathbf{X}(s))\,ds - \mathbf{F}_o(\mathbf{y})\right| > \delta\right\} = 0, \tag{3.35}$$

(iii)

$$\sup_t E\left|\mathbf{F}(\mathbf{y}, \mathbf{X}(t))\right| < \infty.$$

Then, for arbitrary $t > 0,\ \delta > 0$

$$\lim_{\varepsilon\to 0} P\left\{\sup_{0\le t\le T} |\mathbf{Z}^\varepsilon(t) - \mathbf{Z}_1(t)| > \delta\right\} = 0 \tag{3.36}$$

Let us notice that stochastic process $\mathbf{Z}^\varepsilon(t)$ in equation (3.32) can be interpreted as the result of random perturbations of a dynamical system governed by (3.34). Condition (3.35) asserts that these random perturbations are small. Condition (3.35) is satisfied for a wide class of stochastic processes $\zeta(\mathbf{y}, s) = \mathbf{F}(\mathbf{y}, \mathbf{X}(s))$, where $\mathbf{y} \in R_n$ plays a role of a parameter. For example, if $\zeta(y, s)$ is weakly stationary, then it is sufficient if the diagonal elements of its correlation function tend to zero as $\tau \to \infty$; in this case $\mathbf{F}_o(\mathbf{y}) = E[\mathbf{F}(\mathbf{y}, \mathbf{X}(s))]$.

The averaging principle expressed here by Theorem 3.7. gives the conditions under which process $\mathbf{Z}^\varepsilon(t)$ governed by "perturbed" system (3.32) converges to the solution $\mathbf{Z}_1(t)$ of the associated averaged equation (3.34) as $\varepsilon \to 0$. However, in the analysis of specific practical problems we often not deal with $\mathbf{Z}^\varepsilon(t)$ for various ε but just with $Z^\varepsilon(t)$ for one fixed ε. Therefore, the problem which occurs is: when can we guarantee that for small ε process $\mathbf{Z}^\varepsilon(t)$ will be close (in some sense) to $\mathbf{Z}_1(t)$ and what will be the order of magnitude of deviation $\mathbf{Z}^\varepsilon(t) - \mathbf{Z}_1(t)$?

In deterministic case, when the "perturbation" $\mathbf{X}(t)$ is a periodic function, the above difference is of order ε. In stochastic case, difference $\mathbf{Z}^\varepsilon(t) - \mathbf{Z}_1(t)$ is of order $\sqrt{\varepsilon}$. With appropriate assumptions, the normalized error $\frac{1}{\sqrt{\varepsilon}}[\mathbf{Z}^\varepsilon(t) - \mathbf{Z}_1(t)]$ converges weakly(*) to a specified Gaussian Markov process (cf. [136]).

21.4.3. Khasminskii limit theorem

Let us come back to equation (3.31). The averaging principle expressed in Theorem 3.7 can also be formulated as: if the conditions (i), (ii),(iii) are satisfied, then

---

(*) A family of stochastic processes $Y_\lambda(t)$, $\lambda \in \Lambda$ is said to *converge weakly*, as $\lambda \to 0$, to the stochastic process $Y_o(t)$ if for arbitrary n, $n = 1, 2\ldots$ and $t_1, t_2, \ldots, t_n$ the joint distribution of $\{Y_\lambda(t_1), Y_\lambda(t_2), \ldots, Y_\lambda(t_n)\}$ converges to the joint distribution of $\{Y_o(t_1), Y_o(t_2), \ldots, Y_o(t_n)\}$ as $\lambda \to 0$.

for arbitrary $\delta > 0$

$$\lim_{\varepsilon \to 0} P \left\{ \sup_{0 \le t \le \frac{T}{\varepsilon}} |\mathbf{Y}^\varepsilon(t) - \mathbf{Z}_1(t)| > \delta \right\} = 0$$

In the case $\mathbf{F}_0(\mathbf{y}) \equiv 0$ the above assertion implies that $\mathbf{Z}_1(t)$ is constant and, therefore, process $\mathbf{Y}^\varepsilon(t)$ will not move appreciably within time $[0, \frac{T}{\varepsilon}]$ from the initial state.

Stratonovich in 1961 suggested that in such a situation stochastic effects will become important for times t of the order $\frac{1}{\varepsilon^2}$. He also enuncited a limit theorem asserting that on intervals with length of order $\frac{1}{\varepsilon^2}$ process $\mathbf{Y}^\varepsilon(t)$ will approach a Markov diffusion process. A rigorous mathematical formulation and proof of this theorem was provided by Khasminskii in his paper [102]. Later on some improvement and extensions were provided by Papanicolau and others (cf. [119], [120]).

The essence of the Khasminskii theorem is as follows. Let the system of equations have the form

$$\frac{d\mathbf{Y}(t)}{dt} = \varepsilon \mathbf{F}\left[\mathbf{Y}(t), \mathbf{X}(t), t\right],$$

$$\mathbf{Y}(0) = \mathbf{Y}_0 \tag{3.37}$$

where the stochastic process $\mathbf{Y}(t)$ takes its values in R_n, and $\mathbf{X}(t)$ in R_l. Function $\mathbf{F}$ is a mapping of $R_n \times R_l \times [0, \infty)$ into R_n satisfying some conditions such as boundedness with the first partial derivatives. Process $\mathbf{X}(t)$ with zero mean satisfies the so called strong mixing condition which means, roughly speaking, that the values of the process become asymptotically independent when the distance between the corresponding time instants increases (for example, the stationary and Gaussian process with exponentially decaying correlation function satisfies this condition).

The Khasminskii theorem states that the process $\mathbf{Y}^\varepsilon(\tau)$ defined as

$$\mathbf{Y}^\varepsilon(\tau) = \mathbf{Y}\left(\frac{\tau}{\varepsilon^2}\right), \quad \tau = \varepsilon^2 t$$

converges weakly as $\varepsilon \to 0$ to a Markov diffusion process whose drift and diffusion coefficients are defined by the following limits

$$a_i(\mathbf{y}) = \lim_{T \to \infty} \frac{1}{T} \int_{t_0}^{t_0+T} \int_{t_0}^{s} E\left\{ \sum_{j=1}^{n} \frac{\partial F_i(\mathbf{y}, \mathbf{x}(s), s)}{\partial y_j} F_j(\mathbf{y}, \mathbf{x}(\eta), \eta) \right\} d\eta ds$$

(3.38)

$$b_{ij}(\mathbf{y}) = \lim_{T \to \infty} \frac{1}{T} \int_{t_0}^{t_0+T} \int_{t_0}^{s} E\left\{ F_i(\mathbf{y}, \mathbf{x}(s), s) F_j(\mathbf{y}, \mathbf{x}(\eta), \eta) \right\} d\eta ds$$

The above theorem characterizes the behaviour of the process governed by equation (3.37) in the limit $\varepsilon \to 0$ and $t \to \infty$ but $\varepsilon^2 t =$const. It has also been shown that moments of any order of the process $Y^\varepsilon(\tau)$ converge to the corresponding moments of the limiting diffusion process.

The limit theorem outlined above can also be formulated in another, more general, way. We mean here the setup presented in the original paper of Khasminskii [102] as well as that in the paper by Papanicolan and Kohler [120]. Since such a formulation is suitable for analysis of numerous applied problems we will present it here as well.

Instead of (3.37) let us consider a system

$$\frac{d\mathbf{Y}}{dt} = \varepsilon \mathbf{F}(\mathbf{Y}, t, \gamma, \varepsilon),$$
$$\mathbf{Y}(0) = \mathbf{Y}_\circ \tag{3.39}$$

where $\gamma \in \Gamma$, ε is a small parameter and $\mathbf{F}(\mathbf{y}, t, \gamma, \varepsilon)$ is a random function with values in R_n. Let

$$\mathbf{F}(\mathbf{y}, t, \gamma, \varepsilon) = \mathbf{F}_1(\mathbf{y}, t, \gamma) + \varepsilon \mathbf{F}_2(\mathbf{y}, t, \gamma) \tag{3.40}$$

where $\mathbf{F}_1$ and $\mathbf{F}_2$ are measurable stochastic processes for fixed $\mathbf{y}$ and

$$|F_i(y, t, \gamma)| < C, \qquad \left|\frac{\partial F_i}{\partial y_j}\right| < C, \qquad \left|\frac{\partial^2 F_i}{\partial y_j \partial y_k}\right| < C$$

for $i = 1, 2$ and $j, k = 1, \ldots, n$. Functions $\mathbf{F}_1$ and $\mathbf{F}_2$ satisfy also a certain form of a mixing condition.

The behaviour of a solution of (3.39)–(3.40) as $\varepsilon \to 0$ and $t \to \infty$ with $\tau = \varepsilon^2 t$ remaining fixed is characterized by the following assertion:
process $\mathbf{Y}^\varepsilon(\tau)$ converges weakly as $\varepsilon \to 0$ to Markov diffusion process $\mathbf{Y}^\circ(\tau)$ whose coefficients of drift and diffusion are determined through the formulae

$$a_i(\mathbf{y}) = \lim_{T\to\infty} \frac{1}{T} \int_{t_\circ}^{t_\circ+T} E\left\{F_{2,i}(\mathbf{y}, t, \gamma)\right\} dt$$
$$+ \lim_{T\to\infty} \frac{1}{T} \int_{t_\circ}^{t_\circ+T} \int_{t_\circ}^{s} \sum_{j=1}^{n} E\left\{\frac{\partial F_{1,i}(\mathbf{y}, s, \gamma)}{\partial y_j} F_{1,j}(\mathbf{y}, s, \gamma)\right\} ds dt, \tag{3.41}$$

$$b_{ij}(\mathbf{y}) = \lim_{T\to\infty} \frac{1}{T} \int_{t_\circ}^{t_\circ+T} \int_{t_\circ}^{s} E\left\{F_{1,i}(\mathbf{y}, t, \gamma) F_{1,j}(\mathbf{y}, t, \gamma)\right\} ds dt \tag{3.42}$$

where $F_{1,i}(\cdot)$ and $F_{2,i}(\cdot)$ denote the i-th component of $\mathbf{F}_1$ and $\mathbf{F}_2$, respectively.

It is seen that if $\mathbf{F}_2(\mathbf{y}, t, \gamma) \equiv 0$ and $\mathbf{F}_1(\mathbf{y}, t, \gamma) = \mathbf{F}(\mathbf{Y}, \mathbf{X}(t), t)$ formulae (3.41)–(3.42) reduce to (3.38). On the other hand, if we set $\mathbf{F}_1(\mathbf{y}, t, \gamma) \equiv 0$ and assume that $\mathbf{F}_2(\mathbf{y}, t)$ is deterministic then the above theorem reduces to the averaging principle for deterministic equations.

Let us add (cf. [102]) that if $E\left[\mathbf{F}_i(\mathbf{y},t,\gamma)\right]$, $(i=1,2)$ are periodic functions of t with period Θ and also the correlation functions occuring in integrals (3.41), (3.42) are periodic (with respect to s and t) with period Θ, for all $s>0$, $t>0$, then for $\mathbf{F}_i(\mathbf{y},t,\gamma)=\mathbf{F}_i(\mathbf{y},\mathbf{X}(t,\gamma))$, where $\mathbf{X}(t,\gamma)$ is stationary:

$$\begin{aligned} a_i(\mathbf{y}) &= \frac{1}{\Theta}\int_0^{\Theta} E\left\{F_{2,i}(\mathbf{y},t,\gamma)\right\} dt \\ &+ \frac{1}{\Theta}\int_0^{\Theta} ds \int_{-\infty}^{0} \sum_{j=1}^{n} E\left\{\frac{\partial F_{1,i}(\mathbf{y},s,\gamma)}{\partial y_j} F_{1,j}(\mathbf{y},s+u,\gamma)\right\} du \end{aligned} \tag{3.41'}$$

(3.42')

$$b_{ij} = \frac{1}{\Theta}\int_0^{\Theta} ds \int_{-\infty}^{+\infty} E\left\{F_{1,i}(\mathbf{y},s,\gamma)F_{1,j}(\mathbf{y},s+u,\gamma)\right\} du$$

By considering a "slow time" $\tau=\varepsilon^2 t$ from the onset, one can reformulate the above theorem as follows. The solution $\mathbf{Z}^{\varepsilon}(\tau)$ of the problem

$$\begin{aligned} \frac{d\mathbf{Z}^{\varepsilon}}{d\tau} &= \frac{1}{\varepsilon}\mathbf{F}\left(\mathbf{Z},\frac{\tau}{\varepsilon^2},\gamma,\varepsilon\right), \quad \mathbf{Z}^{\varepsilon}(0)=\mathbf{Y}_{\circ} \\ &= \frac{1}{\varepsilon}\mathbf{F}_1\left(\mathbf{Z},\frac{\tau}{\varepsilon^2},\gamma,\varepsilon\right)+\mathbf{F}_2\left(\mathbf{Z},\frac{\tau}{\varepsilon^2},\gamma,\varepsilon\right) \end{aligned} \tag{3.43}$$

converges weakly in the interval $\tau\in[0,\tau_{\circ}]$ to the process $\mathbf{Y}^{\circ}(\tau)$ which is Markov diffusion one and its infinitesimal characteristics are given by the formulae (3.41)–(3.42). Anticipating our derivations in the next section we shall say here that the limiting process $Y^{\circ}(\tau)$ is governed by the Itô stochastic differential equation

$$d\mathbf{Y}^{\circ}(\tau)=\mathbf{m}\left[\mathbf{Y}^{\circ}(\tau)\right]d\tau+\boldsymbol{\sigma}\left[\mathbf{Y}_{\circ}(\tau)\right]d\mathbf{W}(\tau) \tag{3.44}$$

where $\mathbf{W}(t)=[W_1(t),\ldots,W_l(t)]$ is an l-dimensional Wiener process for which $E\left[W_i(t)\right]=0$ and $E\left[W_i^2(\tau)\right]=\tau$, and

$$\begin{aligned} m_i(\mathbf{y}) &= a_i(\mathbf{y}), \\ \sigma(\mathbf{y})\sigma^*(\mathbf{y}) &= \{b_{ij}(y)\} \end{aligned} \tag{3.45}$$

with $a_i(\mathbf{y})$ and $b_{ij}(\mathbf{y})$ given by (3.41)–(3.42).

As illustration of the above Khasminskii limit theorem let us consider a system with random parameters

$$\begin{aligned} &\frac{d\mathbf{Z}^{(\varepsilon)}(\tau)}{d\tau} = \\ &= \mathbf{A}\left(\mathbf{Z}^{\varepsilon}(\tau),\tau\right)+\sum_{k=1}^{N}\mathbf{B}_k\left(\mathbf{Z}^{\varepsilon}(\tau),\tau\right)\frac{1}{\varepsilon}\mathbf{X}_k\left(\frac{\tau}{\varepsilon^2}\right) \end{aligned} \tag{3.46}$$

where matrices **A** and **B** are deterministic with bounded elements, and $X_k(t) = X_k(t,\gamma)$ are stationary and bounded zero mean stochastic processes for which

$$R_{kl}(s) = E\{X_k(t_1)X_l(t_2)\}, \quad s = t_2 - t_1$$
$$\rho_{kl} = \int_0^\infty R_{kl}(s)ds, \qquad k,l = 1,2,\ldots,N$$

and the above integral converges. Comparison of system (3.46) with (3.43) indicates that for $i = 1,2,\ldots,n$

$$F_{1,i}\left(\mathbf{z},\frac{\tau}{\varepsilon^2},\gamma,\varepsilon\right) = \sum_{k=1}^{N} B_i^k(\mathbf{z},\tau)X_k\left(\frac{\tau}{\varepsilon^2},\gamma\right)$$

$$F_{2,i}\left(\mathbf{z},\frac{\tau}{\varepsilon^2},\gamma,\varepsilon\right) = A_i(\mathbf{z},\tau) \tag{3.47}$$

According to the Khasminskii limit theorem, the solution of system of stochastic equations (3.46) converges weakly to a Markov diffusion process with drift and diffusion coefficients given by

$$a_j(\mathbf{z},\tau) = \sum_{k,l=1}^{N}\sum_{i=1}^{n} \rho_{kl} B_i^k(\mathbf{z},\tau)\frac{\partial B_j^l(\mathbf{z},\tau)}{\partial z_i} + A_j(\mathbf{z},\tau),$$

$$b_{ij}(\mathbf{z},\tau) = \sum_{k,l=1}^{N} \rho_{kl} B_i^k(\mathbf{z},\tau)B_j^l(\mathbf{z},\tau). \tag{3.48}$$

The above result suggests a possible interpretation of stochastic differential equations with white noise coefficients (see — Sec. 22.2 and compare with the interpretation implying by the Wong and Zakai theorem).

The results of asymptotic analysis presented in this subsection, especially the Khasminskii limit theorem, constitute a basis for the *stochastic averaging method* which has recently been used for deriving approximate solution of random vibration problems. An illustration of this method in the context of stochastic vibratory systems will be given in Chapter IV.

21.5. Stationary solutions

An important problem in theory and applications of stochastic differential equations is associated with the question: given a stochastic equation in vector form

$$\frac{d\mathbf{Y}(t)}{dt} = \mathbf{F}\left[\mathbf{Y}(t),\mathbf{X}(t,\gamma)\right] \tag{3.49}$$

where $\mathbf{X}(t,\gamma)$ is a given R_m-valued strictly stationary process;when is the process $\mathbf{Y}(t)$ (or $[\mathbf{Y}(t),\mathbf{X}(t)]$) again stationary?

An answer to the above question has attracted an attention of a number of authors (cf. [70],[103],[118]). A solution of (3.49) is usually understood as the so called *weak solution* that is a process $\mathbf{Y}(t)$ — defined on a possible larger probability space than that on which the original process $\mathbf{X}(t,\gamma)$ is given and satisfying (3.49) almost surely.

By contrast, a solution $\mathbf{Y}(t)$ is called a *strong solution* if it is defined on the original probability space (and is measurable with respect to the σ-field $\mathcal{F}$ generated by process $\mathbf{X}(t)$). Strong stationary solutions are of interest since they inherit ergodic properties of the process $\mathbf{Y}(t)$, e.g., if $\mathbf{X}(t)$ is ergodic, any strong solution $\mathbf{Y}(t)$ will be ergodic. There are some situations where weak stationary solutions exist but no strong ones.

Khasminskii [103] investigated these questions for quite general functions $\mathbf{F}$. He showed that a necessary and sufficient condition for the existence of a stationary weak solution $[\mathbf{Y}(t),\mathbf{X}(t)]$ of (3.49) is the existence of a weak solution $\mathbf{Y}(t)$ satisfying

$$\frac{1}{T}\int_0^T P\{|\mathbf{Y}(t)|>R\}\,dt \longrightarrow 0 \quad \text{as} \quad R\to\infty$$

uniformly for $T>0$.

For non-linear systems of particular form, when process $\mathbf{X}(t,\gamma)$ occurs linearly

$$\frac{d\mathbf{Y}(t)}{dt} = \mathbf{F}[\mathbf{Y}(t)] + \mathbf{G}[\mathbf{Y}(t)]\,\mathbf{X}(t,\gamma) \tag{3.50}$$

more effective conditions for the existence of a stationary solution can be given (cf. [103]).

Theorem 3.8. *Let $\mathbf{X}(t)$ be a stationary process and let vector function $\mathbf{F}(\mathbf{y})$ and matrix $\mathbf{G}(\mathbf{y})$ satisfy the Lipschitz condition and*

$$\sup_{\mathbf{y}}|\mathbf{G}(\mathbf{y})|<\infty$$

Let for the deterministic system

$$\frac{d\mathbf{Y}(t)}{dt} = \mathbf{F}(\mathbf{y}) \tag{3.51}$$

there exists a Lapunov function $V(\mathbf{y})$ such that

(a) function $V(\mathbf{y})$ is non-negative and

$$V(\mathbf{y}) \longrightarrow \infty \quad \text{as} \quad |\mathbf{y}|\to\infty$$

(b) function $\frac{d^\circ V}{dt}$, the derivative along the trajectories of systems (3.51), is bounded from above and for $|\mathbf{y}|\to\infty$, $\frac{d^\circ V}{dt}\to-\infty$.

Then equation (3.50) has a stationary solution.

As an illustration let us consider the following second-order stochastic equation

$$\frac{d^2Y(t)}{dt} + f[Y(t)]\dot{Y}(t) + g[Y(t)] = \sigma\left[Y(t), \dot{Y}(t)\right] X(t,\gamma) \tag{3.52}$$

where functions f and g satisfy for $|y| > y_o$ and for some integer positive constants n, k, the conditions:

$$0 < \frac{g(y)}{y^{2n+1}} < c, \quad 0 < \frac{f(y)}{y^{2k}} < c, \quad |\sigma(y, \dot{y})| < c,$$
$$g(y)\tilde{f}(y) \to \infty \quad \text{as} \quad |y| \to \infty,$$
$$\tilde{f}(y)\text{sign}y > \delta > 0 \quad \text{for} \quad |y| > y_o,$$

where

$$\tilde{f}(y) = \int_0^y f(z)dz.$$

Let process $X(t,\gamma)$ be stationary (with finite mean). Then the above equation has stationary solution.

The above assertion follows from theorem 3.8 if one represents equation (3.52) in the form of a system of two equations of the first order (denoting $Y_1 = Y, Y_2 = \dot{Y}$) and apply the Lapunov function

$$V(y_1, y_2) = \left\{ \frac{y_2^2}{2} + \left[\tilde{f}(y_1) - p(y_1)\right] y_2 + \tilde{g}(y_1) + \int_0^{y_1} f(z)\left[\tilde{f}(z) - p(z)\right] dz + 1\right]^{\alpha} - C_1$$

where

$$\tilde{g}(y_1) = \int_0^{y_1} g(z)dz, \quad p(y_1) = \beta \text{arctg} y_1$$

and positive constants β, C_1, α are selected in such a way that

$$\min V(y_1, y_2) = 0,$$
$$\frac{d^o V}{dt} \to -\infty \quad \text{for } y_1^2 + y_2^2 \to \infty.$$

It is important to notice that conditions of the above example are satisfied for the Van der Pol equation (very often used in non-linear dynamics) for which $f(y) = y^2 - 1$, $g(y) = y$.

Interesting results associated with existence of a stationary solution of linear systems of the form

$$\frac{d\mathbf{Y}(t)}{dt} = \mathbf{A}\mathbf{Y}(t) + \mathbf{B}\mathbf{X}(t),$$
$$\mathbf{Y}(0) = \mathbf{Y}_o$$

where $\mathbf{A}$ and $\mathbf{B}$ are constant matrices are presented in paper [65] by Arnold and Wihstutz. In this context it is worth noting that weak solutions of the above linear system are unique if and only if $Re\lambda_i(\mathbf{A}) \neq 0$, where λ_i are the eigenvalues of matrix $\mathbf{A}$. If $\lambda_i(\mathbf{A}) = 0$, the solutions are never unique.

22. ITÔ STOCHASTIC DIFFERENTIAL EQUATIONS

22.1. Existence and uniqueness of a solution

22.1.1. Basic theorem

The equation of the form (3.4), i.e. the equation

$$dY(t) = m\,[t, Y(t)]\,dt + \sigma\,[t, Y(t)]\,dW(t) \tag{3.53}$$

is called the *Itô stochastic differential equation* for a scalar process $Y(t)$; in this equation $W(t)$ is the Wiener process and $m(t,y)$, $\sigma(t,y)$ are assumed to be defined and measurable for $t \in [t_o, T]$, $y \in (-\infty, +\infty)$.

Equation (3.53) means that for small values of h the following approximate equality holds

$$Y(t+h) - Y(t) = m\,[t, Y(t)]\,h + \sigma\,[t, Y(t)]\,[W(t+h) - W(t)] \tag{3.54}$$

If the process $Y(t)$ is considered on the interval $[t_o, T]$ and at t_o the initial condition $Y(t_o) = Y_o$ is posed, then equation (3.53) is equivalent to the following *Itô stochastic integral equation*

$$Y(t) = Y_o + \int_{t_o}^{t} m\,[s, Y(s)]\,ds + \int_{t_o}^{t} \sigma\,[s, Y(s)]\,dW(s) \tag{3.55}$$

where the second integral is to be understood as the Itô stochastic integral. As we know from section 16, for the second integral to be well defined the function $\sigma[s, Y(s)]$ has to belong to the class $H_2[t_o, T]$ of non-anticipative functions; roughly speaking, it is required that $\sigma[t', Y(t')]$ for each t' is a random variable independent of the random variables $\{W(t) - W(s),\ s > t'\}$.

We say that process $Y(t), t \in [t_o, T]$ satisfies equation (3.53) with the initial condition Y_o if:

a) $Y(t)$ is $\mathcal{F}_t$-measurable, i.e. non-anticipating for $t \in [t_o, T]$,
b) both integrals in (3.55) exist,
c) equation (3.55) holds for every $t \in [t_o, T]$ with probability one.

It is clear from previous sections, that if $Y(t)$ is a solution of (3.55), then every process stochastically equivalent to $Y(t)$ is also a solution.

Equation (3.55) is a special case of the Itô stochastic equation for multi-dimensional process $\mathbf{Y}(t) = [Y_1(t), \ldots, Y_n(t)]$:

$$d\mathbf{Y}(t) = \mathbf{m}\,[t, \mathbf{Y}(t)]\,dt + \boldsymbol{\sigma}\,[t, \mathbf{Y}(t)]\,d\mathbf{W}(t) \tag{3.56}$$

where

$$\mathbf{m}(t,\mathbf{y}) = [m_1(t,\mathbf{y}),\ldots,m_n(t,\mathbf{y})]$$
$$\boldsymbol{\sigma}(t,\mathbf{y}) = \{\sigma_{ij}(t,\mathbf{y})\}_{i,j=1}^{n,m}$$

are defined and measurable in $(t,y) \in [t_o,T] \times R_n$; $\mathbf{W}(t) = [W_1(t),\ldots,W_m(t)]$ is a m-dimensional Wiener process, that is a collection of mutually independent scalar Wiener processes $W_j(t)$, $j = 1,\ldots,m$. Of course, system (3.56) can also be represented as

$$dY_i(t) = m_i\,[t,\mathbf{Y}(t)]\,dt + \sum_{j=1}^{m} \sigma_{ij}\,[t,\mathbf{Y}(t)]\,dW_i(t) \tag{3.56'}$$

If $\mathbf{Y}(t_o) = \mathbf{Y}_o$ is an initial condition, then equation (3.56) is equivalent to the following integral equation

$$\mathbf{Y}(t) = \mathbf{Y}_o + \int_{t_o}^{t} \mathbf{m}\,[s,\mathbf{Y}(s)]\,ds + \int_{t_o}^{t} \boldsymbol{\sigma}\,[s,\mathbf{Y}(s)]\,d\mathbf{W}(s) \tag{3.56''}$$

Now, we shall formulate the basic theorem concerning the Itô equation (cf. [61],[90], [91]). It contains the conditions under which the solution exists and is unique; it also determines some important properties of the solution process.

Theorem 3.9. *Let us assume that the following conditions are satisfied:*

a) *vector-valued function* $\mathbf{m}(t,\mathbf{y})$ *and the* $(n \times m)$*-matrix valued function* $\boldsymbol{\sigma}(t,\mathbf{y})$ *are defined and measurable for* $t \in [t_o,T], \mathbf{y} \in R_n$,

b) *there exists a constant* K *such that for* $t \in [t_o,T]$ *and* $\mathbf{y} \in R_n$ *the following conditions hold:*

   A. uniform Lipschitz condition

$$|\mathbf{m}(t,\mathbf{y}_1) - \mathbf{m}(t,\mathbf{y}_2)| + |\boldsymbol{\sigma}(t,\mathbf{y}_1) - \boldsymbol{\sigma}(t,\mathbf{y}_2)| \leq K\,|\mathbf{y}_1 - \mathbf{y}_2| \tag{3.57}$$

   B. growth condition

$$|\mathbf{m}(t,\mathbf{y})| + |\boldsymbol{\sigma}(t,\mathbf{y})| \leq K\,(1 + |\mathbf{y}|) \tag{3.58}$$

   where $|\mathbf{m}(t,\mathbf{y})|$ *denotes the norm in* R_n, *whereas* $|\boldsymbol{\sigma}(t,\mathbf{y})| = \text{trace}\,\boldsymbol{\sigma}\boldsymbol{\sigma}^T$, *where* $\boldsymbol{\sigma}^T$ *is a transposition of* $\boldsymbol{\sigma}$;

c) $\mathbf{Y}(t_o) = \mathbf{Y}_o$ *is a random variable independent of* $\mathbf{W}(t) - \mathbf{W}(t_o)$ *for* $t \geq t_o$; *then:*

1) *equation (3.56) has on* $[t_o,T]$ *a solution* $\mathbf{Y}(t)$ *satisfying for* $t = t_o$ *the initial condition* $\mathbf{Y}_o$,

2) *almost all realizations of* $\mathbf{Y}(t)$ *are continuous on* $[t_o,T]$,

3) *a solution $\mathbf{Y}(t)$ is unique, that is, if $\mathbf{Y}_1(t)$ and $\mathbf{Y}_2(t)$ are two solutions (continuous with probability one) of (3.56) satisfying the same initial condition then*

$$P\left\{\sup_{t_0 \le t \le T} |\mathbf{Y}_1(t) - \mathbf{Y}_2(t)| = 0\right\} = 1,$$

4) *the unique solution $\mathbf{Y}(t)$ is a Markov process on the interval $[t_0, T]$ whose initial probability distribution at $t = t_0$ is the distribution of $\mathbf{Y}_0$ and whose transition probability is given by*

$$P(s, \mathbf{y}, t, A) = P\left\{\mathbf{Y}(t) \in A \middle| \mathbf{Y}(s) = \mathbf{y}\right\},$$

5) *if, additionally, the functions $\mathbf{m}(t, \mathbf{y})$ and $\sigma(t, \mathbf{y})$ are continuous with respect to $(t, \mathbf{y})$ then the solution $\mathbf{Y}(t)$ is a diffusion Markov process on $[t_0, T]$ with the following drift vector $\mathbf{A}(t, \mathbf{y})$ and diffusion matrix $\mathbf{B}(t, \mathbf{y})$:*

$$\begin{aligned} \mathbf{A}(t, \mathbf{y}) &= \mathbf{m}(t, \mathbf{y}), \\ \mathbf{B}(t, \mathbf{y}) &= \sigma(t, \mathbf{y})\sigma^T(t, \mathbf{y}); \end{aligned} \tag{3.59}$$

in particular, if the coefficients of Itô equation (3.56) do not depend on t, then the solution $\mathbf{Y}(t)$ is a homogeneous diffusion process, that is its transition probability is homogeneous in time, what means that for each $\tau \in [0, T - t]$:

$$P(s + \tau, \mathbf{y}, t + \tau, A) = P(s, \mathbf{y}, t, A).$$

A detailed proof of this theorem is given in the books [61],[87],[90]; so, we shall not repeat it here. Instead, we will present a brief sketch of the proof of the existence and uniqueness and the comments on the range of validity of the theorem.

A natural way of proving theorem 3.9 is to construct process $\mathbf{Y}(t)$ and show that it has properties 1)–5). Let, for simplicity, $n = 1$, $m = 1$.

We start from the initial condition $Y_0(\gamma)$ and build the following sequence of successive approximations of the solution process $Y(t)$. Since we assumed that $n = m = 1$ we deal with equation (3.53). Let

$$\begin{aligned} Y_0(t, \gamma) &= Y_0(\gamma) \\ Y_{n+1}(t, \gamma) &= Y_0(\gamma) + \int_{t_0}^{t} m\left[s, Y_n(s, \gamma)\right] ds + \int_{t_0}^{t} \sigma\left[s, Y_n(s, \gamma)\right] dW(s, \gamma) \end{aligned} \tag{3.60}$$

for $n \ge 0$.

Now, one shows that such a sequence is well defined. This requires of a proof that the last integral is correctly defined for each n as the stochastic Itô integral. To do that one shows that for each n, $\sigma[t, Y_n(t,\gamma)]$ is a measurable function of t and γ and for each t it is $\mathcal{F}_t$-measurable and that

$$\int_{t_o}^{T} E\{\sigma^2[t, Y_n(t,\gamma)]\} < \infty$$

The proof of the above facts is usually given by induction.

In the next step one proves that sequence of processes $\{Y_n(t,\gamma)\}$, $(n = 0, 1, \ldots)$ is convergent in the m.s. sense, that is the following relation holds

$$\sup_{m \geq 0} E\left\{[Y_{n+m}(t,\gamma) - Y_n(t,\gamma)]^2\right\} \xrightarrow[n\to\infty]{} 0$$

If we denote a limit by $Y(t,\gamma)$, $t \in [t_o, T]$, i.e.

$$\sup_{t_o \leq t \leq T} E\left\{[Y_n(t,\gamma) - Y(t,\gamma)]^2\right\} \xrightarrow[n\to\infty]{} 0$$

we then show that this limiting process $Y(t,\gamma)$ has the properties stated in the theorem.

For example, in order to prove, that process $Y(t,\gamma)$ satisfies equation (3.55) with initial condition $Y(t_o) = Y_o$, we introduce a quantity $\Delta_t = Y_{n+1}(t,\gamma) - Y(t,\gamma)$, i.e.

$$\Delta_t = Y_{n+1}(t,\gamma) - Y_o - \int_{t_o}^{t} m[s, Y(s)]\, ds - \int_{t_o}^{t} \sigma[s, Y(s)]\, dW(s)$$

Making use of (3.60) we can represent Δ_t as

$$\Delta_t = \int_{t_o}^{t} \{m[s, Y_n(s,\gamma)] - m[s, Y(s,\gamma)]\}\, ds + \int_{t_o}^{t} \{\sigma[s, Y_n(s,\gamma)] - \sigma[s, Y(s,\gamma)]\}\, dW(s).$$

Now, one proves (making use of the Lipschitz condition) that each of the two components on the right hand side of the above equality tends to zero in m.s. sense as $n \to \infty$. Therefore, $E(\Delta_t^2) = 0$ and for each $t \in [t_o, T]$

$$Y(t) = Y_o + \int_{t_o}^{t} m[s, Y(s)]\, ds + \int_{t_o}^{t} \sigma[s, Y(s)]\, dW(s)$$

with probability one.

In order to prove the uniqueness we take two processes $Y_1(t)$ and $Y_2(t)$ and assume that they are two continuous solutions of equation (3.55); then we show that

$$E|Y_1(t) - Y_2(t)|^2 = 0, \quad \text{for all} \quad t \in [t_o, T].$$

To gain the goal we make use of the Lipschitz condition as well as other inequalities; for instance, the Bellman-Gronwall inequality (cf.Appendix).

22.1.2. Restrictions of the main theorem

It is clear that the conditions of the theorem 3.9 are rather restrictive.
Let us consider first the Lipschitz condition. It means that the coefficients of the Itô **equation must be somewhere between continuous and differentiable functions, in the sense that:**
$f(x,y)$ continuously differentiable with respect to y for all $(x,y) \in D \Longrightarrow f(x,y)$ satisfies a Lipschitz condition with respect to $y \Longrightarrow f(x,y)$ is continuous with respect to y.

In particular, if $f(x,y)$ has a continuous derivative with respect to y, then by virtue of the mean value theorem

$$f(x,y) - f(x,y') = \frac{\partial f(x,\tilde{y})}{\partial y}(y - y')$$

where $\tilde{y}$ is a point in the interior of the interval $[y, y']$. Clearly, the Lipschitz condition (with respect to y) is satisfied if we choose

$$K = \sup_{(x,y)\in D} \left| \frac{\partial f(x,y)}{\partial y} \right| .$$

Therefore, functions which are discontinuous with respect to y are excluded as coefficients. But also continuous functions of the type $f(x,y) = |y|^\alpha$, $0 < \alpha < 1$ are not admissible as well.This shows the Girsanov example (cf. [92]); the equation

$$Y(t) = \int_0^t |Y(s)|^\alpha \, dW(s)$$

has exactly one solution for $\alpha \geq \frac{1}{2}$, but infinitely many solutions for $0 < \alpha < \frac{1}{2}$.

To improve the situation a more relaxed form of existence and uniqueness theorem has been proven [90]. Namely, the theorem 3.9 remains valid if we replace the Lipschitz condition by the more general one: for every $N > 0$ there exists a constant K_N such that for all $t \in [t_o, T]$ and $|\mathbf{y}_1| \leq N$, $|\mathbf{y}_2| \leq N$

$$|\mathbf{m}(t,\mathbf{y}_1) - \mathbf{m}(t,\mathbf{y}_2)| + |\boldsymbol{\sigma}(t,\mathbf{y}_1) - \boldsymbol{\sigma}(t,\mathbf{y}_2)| \leq K_N \, |\mathbf{y}_1 - \mathbf{y}_2| .$$

In the literature there exist formulations of the existence and uniqueness theorem for the Itô equations under more relaxed conditions. In particular, Skorokhod in his paper [129] assumed that coefficients are only continuous with respect to their arguments and satisfy usual growth conditions and he proved the existence of a solution; the solution was, however, understood in a weaker sense (cf. weak solution in Sec. 22.8); cf. [113].

Another direction of generalization of the Itô theory is concerned with equations whose coefficients are discontinuous or they are generalized functions (distributions) — cf. [76],[83],[124].

The second condition stated in theorem 3.9 imposes the restrictions on growth of $\mathbf{m}(t, \mathbf{y})$ and $\boldsymbol{\sigma}(t, \mathbf{y})$. These functions must be bounded uniformly with respect to $t \in [t_o, T]$ and can increase at most linearly with respect to $\mathbf{y}$. If this condition is not satisfied we have the effect of an "explosion" of the solution to infinity (known from the determinstic theory). The restriction on the growth of $\mathbf{m}$ and $\boldsymbol{\sigma}$ guarantees that,with probability one, the solution does not escape to infinity in the interval $[t_o, T]$. Generally speaking, when the growth condition is violated, equation (3.56) defines the process $\mathbf{Y}(t)$ only on random interval of time $[t_o, \tau(\gamma)]$ where $\tau(\gamma)$ is the explosion time; at this time the process tends to either $+\infty$ or $-\infty$ (cf. Protter [125]). Process $\mathbf{Y}(t)$ for which $P\{\tau(\gamma) = \infty\} = 1$ is termed as a *regular* Markov process. Regularity of the solution $\mathbf{Y}(t)$ under conditions of theorem 3.9 follows from its continuity. Khasminskii has given (cf. [103]) more general sufficient conditions for the regularity of the solution of Itô equation (3.56).

Remark. The existence and uniqueness theorem 3.9 states that almost all sample functions of the solution process are continuous. However, if $\sigma(t, y)$ does not vanish, the solution of the Itô equation (in scalar and vector form) is, like $W(t)$, non-differentiable and has unbounded variation.

Remark. If the state space of process $\mathbf{Y}(t)$ governed by the Itô equation (3.56), or in scalar case (3.53) is finite or semi-finite, then the appropriate boundary conditions have to be imposed. The analytical form of these conditions depends on the behaviour of the process at the boundary (absorption, reflection, etc.). These questions were discussed in Sec. 6.2;they will also be treated in Sec. 27.1.

22.1.3. Stationary solutions

A question which is of great importance in applications can be stated as: when is the Markov diffusion solution $\mathbf{Y}(t)$ of the Itô equation a stationary process?

In Sec. 6.2.3. we provided necessary and sufficient conditions for a Markov process to be stationary. These conditions are

a) the transition probability is homogeneous in time;
b) there exists an invariant distribution P_{inv} in the state space of the process, that is such distribution that for all $A \in \mathcal{F}$

$$P_{\text{inv}}(A) = \int P(t, \mathbf{y}, A) dP_{\text{inv}}(\mathbf{y}) \tag{3.61}$$

If such a unique invariant distribution exists then, for all $A \in \mathcal{F}$

$$\lim_{t \to \infty} P\{\mathbf{Y}(t) \in A\} = P_{\text{inv}};$$

so, the invariant distribution is a stationary limit distribution. The analytical conditions (1.116′),(1.116″) for the existence of stationary diffusion process hold only in one-dimensional case. In multi-dimensional case the following theorem due to Khasminskii [103] holds.

Theorem 3.10. *Let us assume that coefficients of Itô equation (3.56) do not depend on time and they satisfy the conditions of theorem 3.9 in domain* $U_R = \{\mathbf{y} : |\mathbf{y}| < R\}$ *for each* R*; let, additionally, exist a function* $V(\mathbf{y}) \in C_2$ *defined in* R_n *such that:*

$$V(\mathbf{y}) \geq 0, \tag{3.62}$$
$$\sup_{|\mathbf{y}|>R} LV(\mathbf{y}) = -A_R \longrightarrow -\infty, \quad \text{as} \quad R \to \infty$$

where

$$LV(\mathbf{y}) = \sum_{i=1}^{n} m_i(\mathbf{y}) \frac{\partial V(\mathbf{y})}{\partial y_i} + \frac{1}{2} \sum_{i,j=1}^{n} \{\sigma(\mathbf{y})\sigma^T(\mathbf{y})\}_{ij} \frac{\partial^2 V}{\partial y_i \partial y_j};$$

then, there exists a solution of (3.56) being a stationary Markov process.

It is instructive to apply the above theorem to the equation

$$\frac{d^2Y}{dt^2} + f(Y)\frac{dY}{dt} + g(Y) = D\xi(t) \tag{3.63}$$

where $\xi(t)$ is a white noise and D — constant. When written down in the Itô form ($Y = Y_1$, $\frac{dY_1}{dt} = Y_2$)

$$dY_1 = Y_2(t)dt$$
$$dY_2 = [-Y_2 f(Y_1) - g(Y_1)]\, dt + \sigma dW(t)$$

it indicates that the operator L has the form

$$L = y_2 \frac{\partial}{\partial y_1} - [y_2 f(y_1) + g(y_1)] \frac{\partial}{\partial y_2} + \frac{D^2}{2} \frac{\partial^2}{\partial y_2^2}$$

Let us take the following function $V(\mathbf{y})$:

$$V(y_1, y_2) = \frac{y_2^2}{2} + [F(y_1) - p(y_1)]\, y_2 + G(y_1) + \int_0^{y_1} f(u)\,[F(u) - p(u)]\, du + k$$

where

$$F(y_1) = \int_0^{y_1} f(u)du, \quad G(y_1) = \int_0^{y_1} g(u)du, \quad p(y_1) = a \text{arctg } y_1$$

Substitution to conditions (3.62) shows that they are satisfied (for constants k and a) if for $\delta > 0$, $y_1^\circ > 0$ the following relationships hold

$$\left(\text{sign } g(y_1) = \text{sign } y_1, \quad |y_1| > y_1^\circ, \right)$$

$$g(y_1)F(y_1) - \delta\,|g(y_1)| \longrightarrow 0, \text{ as } |y_1| \to \infty,$$

$$G(y_1) + \delta \int_0^{y_1} \frac{F(u)}{1+u^2} du \longrightarrow \infty, \text{ as } |y_1| \to \infty$$

If the above conditions are satisfied, then there exists a stationary solution of equation (3.63). These conditions are satisfied for the Van der Pol equation, i.e. when $f(y) = \varepsilon(y^2 - 1)$ and $g(y) = y$.

For further discussion of stationary solutions the reader is referred to [64],[104].

22.2. Relation to Stratonovich interpretation

Let us consider a Langevin-type equation and assume, for simplicity, that $n = m = 1$ (one-dimensional case)

$$\frac{dY(t)}{dt} = \tilde{m}\,[t, Y(t)] + \tilde{\sigma}\,[t, Y(t)]\,\xi(t) \tag{3.64}$$

where $\xi(t)$ is a Gaussian white noise. Since white noise is not an ordinary stochastic process the above equality is something as a "pre-equation" which should be properly interpreted. The questions which arise are: how should "pre-equation" (3.64) be interpreted? how is the possible interpretation of (3.64) related to the Itô stochastic differential equation?

To answer the above questions it is natural to start from the interpretation of (3.64) in practice. What one really means by (3.64) in practice is most likely an equation driven by a stationary Gaussian process with spectral density which is flat over a sufficiently wide band of frequences. If we take $\xi(t)$ to be such a process in (3.64), then (3.64) can be interpreted as an equation for sample functions (provided that the spectral density of $\xi(t)$ tends to zero rapidly enough to assure regularity of the sample functions). However, such an interpretation does not lead to an effective probabilistic theory. The interpretation of (3.64) which is physically natural but which also turns out to be mathematically attractive is the following one. We take a sequence of Gaussian processes $\{\zeta_n(t)\}$, $(n = 1, 2 \ldots)$ which converge in some sense to a white Gaussian noise and for each n the processes $\zeta_n(t)$ have regular sample functions. For each n the equation

$$\frac{dY_n(t)}{dt} = \tilde{m}\left[t, Y_n(t)\right] + \tilde{\sigma}\left[t, Y_n(t)\right] \zeta_n(t) \tag{3.65}$$

can be solved (we assume that the functions $\tilde{m}$ and $\tilde{\sigma}$ are such that the sample function solution exists and is unique). In this way we obtain a sequence of processes $Y_n(t)$. Suppose now that a sequence of regular processes $\{\zeta_n(t)\}$ converges in some sense to Gaussian white noise as $n \to \infty$, and the corresponding sequence $\{Y_n(t)\}$ converges to a process $Y(t)$. It is natural to expect that $Y(t)$ is governed by the equation (3.64).

The above is a heuristic answer to the first of our questions. To obtain an answer to the second question we should perform precisely the limiting procedures indicated above (along with the discussion of conditions for the convergence of $Y_n(t)$ to $Y(t)$) and to obtain an equation for the limiting process $Y(t)$. Such an analysis has been provided by Wong and Zakai in their paper [139] — cf. also [39],[59]) and the result is as follows.

If we interpret a white-noise driven "equation" (3.64) by a sequence of equations (3.65), then "equation" (3.64) is equivalent to the following stochastic differential equation:

$$dY(t) = \tilde{m}\left[t, Y(t)\right] dt + \frac{1}{2}\tilde{\sigma}\left[t, Y(t)\right] \frac{\partial \tilde{\sigma}(t, Y)}{\partial Y} dt + \tilde{\sigma}\left[t, Y(t)\right] dW(t) \tag{3.66}$$

We see that this equation differs from the Itô equation (3.53) by the extra term:

$$\frac{1}{2}\tilde{\sigma}\left[t, Y(t)\right] \frac{\partial \tilde{\sigma}(t, Y)}{\partial Y} \tag{3.67}$$

Taking into account formula (2.100) for the Stratonovich integral we conclude that equation (3.66) is the stochastic Stratonovich equation

$$(S) \quad dY(t) = \tilde{m}\left[t, Y(t)\right] dt + \tilde{\sigma}\left[t, Y(t)\right] dW(t) \tag{3.68}$$

since it is equivalent to the following stochastic integral equation

$$(S) \quad Y(t) = Y_0 + \int_{t_0}^{t} \tilde{m}\left[s, Y(s)\right] ds + \int_{t_0}^{t} \tilde{\sigma}\left[s, Y(s)\right] dW(s) \tag{3.69}$$

where the second integral is the Stratonovich stochastic integral (2.100) in which $\Phi(s, Y(s)) = \tilde{\sigma}\left[s, Y(s)\right]$.

It is easy to notice that equation (3.68) is equivalent to the Itô stochastic equation

$$(\text{Itô}) \qquad dY(t) = m\left[t, Y(t)\right] dt + \sigma\left[t, Y(t)\right] dW(t) \tag{3.70}$$

in which

$$m\,[t, Y(t)] = \tilde{m}\,[t, Y(t)] + \frac{1}{2}\tilde{\sigma}\,[t, Y(t)]\,\frac{\partial \tilde{\sigma}(t, Y)}{\partial Y}$$

$$\sigma\,[t, Y(t)] = \tilde{\sigma}\,[t, Y(t)] \tag{3.71}$$

Conclusion: In order to obtain satisfactory mathematical interpretation of the Langevin-type equation (white-noise driven equation) we can proceed in two ways:

1) we make a limiting shift from a "real process" to the white noise in the original equation and solve this equation as a stochastic Itô equation (with coefficients $\tilde{m}$ and $\tilde{\sigma}$);
2) we construct a sequence of equations, as described above and obtain $Y_n(t)$ as a sample function solution of equation disturbed by $\zeta_n(t)$; then we make a shift (from $\zeta(t)$ to $\xi(t)$) in the solution; the limiting process $Y(t)$ is governed by a Stratonovich equation.

These two procedures lead to different results; although both equations (Itô and Stratonovich) define a diffusion process, their drift coefficients are different. Both procedures are mathematically correct, but the question which of them is "better" is a matter of modelling and can be answered in analysis of specific practical problems.

The analysis and conclusion presented above hold also for vector Langevin-type equation for process $\mathbf{Y}(t) = [Y_1(t), \ldots, Y_n(t)]$ (cf. [105])

$$(I) \quad \frac{d\mathbf{Y}(t)}{dt} = \tilde{\mathbf{m}}\,[t, \mathbf{Y}(t)] + \tilde{\boldsymbol{\sigma}}\,[t, \mathbf{Y}(t)]\,\xi(t) \tag{3.72}$$

According to the procedure described above (procedure 2) in the conclusion) the above equation is interpreted as the following Stratonovich equation

$$(S) \quad \mathbf{Y}(t) = \tilde{\mathbf{m}}\,[t, \mathbf{Y}(t)]\,dt + \tilde{\boldsymbol{\sigma}}\,[t, \mathbf{Y}(t)]\,d\mathbf{W}(t) \tag{3.73}$$

where $\mathbf{W}(t)$ is an m-dimensional Wiener process. This equation is equivalent to the Itô equation in which components of drift vector $\mathbf{m}\,[t, \mathbf{Y}(t)]$ and components of diffusion matrix $\boldsymbol{\sigma}\,[t, \mathbf{Y}(t)]$ are

$$m_i\,[t, \mathbf{Y}(t)] = \tilde{m}_i\,[t, \mathbf{Y}(t)] + \frac{1}{2}\sum_{k=1}^{m}\sum_{j=1}^{n}\tilde{\sigma}_{jk}\,[t, \mathbf{Y}(t)]\,\frac{\partial \tilde{\sigma}_{jk}(t, \mathbf{Y})}{\partial Y_j},$$

$$\sigma_{ij}\,[t, \mathbf{Y}(t)] = \tilde{\sigma}_{ij}\,[t, \mathbf{Y}(t)]\,. \tag{3.74}$$

22.3. State transformations and simple solutions

Let us consider the scalar Itô equation (3.53), i.e.

$$dY(t) = m\left[t, Y(t)\right] dt + \sigma\left[t, Y(t)\right] dW(t) \tag{3.53}$$

assuming that the conditions of the existence and uniqueness theorem are satisfied. An important problem which naturally arises is connected with the solving equation (3.53) when the initial condition $Y(0)$ independent of $W(t)$ is given.

In order to solve equation (3.53) it is useful first to reduce it to a simpler equation. This can be done by the transformation of an unknown process $Y(t)$ into another process and by use of the Itô's formula (cf. [90]).

Let $Y(t)$ be a solution of (3.53) and let $f(t,x)$ be a monotonous in x, continuous function defined for $t \in [0,T]$, $x \in (-\infty, +\infty)$ for which the derivatives $f'_t(t,x), f'_x(t,x), f''_{xx}(t,x)$ exist and are continuous. For each $\mathbf{Y}(t) \in [0,T]$ there exists a function $g(t,x)$ inverse to $f(t,x)$, that is such that $f\left(t, g(t,x)\right) = x$, $g\left(t, f(t,x)\right) = x$.

Let

$$Z(t) = f\left(t, Y(t)\right) \tag{3.75}$$

then $Y(t) = g\left(t, Z(t)\right)$ and by use of the Itô formula we have

$$\begin{aligned} dZ(t) = \Big[& f'_t\left(t, Y(t)\right) m\left(t, Y(t)\right) + \\ & + f'_x\left(t, Y(t)\right) m\left(t, Y(t)\right) + \\ & + \frac{1}{2} f''_{xx}\left(t, Y(t)\right) \sigma^2\left(t, Y(t)\right) \Big] dt + \\ & + f'_x\left(t, Y(t)\right) \sigma\left(t, Y(t)\right) dW(t) \end{aligned}$$

So, the process $Z(t)$ satisfies the equation

$$dZ(t) = \hat{m}\left(t, Z(t)\right) dt + \hat{\sigma}\left(t, Z(t)\right) dW(t) \tag{3.76}$$

where

$$\begin{aligned} \hat{m}(t,x) &= f'_t\left(t, g(t,x)\right) m(t, g(t,x)) + f'_x\left(t, g(t,x)\right) m\left(t, g(t,x)\right) \\ &\quad + \frac{1}{2} f''_{xx}\left(t, g(t,x)\right) \sigma^2\left(t, g(t,x)\right), \\ \hat{\sigma}(t,x) &= f'_x\left(t, g(t,x)\right) \sigma\left(t, g(t,x)\right) \end{aligned} \tag{3.77}$$

To reduce our original Itô equation (3.53) to a simpler one we consider few cases:

1) $\hat{\sigma}(t,x) \equiv 1$

In this case the transformation should be such that

$$f'_x(t,x)\sigma(t,x) = 1$$

that is

$$f'_x(t,x) = \frac{1}{\sigma(t,x)}, \quad f(t,x) = \int_0^x \frac{1}{\sigma(t,\xi)} d\xi \tag{3.78}$$

Such a transformation is always possible if $\sigma(t,x) > 0$.

2) $\hat{m}(t,x) \equiv 0$. Now, the function $f(x,t)$ should satisfy the equation

$$m(t,x)\frac{\partial f(t,x)}{\partial t} + m(t,x)\frac{\partial f(t,x)}{\partial x} + \frac{1}{2}\sigma^2(t,x)\frac{\partial^2 f(t,x)}{\partial x^2} = 0 \tag{3.79}$$

Let us assume that: $m(t,x) = m(x)$ and $\sigma(t,x) = \sigma(x)$. The function f is then also independent of t and

$$m(x)f'(x) + \frac{1}{2}\sigma^2(x)f''(x) = 0$$

implying

$$f(x) = C_1 + C_2 \int_0^x exp\left\{-\int_0^u \frac{2m(\xi)}{\sigma^2(\xi)} d\xi\right\} du \tag{3.80}$$

It is of interest to find the conditions under which the coefficients of the transformed equation (3.76) do not depend on the unknown process $Z(t)$.

3) $\hat{m}(t,x) = \hat{m}(t), \quad \hat{\sigma}(t,x) = \hat{\sigma}(t)$.

Simple reasoning leads to the result which in the general case (when the coefficients of the original equation are functions of both x and t) looks a little complicated. If, however, $m(t,x) = m(x)$, $\sigma(t,x) = \sigma(x)$ then the transformation

$$f(t,x) = e^{ct} \int_0^x \frac{1}{\sigma(\xi)} d\xi \tag{3.81}$$

reduces equation (3.53) to

$$dZ(t) = \hat{m}(t)dt + \hat{\sigma}(t)dW(t) \tag{3.82}$$

The solution of the above equation

$$Z(t) = Z(0) + \int_0^t \hat{m}(s)ds + \int_0^t \hat{\sigma}(s)dW(s) \tag{3.83}$$

is a process with independent Gaussian increments since the second integral is normally distributed. The mean and variance are

$$E\left[Z(t) - Z(0)\right] = \int_0^t \hat{m}(s)ds,$$

(3.84)
$$\mathrm{Var}\left[Z(t) - Z(0)\right] = \int_0^t \hat{\sigma}^2(s)ds$$

Another important case is the one in which the coefficients of the original equation can be reduced to the linear form. Let us assume, again, that $m(t,x) = m(x)$, $\sigma(t,x) = \sigma(x)$.

4)

$$\hat{m}(t,x) = \hat{m}(x) = \alpha_1 + \alpha_2 x$$

(3.85)
$$\hat{\sigma}(t,x) = \hat{\sigma}(x) = \beta_1 + \beta_2 x$$

we take the transformation

$$Z(t) = f\left(Y(t)\right)$$

and $g(x)$ is the inverse function to $f(x)$. The general relations (3.77) take the form

$$m\left(g(x)\right) f'_x\left(g(x)\right) + \frac{1}{2}\sigma^2\left(g(x)\right) f''_{xx}\left(g(x)\right) = \alpha_1 + \alpha_2 x$$

(3.86)
$$\sigma\left(g(x)\right) f'_x\left(g(x)\right) = \beta_1 + \beta_2 x$$

The analysis performed in [90] leads to the following condition for reducing the Itô equation with time-independent coeffiecients to the linear case:

(3.87)
$$\frac{d}{dx}\left[\frac{\frac{d}{dx}\left\{\sigma(x)\Lambda'(x)\right\}}{\Lambda'(x)}\right] = 0$$

where $\Lambda(x) = \frac{m(x)}{\sigma(x)} - \frac{1}{2}\sigma'(x)$. When the above condition is satisfied, the appropriate transformation is

(3.88)
$$f(x) = Cexp\left\{\beta_2 B(x)\right\}$$

where C is a constant and

(3.89)
$$\beta_2 = -\frac{\frac{d}{dx}\left\{\sigma(x)\Lambda'(x)\right\}}{\Lambda'(x)}, \quad B(x) = \int_0^x \frac{1}{\sigma(\xi)}d\xi.$$

As an illustration of the method presented above let us consider the **linear equation**

(3.90)
$$dY(t) = \alpha(t)Y(t)dt + \beta(t)Y(t)dW(t)$$

We assume that $Y(0) > 0$. Then because of continuity of $Y(t)$, $Y(t) > 0$ on a certain interval. Let

$$Z(t) = lnY(t) \tag{3.91}$$

Using the Itô formula we obtain

$$dZ(t) = \frac{1}{Y(t)}\alpha(t)Y(t)dt - \frac{1}{2}\frac{1}{Y^2(t)}\beta^2(t)Y^2(t)dt + \frac{1}{Y(t)}\beta(t)Y(t)dW(t)$$

Therefore,

$$dZ(t) = \left[\alpha(t) - \frac{1}{2}\beta^2(t)\right] dt + \beta(t)dW(t) \tag{3.92}$$

So, the solution is:

$$Z(t) = Z(0) + \int_0^t \left[\alpha(s) - \frac{1}{2}\beta^2(s)\right] ds + \int_0^t \beta(s)dW(s) \tag{3.93}$$

and the original process is

$$Y(t) = Y(0)exp\left\{ \int_0^t \left[\alpha(s) - \frac{1}{2}\beta^2(s)\right] + \int_0^t \beta(s)dW(s)\right\} \tag{3.94}$$

It should be noticed that the above formula represents the solution of equation (3.90) as long as this solution does not vanish. Since the right hand side of (3.93) does not vanish for $t > 0$, (3.94) gives an arbitrary solution of equation (3.90) when $Y(0) > 0$. Analogous argument applied to $Y(t)$ with $Y(0) < 0$ leads to the conclusion that (3.94) holds if $Y(0) \neq 0$.

In the particular case, where equation (3.90) has constant coefficients $\alpha(t) = \alpha_o$, $\beta(t) = \beta_o$ the solution is

$$\begin{aligned} Y(t) &= Y(0)exp\left\{\int_0^t \left[\alpha_o - \frac{\beta_o^2}{2}\right] ds + \int_0^t \beta_o dW(s)\right\} \\ &= Y(0)exp\left\{\left(\alpha_o - \frac{\beta_o^2}{2}\right) t + \beta_o W(t)\right\} \end{aligned} \tag{3.95}$$

Remark: If $\beta(t)$ is differentiable on the considered interval then the second integral occuring in solution (3.94) is $\left(\dot\beta(t) = \frac{d\beta}{dt}\right)$:

$$\int_0^t \beta(s)dW(s) = \beta(t)W(t) - \int_0^t \dot\beta(s)W(s)ds \tag{3.96}$$

Remark: It should be noticed that when equation (3.90) is regarded as the Stratonovich equation then its solution is the following

$$Y(t) = Y(0) exp\left\{\int_0^t \alpha(s)ds + \int_0^t \beta(s)dW(s)\right\} \tag{3.97}$$

Formula (3.94) can be used for various purposes. Making use of this formula (due to the properties of the Itô integral) we obtain directly the expression for the moments of arbitrary order p. For variable coefficients:

$$E|Y(t)|^p = E|Y(0)|^p exp\left\{p\int_0^t \alpha(s)ds + \frac{p^2}{2}\int_0^t \beta^2(s)ds\right\} \tag{3.98}$$

and for constant coefficients:

$$E|Y(t)|^p = E|Y(0)|^p exp\left\{p\left[\alpha_0 + \frac{p^2}{2}\beta_0^2\right]t\right\} \tag{3.99}$$

The reader will easily notice that in order to obtain the above formulae the following fact connected with normal distribution is used: if $X(\gamma)$ is a Gaussian random variable with mean m_x and variance σ_x^2, then for every $p > 0$

$$E\left[e^{X(\gamma)}\right]^p = e^{pm_X + \frac{1}{2}p^2\sigma_X^2} \tag{3.100}$$

Remark: The solutions of the linear vector Itô equations are more complex; the appropriate analysis can be found in book [61]. Applying the Itô formula for the function $f(y) = y_i y_j$ (or making use of the corresponding Fokker-Planck-Kolmogorov equation) one can derive a system of linear deterministic differential equations for moments of different orders. The equations for moments shall be considered in the next chapter.

Remark: In various situations, when the Itô stochastic differential equation constitutes a model of a complex system the transformations of the state variable might be non-applicable (the transformation shown above are connected only with scalar equations). In such cases the approximative and numerical methods of solving the stochastic Itô equations will be helpful. Of course, various approximative schemes have to take into account specific features of the Itô integral. A detailed discussion of numerical methods for stochastic equations is presented in Chapter V of this book.

22.4. Asymptotic properties

Just as in the analysis of deterministic differential equations an interesting class of problems is concerned with the behaviour of a solution of stochastic differential equation for large values of t, that is when $t \to \infty$.

Although we do not intend to pay here much attention to the asymptotic behaviour of solutions, it seems, nevertheless, to be advisable to give the reader an idea about the problems and some results.

A) First, we shall mention the problem associated with the **ergodic-type theorems**. They can be at least of two kinds. The first kind includes the theorems associated with limiting properties of averages

$$\frac{1}{T}\int_0^T f(Y(s))\,ds$$

as $T \to \infty$ and where $Y(t)$ is a diffusion process governed by a stochastic equation and f is a function belonging to a sufficiently wide class (e.g. f is a continuous function). To the second class of ergodic theorems belong all statements concerning the existence of a limiting distribution for $Y(t)$ when $t \to \infty$.

There exists an interesting connection between these two ergodic problems (cf. [90]). Namely, for the most common cases

$$\lim_{t\to\infty}\frac{1}{T}\int f(Y(s))\,ds = \int f(y)dG(y) \tag{3.101}$$

where G is the limiting distribution of $Y(t)$ for $t \to \infty$. Such a distribution is called an *ergodic distribution* for the process $Y(t)$.

An ergodic distribution can only exist when $|Y(t)|$ remains finite with probability one; this will be the case if the function

$$\varphi(x) = \int_0^x exp\left\{-\int_0^u \frac{2m(\xi)}{\sigma^2(\xi)}d\xi\right\}du \tag{3.102}$$

is such that: $lim_{x\to\infty}\varphi(x) = +\infty$ and $\lim_{x\to-\infty}\varphi(x) = -\infty$.

Let us consider the scalar Itô equation with time-independent coefficients, i.e. $m(t,x) = m(x)$, $\sigma(t,x) = \sigma(x)$. As we know,the appropriate transformation $Z(t) = f(Y(t))$ reduces the original Itô equation to the equation: $dZ(t) = \hat{\sigma}(Z(t))\,dW(t)$. Since f is monotonous the limiting process associated with $Z(t)$ and $Y(t)$ exist simultaneously. Hence, we can consider equation

$$dY(t) = \sigma[Y(t)]\,dW(t) \tag{3.103}$$

The following theorems hold [90].

Theorem 3.11. *Let g be a bounded measurable function, then for arbitrary y*

$$\lim_{T\to\infty}\frac{1}{T}\int_0^T E[g(Y(t))]\,dt = D\int_{-\infty}^{+\infty}\frac{g(y)}{\sigma^2(y)}dy \tag{3.104}$$

where

$$D = \left(\int_{-\infty}^{+\infty}\frac{1}{\sigma^2(\xi)}d\xi\right)^{-1} \tag{3.105}$$

Theorem 3.12. *Let $\sigma(y)$ satisfy a Lipschitz condition and let*

$$\int_{-\infty}^{+\infty} \frac{1}{\sigma^2(\xi)} d\xi < \infty$$

and

$$F_t(y) = P\{Y(t) < y\}$$

Then

$$\lim_{t \to \infty} F_t(y) = D \int_{-\infty}^{y} \frac{1}{\sigma^2(\xi)} d\xi \tag{3.106}$$

The next theorem gives the conditions under which there exists limting distribution for the process $\frac{Y(t)}{\alpha(t)}$, where $\alpha(t)$ is a suitable deterministic function tending to infinity as $t \to \infty$. It turns out that the limiting distribution of $Y(t)$ normed in this way is Gassian.

Theorem 3.13. *Assume that the coefficients of equation*

$$dY(t) = m[Y(t)]\,dt + \sigma[Y(t)]\,dW(t)$$

fulfill the conditions

a) $m(y)$, $\sigma(y)$ *and* $\sigma'(y)$ *satisfy the Lipschitz condition;*
b) $\sigma(y) > 0$ *and* $\lim_{|y| \to \infty} \sigma(y) = \frac{1}{s_0} > 0$ *exists;*
c) $\int_{-\infty}^{+\infty} \frac{m(\xi)}{\sigma^2(\xi)} d\xi = 0$.

Then

$$\lim_{t \to \infty} P\left\{\frac{Y(t)}{\sqrt{t}} < \frac{y}{s_0}\right\} = \\ = \frac{1}{\sqrt{2\pi}} \int_{-\infty}^{y} e^{-\frac{u^2}{2}} du \tag{3.107}$$

B) Another important class of asymptotic problems in the theory of stochastic differential equations is associated with *stability* of stochastic dynamic systems.

Stability of dynamic system (or, of a solution of differential equation) is usually understood as insensitivity of the state of the system within unbounded time interval $[t_0, \infty)$ to small changes in the initial state or in the parameter of the system. As it is known there are a number of stability concepts in the study of deterministic differential equations. In the stochastic case a number of different stability notions is even greater because of larger variety of the concepts of stochastic convergence. The concepts of stochastic stability which are studied most often are: the stability in probability, stability with probability one and stability of moments.

Let $Y(t)$ be a global solution on $[t_0, \infty)$ of a stochastic equation in question (which has not to be necessarily the Itô equation). Without loss of generality the trivial (or equilibrium) solution $Y(t) \equiv 0$ can be studied.

Definition 3.5. *A trivial solution of a stochastic differential equation is said to be:*

a) **stable in probability,** *if for arbitrary (sufficiently small) numbers* $\varepsilon > 0$, $\delta > 0$ *there exists such number* $r > 0$, *that for arbitray solution* $Y(t)$ *for which at* $t = t_\circ$

$$|y^\circ| = |Y(t_\circ)| < r$$

the following inequality holds for all $t \geq t_\circ$

$$P_t\{|Y(y^\circ, t_\circ; t)| > \varepsilon\} < \delta \tag{3.108}$$

where $P_t\{-\}$ *denotes the probability that at time* $t \geq t_\circ$ *the event* $\{-\}$ *occurs; according to Khasminskii [103]* $Y(t)$ *is stable in probability if for every* $\varepsilon > 0$

$$\lim_{|y^\circ| \to 0} P\left\{\sup_{t \geq t_\circ} |Y(y^\circ, t_\circ; t)| > \varepsilon\right\} = 0; \tag{3.109}$$

a′) **asymptotically stable in probability** , *if it is stable in probability and*

$$\lim_{t \to \infty} P_t\{|Y(y^\circ, t_\circ; t)| < \varepsilon\} = 1 \tag{3.110}$$

a″) **asymptotically stable in probability in the large** *if the above equality holds for all* $y^\circ \in R_1$ *(or* $\mathbf{y}^\circ \in R_n$, *in the case of vector equations);*

b) **stable with probability one** *(or, almost surely), if for arbitrary solution* $Y(t)$ *such that* $|Y(t_\circ| < r$ *almost all trajectories of* $Y(t)$ *are stable;*

b′) **asymptotically stable with probability one,** *if*

$$|Y(y^\circ, t_\circ; t)| \underset{t \to \infty}{\longrightarrow} 0 \quad \textit{almost surely}; \tag{3.111}$$

c) **stable in p-th mean** *(where* $p > 0$*) if for every* $\varepsilon > 0$, *there exists* $r > 0$ *such that* $|y^\circ| < r$ *implies*

$$E\{|Y(y^\circ, t_\circ; t)|^p\} < \varepsilon, \quad t \geq t_\circ; \tag{3.112}$$

c′) **asymptotically stable in p-th mean** *if it is stable in p-th mean and if*

$$\lim_{t \to \infty} E\{|Y(y^\circ), t_\circ; t)|^p\} = 0; \tag{3.113}$$

c″) **exponentially stable in p-th mean** *if there exist positive constants* c_1 *and* c_2 *such that for sufficiently small* y°

$$E\{|Y(y^\circ, t_\circ; t)|^p\} \leq c_1 |y^\circ|^p e^{-c_2(t - t_\circ)}; \tag{3.114}$$

in the case $p = 1$ *or* $p = 2$ *we speak of (asymptotic, exponential)* stability in mean *or in* mean square, *respectively.*

Since the function $E\{|Y|^p\}^{1/p}$ is monotonically increasing when $p > 0$, the stability in p-th mean implies the same in q-th mean when $0 < q < p$. It follows from the Markov inequality (cf. (1.8)) that stability in p-th mean implies the stability in probability.

In the last decades the problem of stability of stochastic systems has generated a great deal of interest. In the early period of the development most of classical methods of stability theory was extended to stochastic systems. Essential advances are concerned with elaboration of stochastic Lapunov function method. Very recently, the asymptotic behaviour (when $t \to \infty$) of stochastic systems, including stability, is studied by use of the Lapunov exponents (cf. [62],[66],[114]). Since the thecrey of stochastic stability has at present an extensive literature including a number of books (cf. [61],[103],[112]) and review papers (cf. [109],[110]) we leave these problems out of our attention here. In the next chapter and in chapter VI we shall present appropriate examples and applications.

22.5. Equations with Markov coefficients

Let us consider the stochastic equation (in a vector form) for an n-dimensional process $\mathbf{Y}(t)$:

$$\frac{d\mathbf{Y}(t)}{dt} = F\left[\mathbf{Y}(t), \mathbf{X}(t)\right], \quad \mathbf{Y}(0) = \mathbf{Y}_0 \tag{3.115}$$

where $\mathbf{X}(t)$ is an R_m-valued Markov diffusion process. Let us assume additionally here that $\mathbf{X}(t)$ is a time homogeneous process whose drift and diffusion coefficients are $a_i(\mathbf{x})$ and $b_{ij}(\mathbf{x})$, $i, j = 1, 2 \ldots, m$, respectively.

According to the theory of Itô stochastic differential equations (see Sec. 21.1) the process $\mathbf{X}(t)$ can be represented as a solution of the vector Itô's stochastic equation ($i = 1, \ldots, m$)

$$dX_i(t) = a_i\left[\mathbf{X}(t)\right] dt + \sum_{j=1}^{m} \sigma_{ij}\left[\mathbf{X}(t)\right] dW_j(t) \tag{3.116}$$

where $W_1(t), \ldots, W(m(t)$ are mutually independent Wiener process and the vectors

$$\boldsymbol{\sigma}_j = \left[\sigma_{1j}(x), \ldots, \sigma_{mj}(x)\right], \quad j = 1, 2, \ldots, m \tag{3.117}$$

are the eigenvectors of the matrix $\boldsymbol{\sigma}^2 = \{b_{ij}(x)\}$. If equation (3.115) is written for components and in differential form

$$dY_k(t) = F_k\left[\mathbf{Y}(t), \mathbf{X}(t)\right] dt, \quad k = 1, 2, \ldots, n \tag{3.118}$$

then the system of equations (3118), (3.116) is a system of Itô stochastic equations for the vector process $[\mathbf{Y}(t), \mathbf{X}(t)]$. This $(n+m)$-dimensional process is a diffusion Markov process with the following drift vector and diffusion matrix

$$\begin{gathered} [F_k(\mathbf{y}, \mathbf{x}), a_i(\mathbf{x})], \quad k = 1, 2 \ldots, n \\ \{b_{ij}(\mathbf{x})\}, \quad i, j = 1, 2, \ldots, m \end{gathered} \tag{3.119}$$

The transition probability density of the process $[\mathbf{Y}(t), \mathbf{X}(t)]$ $p(\mathbf{y}_\circ, \mathbf{x}_\circ; t, \mathbf{y}, \mathbf{x})$, if it exists, satisfies the following Fokker-Planck-Kolmogorov equation

$$\begin{gathered} \frac{\partial p}{\partial t} + \sum_{k=1}^{n} \frac{\partial}{\partial y_k} [F_k(\mathbf{y}, \mathbf{x})p] + \sum_{i=1}^{m} \frac{\partial}{\partial x_i} [a_i(\mathbf{x})p] \\ - \frac{1}{2} \sum_{ij=1}^{m} \frac{\partial}{\partial x_i, \partial x_j} [b_{ij}(\mathbf{x})p] = 0, \\ p(\mathbf{y}_\circ, \mathbf{x}_\circ; 0, \mathbf{y}, \mathbf{x}) = \delta(\mathbf{x} - \mathbf{x}_\circ)\delta(\mathbf{z} - \mathbf{z}_\circ). \end{gathered} \tag{3.120}$$

In the scalar case, when $n = m = 1$, equations (3.118) and (3.116) form a system of two equations

$$\begin{aligned} dY(t) &= F[Y(t), X(t)]\, dt \\ dX(t) &= a[X(t)]\, dt + \sqrt{b[X(t)]} dW(t) \end{aligned} \tag{3.121}$$

and the corresponding Fokker-Planck-Komogorov equation is

$$\frac{\partial p}{\partial t} \frac{\partial}{\partial y} [F(y, x)p] + \frac{\partial}{\partial x} [a(x)p] - \frac{1}{2} \frac{\partial^2}{\partial x^2} [b(x)p] = 0 \tag{3.122}$$

where, of course, $a(x)$ and $b(x)$ are the drift and diffusion coefficients of a given one-dimensional diffusion process $X(t)$.

In the next chapter we shall illustrate this approach.

22.6. Equations with jump processes

The Itô stochastic differential equations considered so far consitute a model of dynamical systems whose displacement within a small interval h can be represented by formula (3.54). More general representation of an increment of a process $\mathbf{Y}(t)$ — written for a vector case — is

$$d\mathbf{Y}(t) = \mathbf{Y}(t + \Delta t) - \mathbf{Y}(t) = \mathbf{A}[t, \mathbf{Y}(t)]\Delta t + \boldsymbol{\eta}[t, \mathbf{Y}(t), \Delta t] \tag{3.123}$$

where $\boldsymbol{\eta}[t, \mathbf{Y}(t), \Delta t]$ consists of two components; the first component is a continuous function of Δt and the second one describes the changes by jumps, i.e.

$$\begin{aligned} &\boldsymbol{\eta}[s,t,\mathbf{y}] = \\ (3.124) \qquad &= \mathbf{B}(t,\mathbf{y})\,[\mathbf{W}(t+s)-\mathbf{W}(t)] + \int_{R_n} h(t,\mathbf{y},\mathbf{u})N(t,t+s,d\mathbf{u}) \end{aligned}$$

where $\mathbf{B}(t,\mathbf{y}) = \{b_{jk}(t,\mathbf{y})\}, (j = 1,2,\ldots,m; k = 1,2,\ldots,m)$ is a known diffusion matrix and $\mathbf{W}(t) = [W_1(t),\ldots,W_m(t)]$ is an m-dimensional Wiener process. The meaning of the second term is as follows.

We assume that a random measure N with integer values is defined on the σ-algebra $\mathcal{B}$ of measurable sets of space $[0,\infty) \times R_n$; this measure has the property that its values are independent on non-overlapping sets belonging to $\mathcal{B}$ and for all sets of the form $[t_1,t_2] \times A$ (where A is measurable set in R_n) the random variable $N([t_1,t_2] \times A)$ has the Poisson distribution with the intensity parameter

$$(3.125) \qquad \int_{t_1}^{t_2} \mu(t,A)dt$$

where $\mu(t,A)$ for every t is a measure in R_n and for each A it is a measurable function of t, $t \geq 0$. Introducing the denotation

$$\begin{aligned} N([T_1,t_2] \times A) &= N(t_1,t_2,A) \\ (3,126) \qquad N(0,t;A) &= N(t,A) \end{aligned}$$

we interpret $N(t_1,t_2;A)$ as a quantity which characterizes a number of events which occured within time interval $[t_1,t_2]$ and took place in $A \subset R_n$. Let us assume that if an event occured in time t and at point $\mathbf{u} \in R_n$ then a system which at time t was at $\mathbf{y}$ undergoes a displacement of "length" $h(t,\mathbf{y},\mathbf{u})$. It is clear that an overall displacement of a system due to events occured in $[t_1,t_2]$ which took place in R_n is

$$(3.127) \qquad \int_{t_1}^{t_2}\int_{R_n} h(t,\mathbf{y}(t),\mathbf{u})N(dt,d\mathbf{u})$$

The second component in (3.124) represents a part of the above quantity related to a small time interval $(t,t+s)$.

Relations (3.123), (3.124) can be written in the form of the following stochastic equation

$$\begin{aligned} &d\mathbf{Y}(t) = \mathbf{A}[t,\mathbf{Y}(t)]\,dt + \mathbf{B}[t,\mathbf{Y}(t)]\,d\mathbf{W}(t) + \int_{R_n} h[t,\mathbf{Y}(t),\mathbf{u}]\,N(dt,d\mathbf{u}) \\ (3.128) \qquad & \\ &\mathbf{Y}(t_o) = \mathbf{Y}_o \end{aligned}$$

The above equation is equivalent to the following integral equation

$$\mathbf{Y}(t) = \mathbf{Y}(t_o) + \int_{t_o}^{t} \mathbf{A}(s, \mathbf{Y}(s))\, dt$$

$$(3.129) \qquad + \int_{t_o}^{t} \mathbf{B}(s, \mathbf{Y}(s))\, d\mathbf{W}(s) + \int_{t_o}^{t} \int_{R_n} h(s, \mathbf{Y}(s), \mathbf{u})\, N(ds, du)$$

The third integral in the above equation is the *counting integral* (with respect to the Poisson random measure) — cf. Section 18.

Let us formulate now the conditions for the existence and uniqueness of the solution of the initial value problem (3.128) — cf. [91].

Suppose that the functions $\mathbf{m}(t, \mathbf{y})$, $\sigma(t, \mathbf{y})$ and $\mathbf{h}(t, \mathbf{y}, \mathbf{u})$ satisfy the following conditions.

1) there exists such a constant L, that

$$(3.130) \qquad |\mathbf{A}(t,\mathbf{y})|^2 + |\mathbf{B}(t,\mathbf{y})|^2 + \int |h(t,\mathbf{y},\mathbf{u})|\, \lambda(d\mathbf{u}) \le L\,(1 + |\mathbf{y}|^2)\,;$$

2) for an arbitrary $R > 0$ there exists a constant C_R such that for $|\mathbf{y}_1| \le R$, $|\mathbf{y}_2| \le R$

$$(3.131) \qquad \begin{aligned} &|\mathbf{A}(t,\mathbf{y}_1) - \mathbf{A}(t,\mathbf{y}_2|^2 + |\mathbf{B}(t,\mathbf{y}_1) - \mathbf{B}(t,\mathbf{Y}_2)|^2 \\ &+ \int |h(t,\mathbf{y}_1,\mathbf{u}) - h(t,\mathbf{y}_2,\mathbf{u})|^2\, \lambda(d\mathbf{u}) \le C_R\, |y_1 - y_2|^2 \end{aligned}$$

3) there exists such a constant L_1 and function $g(\varepsilon) \ge 0$, $g(\varepsilon) \to 0$, as $\varepsilon \to 0$ that

$$(3.132) \qquad \begin{aligned} &|\mathbf{A}(t+\varepsilon,\mathbf{y}) - \mathbf{A}(t,\mathbf{y})|^2 + |\mathbf{B}(t,+\varepsilon,\mathbf{y}) - \mathbf{B}(t,\mathbf{y})|^2 \\ &+ \int |h(t+\varepsilon,\mathbf{y},\mathbf{u}) - h(t,\mathbf{y},\mathbf{u}|^2\, \lambda(d\mathbf{u}) \le L_1\,(1 + |\mathbf{y}|^2)\, g(\varepsilon). \end{aligned}$$

Theorem 3.14. *If the conditions 1), 2), 3) are satisfied, then stochastic differential equation (3.128) has a unique solution whose almost all sample functions are continuous from the right.*

Of course, in addition to the condition 1)-3) we always assume that the initial condition $\mathbf{Y}_o$ does not anticipate the underlying point process on $[t_o, \infty)$, by which we mean that $\mathbf{Y}_o$ is statistically independent of

$$\int_{t_1}^{t_2} \int_{R_n} N(ds, du)$$

for all t_1 and t_2 such that $t_o \le t_1 \le t_2$. This property of the initial condition reproduces itself in time with the important implication that the process $\mathbf{Y}(t)$, $t \ge t_o$

defined by equation (3.128) is a Markov process. This Markovian property of a solution (which can be proven rigorously — cf. [91]) is a reason that processes described by equation (3.129), especially in the case where the diffusion matrix $\mathbf{B}(t, \mathbf{y})$ vanishes are often called the *Poisson driven processes.* The transition probability density satisfies an integrodifferential equation known as a Feller-Kolmogorov equation. While this equation can be solved only in certain very special circumstances, it still provides a useful starting point for the approximate statistical analysis of non-linear systems excited by a compound Poisson process.

Of course, like in the purely diffusion case (where process $\mathbf{Y}(t)$ is governed by the Itô differential equation (3.53)) an important tool in investigating of equation (3.129) is the generalized Itô formula (2.125). To illustrate this point let us consider the following linear scalar equation

$$dY(t) = \alpha(t)Y(t)dt + \beta(t)Y(t)dW(t) + \int h(t,u)N(dt,du) \tag{3.133}$$

Introducing a new unknown process $Z(t) = ln\,|Y(t)|$ and use of the Itô formula (2.125) give

$$dZ(t) = \left[\alpha(t) - \frac{1}{2}\beta^2(t)\right] dt + \beta(t)dW(t) + \int ln\,|1 + h(t,u)|\, N(dt,du)$$

This means that

$$Z(t) = Z_0 + \int_0^t \left[\alpha(\tau) - \frac{1}{2}\beta^2(\tau)\right] d\tau + \int_0^t \beta(\tau)dW(\tau) + \int_0^t \int ln\,|1 + h(t,u)|\, N(dt,du)$$

and

$$Y(t) = Y_0 exp\Big\{ \int_0^t \left[\alpha(\tau) - d\beta^2(\tau)\right] d\tau + \int_0^t \beta(\tau)dW(\tau) + \int_0^t \int ln\,|1 + h(t,u)|\, N(dt,du)$$

The above formula represents the solution process as a linear transformation of the coefficients and processes $W(t)$ and $N(t,u)$ and is convenient for averaging.

22.7. Equations with functional coefficients

Here we wish to present some comments on more general Itô stochastic equations. Namely, we shall assume that the coefficients of equation depend on the history of unknown process, that is upon the trajectories of $\mathbf{Y}(t)$ up to time t. The equation has the form

$$
\begin{aligned}
&d\mathbf{Y}(t) = \mathbf{m}\,[t, \mathbf{Y}(\cdot)]\,dt + \boldsymbol{\sigma}\,[t, \mathbf{Y}(\cdot)]\,d\mathbf{W}(t) \\
&\mathbf{Y}(t_\circ) = \mathbf{Y}_\circ
\end{aligned}
\tag{3.135}
$$

where $\mathbf{m}\,[t, \mathbf{y}(\cdot)]$ is a measurable mapping: $[t_\circ, T] \times C_{[t_\circ,T]}(R_m) \rightarrow R_m$, where $C_{[t_\circ,T]}(R_m)$ is a space of continuous functions defined on $[t_\circ, T]$ with values in R_m; $\boldsymbol{\sigma}[t, \mathbf{y}(\cdot)]$ is a $m \times m$ matrix-valued function measurable with respect to both arguments; for each $t \in [t_\circ, T]$ it is $\mathcal{R}$-measurable, where $\mathcal{R}$ is a σ-algebra of subsets of $C_{[t_\circ,T]}(R_m)$ and for each $\mathbf{y}(\cdot) \in C_{[t_\circ,T]}(R_m)$ it is bounded as function of t.

If coefficients $\mathbf{m}\,[t, \mathbf{y}(\cdot)]$ and $\boldsymbol{\sigma}\,[t, \mathbf{y}(\cdot)]$ do not depend on the whole trajectory $\mathbf{y}(s)$, $s \leq t$ but just on the values of function $\mathbf{y}(\cdot)$ at moment t then equation (3.135) reduces to (3.53). The notion of a solution of equation (3.135) is analogous to that corresponding equation (3.53). The basic existence and uniqueness theorem can be stated as follows.

Theorem 3.15. *Let* $\mathbf{m}\,[t, \mathbf{y}(\cdot)]$ *be such as described above and satisfy the conditions:*

1) *there exists a constant* K, *such that for* $t \leq T$

$$|\mathbf{m}\,[t, \mathbf{y}(\cdot)]|^2 + |\boldsymbol{\sigma}\,[\mathbf{y}(\cdot)]|^2 \leq K \left(1 + ||\mathbf{y}(\cdot)||_t^2\right)$$

where $||\mathbf{y}(\cdot)||_t = \sup_{s \leq t} |\mathbf{y}(s)|$;

2) *for each* $R > 0$ *there exists such a constant* C_R *that for* $t \in [t_\circ, T]$, $||\mathbf{y}_1(\cdot)||_T \leq R$, $||\mathbf{y}_2(\cdot)||_T \leq R$ *the following relation holds*

$$|\mathbf{m}\,[t, \mathbf{y}_1(\cdot)] - \mathbf{m}\,[t, \mathbf{y}_2(\cdot)]|^2 + |\boldsymbol{\sigma}\,[t, \mathbf{y}_1(\cdot)] - \boldsymbol{\sigma}\,[t, \mathbf{y}_2(\cdot)]|^2 \leq C_R\, ||\mathbf{y}_1(\cdot) - \mathbf{y}_2(\cdot)||_t\,.$$

Then there exists a unique solution $\mathbf{Y}(t)$ *of equation (3.135).*

Although the increments of process $\mathbf{Y}(t)$ within small (infinitesimal) time interval shall be approximately equal to increments of the process governed by Itô equation (3.53), in the case of (3.135) coefficients depend, in general, on all previous states of the process, and therefore the solution is not Markovian.

The equations just described include, as a special case, the equations with retarded argument.

22.8. Strong and weak solutions

The notion of a solution of Itô stochastic differential equation defined and used in the present section means a stochastic process $Y(t)$ which is defined on a given probability space and is connected with the Wiener process $W(t)$ by relation (3.53). Such a solution is at arbitrary time t completely determined by the values of a given Wiener process in all earlier than t instants and by the initial condition; according to Sec. 21.1. it is $\mathcal{F}_t$-measurable for each $t \in [t_0, T]$, i.e. non-anticipating.

A solution $Y(t)$ of equation (3.53), or more precisely a pair $[Y(t), W(t)]$ in the above sense is called a strong solution. Also, a uniqueness of a solution defined by statement 3) of theorem 3.9 is usually termed as uniqueness in a strong sense, or a sample function uniqueness.

Although the theorems for existence and uniqueness of a strong solution (3.53) or vector equation (3.60) can be formulated in various forms and the conditions formulated in theorem 3.9 can be relaxed, the Lipschitz condition (3.57) or its counterparts are still too restrictive. Many problems lead to stochastic equations whose coefficients do not satisfy these conditions. This is one of the reasons that a concept of a solution should be extended. Considering a stochastic Itô equation it is reasonable to assume that all what is given are just coefficients of the equation. Then, one should construct a probability space and processes $W(t)$ and $Y(t)$ on it in such a way that these processes are related to each other by equation (3.53) with given coefficients $m(t, y)$ and $\sigma(t, y)$; it is necessary in the same time that the minimal σ-algebra generated by the values of $Y(s)$ for $s \leq t$ and increments $W(t + \tau) - W(t)$, $\tau > 0$, to be independent. This leads to the concept of a weak solution.

Definiton 3.6. *Let us assume that the functions $m(t, y)$ and $\sigma(t, y)$ are given (such as in defintion of a strong solution). If there exists a probability space $(\Gamma, \mathcal{F}, P)$ with a family of σ-algebras $\{\mathcal{F}_t,\ t \in [t_0, T]\}$ and such a pair of processes $[Y(t), W(t)]$ adapted to the family $\mathcal{F}_t$ that process $W(t)$ is a Wiener process and processes $Y(t)$ and $W(t)$ are, with probability one, connected to each other by the integral relation (3.55) for all $t \in [t_0, T]$, then the pair of processes $[Y(t), W(t)]$ is called* a weak solution *of equation (3.53) or (3.55)*

It is clear that every strong solution of equation (3.53) is, in the same time, a weak solution. If arbitrary two solutions $[Y_1(t), Y_2(t)]$ of equation(3.53) have the same finite-dimensional distributions, then we say a solution is unique in a weak sense (or, unique in probability).

It can be proved that an arbitrary solution of equation (3.53) defined on some probability space is either a strong solution, or none of the solutions defined on any probability space is a strong one and they are not unique in a strong sense. Simple conditions for the existence and uniqueness of a weak solution of a scalar Itô equation with time-independent coefficients are given in paper [126].

More detailed studies of weak solutions of stochastic differential equations an interested reader will find in such monographes as [91],[98].

23. STOCHASTIC ABSTRACT DIFFERENTIAL EQUATIONS

23.1. Introduction; deterministic theory

The formulation and analysis of a wide class of stochastic differential equations can be made systematically and rigorously by use of stochastic differential equations in Banach space. Such equations have been studied in the last years in various contexts. They include, as a special case, not only partial stochastic differential equations, but also ordinary stochastic delay equations and, of course, stochastic ordinary differential equations considered in the previous sections.

A natural approach to the Hilbert space stochastic equations is based on the concepts and theory of semigroups of operators in Hilbert space. The semigroup approach, or more generally, the evolution operator approach is most often used in the recent literature. Such an approach is an especially appealing one for the usage in applications since the abstract formulation and solutions look like those in finite-dimensional case.To make these words clear we shall first give a brief description of the evolution of a dynamical system.

Let us consider a dynamical system whose states are the elements of a Banach space B; the initial state of a system at time $t = 0$ is denoted by y_o, which is an element of B. The states of the system at subsequent times $t > 0$ are characterized by $y(t)$. Let us assume that our system is linear and autonomous. We can introduce an operator T_t, $t \geq 0$ such that

$$y(t) = T_t y_o \tag{3.136}$$

$$T_t : B \longrightarrow B \ , \ T_o = I \tag{3.137}$$

where I is the identity operator on B.

If the evolution of the system is continuous (it depends continuously on the initial state) then T_t is bounded.

If we assume that $y(t+s)$ is the same state as that reached first during time s and then by going from the state $y(s)$ at time s to the state $y(t)$ at time t we obtain the basic semigroup property of our dynamical system:

$$y(t+s) = T_{t+s} y_o = T_t y(s) = T_t T_s y_o \tag{3.138}$$

or

$$T_{t+s} = T_t T_s \tag{3.139}$$

If, additionally $y(t) \longrightarrow y_o$ as $t \rightarrow 0$ for all $y_o \in B$ we can simply say that T_t is a continuous semigroup of operators on B. The definition is as follows (cf. [7],[20]).

Definition 3.7. *A family T_t of operators from B to B is called* a strong continuous semigroup *if*

$$T_{t+s} = T_t T_s \qquad o \leq s \leq t$$

$$T_o = I$$

$$\|T_t y_o - y_o\| \longrightarrow 0 \quad \text{as } t \rightarrow 0, \text{ for each } y_o \in B.$$

An interesting fact is that

$$\|T_t\| \leq M e^{\omega t} \tag{3.141}$$

where ω and $M \geq 0$ are constants.

It is clear that we can interpret T_t as an operator-valued function (which maps R^+ into the space $\mathcal{L}(B)$ of operators from B to B). So, it is natural to introduce an operator A analogous to the derivative

$$Ay_o = \lim_{h \to 0} \frac{1}{h} [T_h y_o - y_o] \tag{3.142}$$

The above operator is called the *infinitesimal generator* of T_t and (cf. [7])

$$\frac{d}{dt}(T_t y_o) = AT_t y_o = T_t A y_o \qquad \text{for all} \quad y_o \in \mathcal{D}(A) \tag{3.143}$$

Therefore, the evolution of a dynamical system can be respresented as

$$\frac{d}{dt} y(t) = Ay(t), \qquad y(0) = y_o \tag{3.144}$$

The unique solution of (3.144) is: $y(t) = T_t y_o$.

It is clear that for a system of ordinary differential equations

$$\dot{\mathbf{y}} = \mathbf{A}\mathbf{y},$$
$$\mathbf{y}(0) = \mathbf{y}_o, \quad \mathbf{y}, \mathbf{y}_o \in R_n$$

the semigroup is: $T_t = e^{\mathbf{A}t}$.

Let us take a "system" governed by the equation (heat conduction)

$$\frac{\partial y}{\partial t} = \frac{\partial^2 y}{\partial x^2}, \qquad \begin{array}{c} y(0,t) = y(1,t) = 0 \\ y(x,0) = y_o(x) \end{array} \tag{3.145}$$

If we take $B = H = L_2(0,1)$ and the operator A on H as

$$A = \frac{\partial^2}{\partial x^2} \tag{3.146}$$

where the domain $\mathcal{D}(A)$ of A is: $\left\{ z \in H : \frac{\partial z}{\partial x}, \frac{\partial^2 z}{\partial x^2} \in H;\ z(0) = z(1) = 0 \right\}$ then the initial-boundary problem (3.145) is equivalent to the following abtract initial problem

$$\frac{dy}{dt} = Ay, \quad y(0) = y_o \tag{3.147}$$

with A given by (3.146). Operator A is a generator of the strongly continuous semigroup given by

$$T_t z(x) = \sum_{k=1}^{\infty} 2e^{-k^2\pi^2 t} \sin k\pi x \int_0^1 \sin k\pi x z(x) dx \tag{3.148}$$

So, the solution of (3.145) is: $y(t) = T_t y_0$.

In general, many linear partial differential equations can be expressed as an abstract evolution equation of type (3.147) for a suitable choice of operator A and Banach space B, and their solutions are then given in terms of the strongly continuous semigroup T_t generated by A.

Let us take the inhomogeneous equation

$$\frac{dy}{dt} = Ay + f, \quad y(0) = y_0 \tag{3.149}$$

where f takes its values in B. It can be shown (cf. [7]) that if A generates a strongly continuous semigroup T_t on a Banach space B, and if

(i) f is continuously differentiable on $[0, T]$,

(ii) $y_0 \in \mathcal{D}(A)$ then

$$y(t) = T_t y_0 + \int_0^t T_{t-s} f(s) ds \tag{3.150}$$

is continuously differentiable on $[0, T]$ and is a solution of (3.149).

Of course, the assumption (i) posed above on f is often too restrictive. However, if we accept solutions satisying equation (3.149) only almost everywhere, the hypothesis on f can be weakened. Condition (i) can be replaced by:

(a) $T_{t-s} f(s) \in \mathcal{D}(A)$ for all $t > s \in [0, T]$

(b) $f \in L_1(T; B)$, $\quad AT_{t-s} f(s) \in L_1(T; B)$,

where $L_1(T; B)$ is a space of integrable functions defined on T with values in B.

More essential weakening of the above assertions leads to the concept of a mild solution of equation (3.149) by which we mean that (if $f \in L_p(T; B)$, $p \geq 1$) (3.150) holds.

A more general class of abstract evolution equations takes the form

$$\frac{dy}{dt} = [A + F(t)] y, \quad y(0) = y_0 \tag{3.151}$$

where $F(t)$ belongs to the class of strongly measurable operator-valued functions which map $(0, T)$ into space $\mathcal{L}(B)$ and are bounded. To express the solution of (3.151) we need the concept of a mild evolution operator (cf. [7]).

Let $\Delta(T)$ be defined as the following set of pairs $\{(t, s) : 0 \leq s \leq t \leq T\}$.

Definition 3.8. $U(t,s)$ *is a* mild evolution operator *defined on* $\Delta(T)$ *with values in* $\mathcal{L}(B)$ *if*

a) $U(t,t) = I$, $t \in [0,T]$,

b) $U(t,r)U(r,s) = U(t,s)$, $0 \le s \le r \le t \le T$,

c) $U(t,s)$ *is strongly continuous in* s *on* $[0,T]$ *and in* t *on* $[s,T]$.

A special case of mild evolution operators is formed by $U(t,s) = T_{t-s}$. In general, $U(t,s)$ is not differentiable in t, but if $F(t) \in C^1$ in t (in the uniform operator toplogy) then $U(t,s)$ is differentiable and is termed as a **evolution operator**. Such operator has the property

$$\frac{\partial}{\partial t}[U(t,s)h] = (A + F(t))\,U(t,s)h \qquad \text{for} \quad h \in \mathcal{D}(A)$$

This means that

$$y(t) = U(t,0)y_0 \tag{3.152}$$

is differentiable in t when $y_0 \in \mathcal{D}(A)$ and it is a (strong) solution of (3.151).

Otherwise (when $U(t,s)$ is not differentiable) $U(t,0)y_0$ can be regarded as a mild solution of (3.151) by which we mean that $y(t)$ satisfies the integral equation

$$y(t) = y_0 + A\int_{t_0}^{t} y(s)ds + \int_0^t F(s)y(s)ds. \tag{3.153}$$

Let us consider inhomogeneous equation

$$\frac{dy}{dt} = [A + F(t)]\,y + f(t), \quad y(0) = y_0 \tag{3.154}$$

If

a) $A + F(t)$ generates a strong evolution operator,

b) $Y_0 \in \mathcal{D}(A)$,

c) $F(t) \in C^1([0,T];B)$

then (cf. [7]) equation (3.154) has the strong solution

$$y(t) = U(t,0)y_0 + \int_0^t U(t,s)f(s)ds \tag{3.155}$$

If any of the conditions a), b), c) fail, $y(t)$ may not be differentiable and (3.155) is regarded as the mild solution of (3.154). It is worth noting that in the analysis of partial differential equations via abstract evolution equations the mild solution is a weak solution defined in terms of distributions.

23.2. Itô stochastic equations

23.2.1. Linear equations; basic theorems

Let us consider the following linear stochastic equation for the process $Y(t), t \in [t_0, T]$

$$dY(t) = AY(t)dt + B(t)dW(t),$$
$$Y(t_0) = Y_0 \tag{3.156}$$

Let us remember that in the "classical" case where A and B are $(n \times n)$ and $(n \times m)$ matrices, respectively, Y_0 is an n-dimensional random vector independent of the k-dimensional Wiener process $W(t)$, the unique solution of (3.156) is

$$Y(t) = e^{At}Y_0 + \int_{t_0}^{t} e^{a(t-s)}B(s)dW(s) \tag{3.157}$$

Now, let A be a generator of a strongly continuous semigroup T of bounded linear operators from the Hilbert space H to H; $B(t)$ is a family of bounded operators from the Hilbert space K into H and $W(t)$ is the Wiener process with values in K. Equation (3.156) is equivalent to the integral equation

$$Y(t) = Y_0 + \int_{t_0}^{t} AY(s)ds + \int_{t_0}^{t} B(s)dW(s) \tag{3.158}$$

where the second integral is the Itô integral with respect to the K-valued Wiener process $W(t)$ — cf. Sec. 19.

Definition 3.9. *A H-valued stochastic process $Y(t)$ is called* **a strong solution of equation** *(3.156) or (3.158) if: $Y(t) \in \mathcal{D}(A)$ with probability one, $Y(t)$ satisfies (3.158) almost everywhere on $[t_0, T] \times \Gamma$ and $Y(t)$ has continuous sample functions.* Solution $Y(t)$ is regarded to be unique if whenever $Y_1(t)$ is another solution

$$P\left\{ \sup_{t_0 \leq t \leq T} \|Y(t) - Y_1(t)\|_H \neq 0 \right\} = 0 \tag{3.159}$$

In general (without additional assumptions) equation (3.156) has not a strong solution. Nevertheless, it describes a well-defined stochastic process which is continuous in the mean square sense and is expressed by the formula

$$\begin{aligned} Y(t) &= T_{t-t_0}Y_0 + \int_{t_0}^{t} T_{t-s}B(s)dW(s) \\ &= T_{t-t_0}Y_0 + \sum_{k=1}^{\infty} \int_{t_0}^{t} T_{t-s}B(s)e_k\, d\beta_k \end{aligned} \tag{3.160}$$

where $\{e_k\}_{k=1}^{\infty}$ form a complete orthogonal basis in H and occur in defintion (1.34) of the H-valued Wiener process; β_k, $k = 1, 2, \ldots$ are real Wiener processes.

Theorem 3.16. *If $T_{t-t_o}Y_o \in \mathcal{D}(A)$ with probability one, $T_{t-s}B(s)e_k \in \mathcal{D}(A)$ for all k and*

$$\sum_{k=1}^{\infty} \lambda_k \int_{t_o}^{t} \|AT_{t-s}B(s)e_k\|_H \, ds < \infty \tag{3.161}$$

then $y(t)$ given by (3.160) is the unique strong solution of (3.156) or (3.158).

If the conditions of the above theorem are not satisfied the process $Y(t)$ represented by (3.160) is called the *mild solution*.

To illustrate the above notions let us consider a stochastic heat conduction equation, i.e. equation (3.156) in which

$$A = \frac{\partial^2 Y}{\partial x^2}, \quad B(t) \equiv 1$$

Associated initial and boundary conditions are the same as in (3.145). The same is also the domain $\mathcal{D}(A)$. We represent H-valued Wiener process $W(t)$ as

$$W(t) = \sum_{k=1}^{\infty} \beta_k(t) e_k$$

where $\beta_k(t)$ are mutually independent real Wiener processes with variances λ_k and $e_k(x) = \sin \pi k x$. A strongly continuous semigroup generated by operator A is given by (3.148).The process given by formula (3.160) is the well-defined mild solution of our equation. It is strong solution provided that

$$\sum_{k=1}^{\infty} \lambda_k \int_{0}^{t} \|AT_{t-s} \sin k\pi(\cdot)\|_{L_2(0,1)} \, ds < \infty$$

But

$$AT_{t-s} \sin k\pi(\cdot) = -\pi^2 k^2 e^{-k^2\pi^2(t-s)} \sin k\pi(\cdot)$$

So, the condition is

$$\sum_{k=1}^{\infty} \lambda_k k^2 < \infty$$

which is a serious restriction on the intensity of the noise process.

Analogous assertions hold for a more general abstract stochastic equation cf. [77])

$$\begin{aligned} &dY(t) = AY(t)dt + B(t)dW(t) + g(t)dt, \\ &Y(t_o) = Y_o \end{aligned} \tag{3.162}$$

which can be written down in the integral form

$$Y(t) = Y_o + \int_{t_o}^{t} [AY(s) + g(s)]\, ds + \int_{t_o}^{t} B(s)dW(s) \tag{3.162'}$$

where g is an integrable H-valued function defined on $[t_o, T]$.

The mild solution of (3.162) has form

$$Y(t) = T_{t-t_o}Y_o + \int_{o}^{t} T_{t-s}g(s)ds + \int_{t_o}^{t} T_{t-s}B(s)dW(s) \tag{3.163}$$

If, an addition to the assumptions stated in theorem 3.16.

$$\int_{t_o}^{t} \|AT_{t-s}g(s)\|\, ds < \infty \tag{3.164}$$

then $Y(t)$ given by (3.163) is the unique strong solution of (3.162).

A stochastic counterpart of abstract deterministic equation (3.151) has the form

$$\begin{aligned} &dY(t) = [A + F(t)]\, Y(t)dt + B(t)dW(t), \\ &Y(t_o) = Y_o \end{aligned} \tag{3.165}$$

The strong solution of this equation is defined analogously to definition 3.9 and its mild solution by

$$Y(t) = U(t, t_o)Y_o + \int_{t_o}^{t} U(t, s)B(s)dW(s) \tag{3.166}$$

where $U(t, s)$ is the evolution operator generated by $A + F(t)$.

It can be shown (cf. [77]) that if $A(t) = A + F(t)$ generates a strong evolution operator $U(t, s)$ and $U(t, s)B(s)e_k \in \mathcal{D}(A)$, $U(t, t_o)Y_o \in \mathcal{D}(A)$ with probability one for all $t_o \leq s \leq t \leq T$ and

$$\sum_{i=1}^{\infty} \lambda_k \int_{t_o}^{t} \|A(t)U(t, s)B(s)e_k\|_H^2\, ds < \infty \tag{3.167}$$

then $Y(t)$ given by (3.166) is the unique strong solution of (3.165).

The results presented above, though general in nature, have direct applicatory signifance in analysis of linear systems with distributed parameters (for example, heat and mass transfer processes, wave propagation, turbulence etc.) disturbed by additive noises. It is clear, that solutions (3.163), (3.166) give explicite expressions for the solution process in terms of given elements (Y_o, $W(t)$, $B(s)$) and the evolution operator $U(t, s)$ or T_{t-s}. If Y_o is Gaussian then $Y(t)$ is Gaussian as well. Formulae

(3.163), (3.166) along with the properties of Itô integral with respect to $W(t)$ can be used for calculating the moments.

Analogously as in finite-dimensional case an important class of problems is concerned with such properties as: existence of stationary measures for the solution process, stability of the solutions, asymptotic (as $t \to \infty$) behaviour, etc. Such problems have been carefully studied in a number of papers — cf. Curtin [79], Haussmann [93], Ichikawa [95], Zabczyk [142], [143]. Following [95] we may say that solution (3.166) is stable in the mean square sense if

$$\lim_{T\to\infty} \frac{1}{T} \int_0^T E\{\|Y(t)\|\}\, dt < \infty \tag{3.168}$$

However, the analysis of moments of solution (3.166) gives the upper bounds for $E\left\{\|Y(t)\|^2\right\}$ in terms of $\|U(t,s)\|$, $\|B(s)\|$ and the variance of Y_0 — cf. [63].

Hence, if $U(t,s)$ is exponentially stable, then solution (3.166) is stable in the mean square sense. It is clear, however, that in the semigroup case (when $U(t,s) = T_{t-s}$) — cf. solution (3.163) — we have the bound (3.141) which says that for $\omega < 0$ we have exponential asymptotic stability of the zero solution of equation (3.162) and instability for $\omega > 0$. In other cases one should make use of the estimates of the evolution operator $U(t,s)$ — cf. [7].

The definition of a mild solution does not require its sample continuity. However, the sufficient conditions for such continuity can be established. Once this is guaranteed it is meaningful to consider sample path stability (or stability with probability one — cf. [79], [93].

23.2.2. Equation for characteristic functional

The basic quantity which characterizes a probability measure in infinite-dimensional Hilbert space is the characteristic functional. As we have already said in Section 11, the characteristic functional is a direct generalization of the concept of characteristic function. According to the definition 1.31

$$F(t,\lambda) = E\left\{e^{i(X_t,\lambda)_H}\right\} \tag{3.169}$$

where $t \in [0,T]$ is the parameter of the H-valued stochastic process $X(t)$, $\lambda \in H$ and $(\cdot,\cdot)_H$ is the inner product in H.

It is of interest to try to construct characteristic functional for the processes governed by stochastic differential equations. First attempt in this direction was made by Hopf who derived differential equations (in functional derivatives) for the characteristic functional of the solution of Navier-Stokes equations of turbulence theory. Recently, Chow (cf. [75]), Curtin [78] and Kotulski [106] derived the equations for the characteristic functional of the solution of the Itô-type stochastic abstract equation. Here, we shall report the equation derived in the thesis [106], (cf. also [107]).

Let us consider the following abstract equation

$$dY(t) = AY(t)dt + [BY(t)]\,dW(t),$$
$$Y(0) = Y_o \in H \tag{3.170}$$

where, as previously, $A : \mathcal{D}(A) \to H$, $(\mathcal{D}(A) \subset H)$ is a generator of a strongly continuous semigroup T_t, $t \geq 0$, of bounded linear operators from the Hilbert space H into H. $B : H \to \mathcal{L}(H_1, H)$ where $\mathcal{L}(H_1, H)$ is the space of linear bounded operators from H_1 to H; $W(t)$ is an H_1-valued Wiener process, which is Gaussian with zero mean and the covariance operator

$$E\,[W(t) \otimes W(s)] = Q \min(t, s) \tag{3.171}$$

where $\otimes$ denotes here the tensor product of elements of H_1. If we denote by α_k, $k = 1, 2, \ldots$, the eigenvalues of the operator Q and by e_k — its eigenvectors, i.e.

$$Qe_k = \alpha_k e_k \qquad k = 1, 2, \ldots$$

then

$$\alpha_k > 0, \qquad \sum_{k=1}^{\infty} \alpha_k < \infty$$

and, moreover, $\{e_k\}$, $k = 1, 2, \ldots$, constitutes an orthonormal basis in H_1. The Wiener process $W(t)$ has the following representation

$$W(t) = \sum_{k=1}^{\infty} \sqrt{\alpha_k}\, e_k \beta_k(t) \tag{3.172}$$

where $\beta_k(t)$ are real independent Wiener processes with unit intensities. Equation (3.170) can be written down in the following integral form

$$Y(t) = Y_o + \int_0^t AY(s)ds + \sum_{k=1}^{\infty} \sqrt{\alpha_k} \int_0^t [BY(s)]\, e_k d\beta_k(s) \tag{3.173}$$

In [106],[107] it has been shown that under rather general conditions the characteristic functional $F(t, \lambda)$ of the mild solution of equation (3.170), i.e. of the solution of (3.173), satisfies the following differential equation (in Fréchet derivatives)

$$\frac{\partial}{\partial t} F(t, \lambda) = \left(\frac{\delta F(t, \lambda)}{\delta \lambda}, A^* \lambda \right)_H$$
$$+ \frac{1}{2} \sum_{k=1}^{\infty} \alpha_k \left(\left[B \frac{\delta}{\delta \lambda} \left(\left[B \frac{\delta F(t, \lambda)}{\delta \lambda} \right] e_k, \lambda \right)_H \right] e_k, \lambda \right)_H \tag{3.174}$$

with the following initial and normalization conditions

$$F(0,\lambda) = e^{i(Y_o,\lambda)_H}$$

$$F(t,0) = 1 \tag{3.175}$$

where the initial value Y_o has been assumed to be deterministic. In (3.174) operator A^* is adjoint to A. Obtaining the solution of equation (3.174) is not an easy task. As a matter of fact, the existing theory of differential equations in functional derivatives do not yet permit its consistent utilization in solving the equations of type (3.174). However, equation (3.174) constitutes a basis for deriving the equations for moments.

The moment $m_k(t)$ of the H-valued process $Y(t)$ is defined as the following tensor-valued function

$$m_k(t) = E\left\{X(t)\otimes X(t)_{\underset{k\ \text{times}}{\cdots}}\otimes X(t)\right\},\ k=1,2,\ldots \tag{3.176}$$

Looking for the solution of equation (3.174) in the form of the following series

$$F(t,\lambda) = \sum_{k=1}^{\infty}\frac{(i)^k}{k!}m_k(t)\cdot\lambda\otimes\underset{k\ \text{times}}{\cdots}\otimes\lambda \tag{3.177}$$

and substituting it into (3.174), after indicated differentiations and comparison of the terms of the same power in λ, one obtains the following equations for the moments (cf. [107])

$$\frac{d}{dt}m_k(t) = \sum_{i=1}^{k}A_i m_k(t)$$

$$+\frac{1}{2}\sum_{l=1}^{\infty}\sum_{i,j=1}^{k}\alpha_l\left[B_i\left[B_j m_k(t)\right]e_l^j\right]e_l^i \tag{3.178}$$

with initial conditions

$$m_k(0) = Y_o\otimes\cdots\otimes Y_o$$

$k=1,2,\ldots.$

The operators A_i and $[B_i\cdot]e^i$ act on i-th coordinate of simple tensors of the elements of Hilbert space.

In particular, the equation for the mean of the solution of equation (3.173) takes the form

$$\frac{d}{dt}m_1(t) = Am_1(t)+\frac{1}{2}\sum_{l=1}^{\infty}\alpha_l\left[B\left[Bm_1(t)\right]e_l\right]e_l \tag{3.179}$$

In paper [107] the sufficient conditions for the existence and uniqueness of the solution of equations (3.178) have been derived.

32.2.3. Non-linear equations; remarks

Recently, an increasing amount of attention has been focussed on non-linear stochastic Itô equations. An effort has been made to extend the results known for the theory of stochastic equations in finite-dimensional case to the abstract stochastic equations.

For instance, Arnold, Curtain and Kotelenez [63] gave the sufficient conditions for the existence of mild and strong solutions of non-linear stochastic equations of the form

$$dY(t) = A(t)Y(t)dt + B(t, Y)dW(t) \tag{3.180}$$

Their work also contains the theorems on continuous dependence of the unique mild solution on its initial state as well as on the Markovian property of the mild solution.

Boundedness and stability of the Markov solutions of the so called semi-linear equations of the form

$$dY(t) = [AY + f(Y)]\, dt + B(Y)dW(t) \tag{3.181}$$

are studied by Ichikawa [96], [97] by use of the Lapunov method. In equation (3.181), operator A is the infinitesimal generator of a strongly continuous semigroup T_t of linear bounded operators on a given Hilbert space, whereas in equation (3.180) time dependent operator $A(t)$ generates a strong evolution operator $U(t, s)$ on a Hilbert space. The mild solutions of non-linear abstract stochastic equations with bounded operators are investigated by Makhno in paper [115].

The results mentioned above have been obtained by use of a functional analytic approach based on the theory of semigroups and their two-parameter version, evolution operator. Another approach (variational in nature) originates in early paper by Bensoussan and Temam [67] and in the thesis of Pardoux [121], cf. also [122]. A martingale approach introduced earlier by Stroock-Varadhan in finite-dimensional case was extended to abstract equations by Viot [137]. In the papers just mentioned the equation has the form

$$dY(t) = A(Y)dt + B(Y)dW(t) \tag{3.182}$$

where the operators A, B (from Hilbert space H to H_1) are bounded and continuous. The problems of boundedness and asymptotic stability of sample paths of solutions of equation (3.182) were investigated by Chow [74] by use of Itô formula and Lapunov functionals.

Closing this section it is worth noting that in addition to the notions of strong and mild solutions defined above one can introduce also other types of interpretation of the solution of stochastic abstract differential equations. For equations in Hilbert space Chojnowska-Michalik [72] defined and investigated four types of solutions (strong, weakened, mild and mild integral one). She showed, however, that three last interpretations (weakened, mild and mild integral) are equivalent, what provides a motivation for preference of the mild solution (in the integral form defined above) which is the most convenvient in analysis.

For other results on abstract stochastic equations of Itô type we refer to the recent papers (cf. [73],[81],[82],[85]).

23.3. Other problems

The theory of abstract stochastic differential equations is at present in the period of its rapid development. As we tried to show in this section, the efforts are focussed mainly on stochastic Itô differential equations in infinite-dimensional spaces. But, one can also consider a class of equations which can be interpreted as regular stochastic equations (by analogy to the terminology introduced in this book for finite-dimensional case.) In order to illustrate a possible approach we shall give here a brief résumé of the idea reported by Kotulski and Sobczyk in paper [108].

Let us consider the following equation

$$\frac{dY}{dt} = AY + B(\eta, t)Y + f(t)$$
$$Y(0) = Y_0 \tag{3.183}$$

where A is the infinitesimal generator of a strongly continuous semigroup $K(t)$ of bounded operators acting on the Hilbert space H, $t \in [0,T]$, $B(\eta,t)$ denotes the family of operators $(H \to H)$ depending on t and some Z-valued stochastic process $\eta(t,\gamma)$, where Z is a Banach space and, as always, $\gamma \in \Gamma$ with $(\Gamma, \mathcal{F}, P)$ being a complete probability space; $Y_0 = Y_0(\gamma)$ is a H-valued random variable, $f(t) = f(t,\gamma)$ is a square-integrable H-valued stochastic process (i.e. $f \in L_2([0,T] \times \Gamma, H)$.

Assume that η and $(Y_0, f(t))$ are statistically independent. Let $\Phi_0[a,b]$ denote the joint characteristic functional of the pair (Y_0, f), i.e.

$$\Phi_0[a,b] =$$
$$= E\left\{exp\left[i(Y_0,a)_H + i\int_0^T (b(t), f(t))_H \, dt\right]\right\} \tag{3.184}$$

where E denotes averaging with respect to the probability measure generated by the pair $(Y_0, f(t))$, $a \in H, b \in L_2([0,T]H)$; $(.,.)_H$ is an inner product on H.

Let us introduce the characteristic functional $\Phi[\lambda]$ of the solution process $Y(t)$, i.e.

$$\Phi[\lambda] = E\left\{exp\left[i\int_0^T (\lambda(t), Y(t))_H \, dt\right\} =\right.$$
$$= \int exp\left[i\int_0^T (\lambda(t), Y(t))_H \, dt\right]\mu(dY) \tag{3.185}$$

where $\mu(\cdot)$ is a probability measure on $L_2([0,T],H)$ generated by $Y(t)$ and $\lambda(t) \in L_2([0,T],H)$.

It is of interest to construct functional $\Phi[\lambda]$ effectively in terms of given functional $\Phi_0[a,b]$ and the probability measure ν associated with process $\eta(t)$. In paper [108] it has been shown that

$$\Phi[\lambda] = \int \Phi_0 \left[\int_0^T K^*(t)S(t)dt, \int_t^T K^*(s-t)S(t)ds \right] \nu(d\eta) \tag{3.186}$$

where $K^*(t-s)$ is the adjoint to $K(t-s)$ and $S(t)$ is a solution of the following integral equation:

$$S(t) - B^*(\eta,t) \int_t^T K^*(s-t)S(s)ds = \lambda(t) \tag{3.187}$$

where $B^*(\eta,t)$ is the adjoint to $B(\eta,t)$. The conditions under which there exists a solution of (3.187) are given in [108].

Therefore, to make the construction fully effective one has to solve equation (3.187) and then to perform integration in (3.186). In the general case, these are not easy problems, but in some special, simplified, cases when randomness appears only in the inhomogeneous term $f(t)$, the construction of the characteristic functional $\Phi[\lambda]$ of the solution process can be performed analytically (cf. [108]). The application to the heat conduction equation with random source will be given in Sec. 29.

Chapter IV

STOCHASTIC DIFFERENTIAL EQUATIONS: ANALYTICAL METHODS

24. INTRODUCTORY REMARKS

In addition to the theory of stochastic differential equations (dealing with such qualitative problems as these presented in the previous chapter) there exists now a large body of research devoted to the development of the effective methods of obtaining solutions. By the term "solution" we understand a stochastic process satisfying the equation together with its probabilistic properties.

Of course, we would be very pleased to have at our disposal the methods which provide the solution in exact and closed mathematical form. Unfortunately, such exact analytical solutions are only possible in the simplest cases. Most often, all that we are able to gain are the approximate solutions which can be represented by analytical formulae or, by numerical schemes. In this chapter we focus our attention on various analytical methods and procedures for stochastic differential equations; the constructive approximation schemes for numerical characterization of the solution process shall be presented in the next chapter.

The effective methods of solution of stochastic differential equations have been most often elaborated during the study of various specific practical problems. Due to this fact, as we have already mentioned in Sec. 20, it is natural to categorize all stochastic equations (and the methods associated) according to the physical origin of randomness.

Stochastic problems associated with differential equations are methodically most transparent in the case when the random elements enter only through initial conditions. As it was indicated in the Introduction, the study of motion of an ensemble of N material particles (considered in statistical physics) leads very naturally to this type of problems. But, of course, many other real processes are adequately described by equations with random initial data. Differential determistic equations with random initial conditions shall be considered first.

If dynamical systems are subjected to random excitation (of internal or external origin) then the governing equations are stochastic ones, and the principles of stochastic calculus constitute in this situation a basis for the analysis. The methods of solving such equations shall be discussed systematically in the subsequent sections of this chapter. For further solutions and applications the reader is referred to the papers given in the bibliography.

25. SYSTEMS WITH RANDOM INITIAL CONDITIONS

25.1. Probability distribution of solution

Let us consider a dynamical system described by the set of equations

$$\frac{dY_i}{dt} = F_i\left[t, Y_1(t), \ldots, Y_n(t)\right] \quad i = 1, 2, \ldots, n \tag{4.1}$$

and the initial conditions

$$Y_i(t_o) = Y_i^o(\gamma) \tag{4.2}$$

where $Y_i^o(\gamma)$ are random variables with unknown joint probability distribution. In the vector form the initial-value problem (4.1), (4.2) is written as

$$\begin{aligned} \frac{d\mathbf{Y}(t)}{dt} &= \mathbf{F}\left[t, \mathbf{Y}(t)\right] \\ Y(t_o) &= Y^o(\gamma) \end{aligned} \tag{4.3}$$

Let us assume that an explicit solution of the above system of equations can be found. This solution has the following general form

$$Y_i(t, \gamma) = g_i\left[t_o, t, Y_1^o(\gamma), \ldots, Y_n^o(\gamma)\right] \tag{4.3'}$$

or, in vector form

$$\mathbf{Y}(t, \gamma) = \mathbf{g}\left[t_o, t, \mathbf{Y}^o(\gamma)\right] \tag{4.3''}$$

Functions (4.3′) or (4.3″) characterize at arbitrary time t the state of a system which at $t = t_o$ was at the state $\mathbf{Y}(\gamma) = [Y_1^o(\gamma), \ldots, Y_n^o(\gamma)]$. In other words, these functions represent (at each time t) a transformation of the random vector $\mathbf{Y}^o(\gamma)$ into random vector $\mathbf{Y}(t, \gamma) = [Y_1(t, \gamma), \ldots, Y_n(t, \gamma)]$. It is clear that the probability distribution of the solution (4.3′)–(4.3″) can be obtained by use of the theorem on transformation of random vectors.

Let us assume that the transformation defined by (4.3″) is continuous with respect to $\mathbf{Y}^o$ (for each t), has continuous partial derivatives with respect to $\mathbf{Y}^o$, and defines a one-to-one mapping. Then, if the inverse transformation is written as

$$\mathbf{Y}^o(\gamma) = \mathbf{h}[t, \mathbf{Y}] \tag{4.4}$$

the probability density $f(\mathbf{y}; t]$ of the solution $\mathbf{Y}(t, \gamma)$ is

$$f(\mathbf{y}; t) = f_o\left(\mathbf{y}^o = \mathbf{h}(t, \mathbf{y})\right) |J| \tag{4.5}$$

where $f_o(y^o)$ is a given probability density of random vector $\mathbf{Y}^o(\gamma)$ and J is the Jacobian of the inverse transformation, i.e.

$$J = \frac{\partial \mathbf{y}^o}{\partial \mathbf{y}} \tag{4.6}$$

Formula (4.5) written for components takes the form

$$\begin{aligned} &f(y_1, y_2, \dots, y_n; t) = \\ &= f_o(y_1^o, y_2^o, \dots, y_n^o) \left| \frac{\partial(h_1, h_2, \dots, h_n)}{\partial(y_1, y_2, \dots, y_n)} \right| \end{aligned} \tag{4.5'}$$

where h_i are the inverse functions with respect to g_i for each t, i.e.

$$y_i^o(\gamma) = h_i[t; y_1, y_2, \dots, y_n] \quad i = 1, 2, \dots, n. \tag{4.4'}$$

Let the system considered be linear. Then instead of (4.3) we have

$$\begin{aligned} &\frac{d\mathbf{Y}(t)}{dt} = \mathbf{A}(t)\mathbf{Y}(t), \\ &\mathbf{Y}(t_o) = \mathbf{Y}(\gamma) \end{aligned} \tag{4.7}$$

If $\mathbf{A}(t)$ is continuous, then there exists a unique solution (in mean-square sense or sample functions) which has representation

$$\mathbf{Y}(t, \gamma) = \mathbf{\Phi}(t, t_o)\mathbf{Y}^o(\gamma) \tag{4.8}$$

where $\mathbf{\Phi}(t, t_o)$ is the fundamental matrix of the system. This is a particular form of (4.3″).
In the case where the matrix $\mathbf{A}(t) = \mathbf{A}_o$ is constant the fundamental matrix $\mathbf{\Phi}(t, t_o)$ has the explicit form

$$\mathbf{\Phi}(t, t_o) = e^{\mathbf{A}_o(t - t_o)} = \sum_{k=0}^{\infty} \frac{(t - t_o)^k \mathbf{A}_o^k}{k!} \tag{4.9}$$

As it is known from the theory of deterministic equations the fundamental matrix $\mathbf{\Phi}(t, t_o)$ satisfies the following conditions

$$\begin{aligned} &\frac{d\mathbf{\Phi}(t, t_o)}{dt} = \mathbf{A}(t)\mathbf{\Phi}(t, t_o), \\ &\mathbf{\Phi}(t, t_o) = \mathbf{I} \end{aligned} \tag{4.10}$$

$$\mathbf{\Phi}(t_3, t_1) = \mathbf{\Phi}(t_3, t_2)\mathbf{\Phi}(t_2, t_1) \quad \text{for all} \quad t_1, t_2, t_3 \tag{4.11}$$

The moments of the solution of stochastic initial-value problem (4.3) can be determined directly from the explicit form of the solution by integration with respect to the intitial condition. The moment of order k of a particular component $Y_i(t,\gamma)$, $i = 1.2.\ldots,n$ is given by

$$E\left[Y_i^k(t,\gamma)\right] = \int_{-\infty}^{+\infty} g_i^k(t_0, t, \mathbf{y}^\circ) f_0(\mathbf{y}^\circ) d\mathbf{y}^\circ \tag{4.12}$$

Let us illustrate the above reasoning by a simple example. Let

$$\begin{aligned} \frac{d^2Y}{dt^2} &= 0 \qquad 0 \le t \le \infty \\ Y(0) &= Y_1^\circ(\gamma), \\ \frac{dY}{dt}(0) &= Y_2^\circ(\gamma) \end{aligned} \tag{4.13}$$

where $Y_1^\circ(\gamma), Y_2^\circ(\gamma)$ are independent Gaussian random variables with zero means and variances $\sigma_{Y_1}^2$ and $\sigma_{Y_2}^2$, respectively. It means that the joint probability density of $Y_1^\circ(\gamma)$ and $Y_2^\circ(\gamma)$ is

$$f_0(y_1^\circ, y_2^\circ) = \frac{1}{2\pi\sigma_{Y_1^\circ}\sigma_{Y_2^\circ}} exp\left\{-\frac{1}{2}\left(\frac{y_1^{\circ 2}}{\sigma_{Y_1^\circ}^2} + \frac{y_2^{\circ 2}}{\sigma_{Y_2^\circ}}\right)\right\}$$

Equation (4.13) can be written in the vector form $(Y(t) = Y_1(t))$

$$\begin{aligned} \frac{dY_1(t)}{dt} &= Y_2(t) \qquad & Y_1(0) &= Y_1^\circ(\gamma) \\ \frac{dY_2(t)}{dt} &= 0 & Y_2(0) &= Y_2^\circ(\gamma) \end{aligned}$$

The solution is

$$\begin{aligned} Y_1(t,\gamma) &= Y_1^\circ(\gamma) + tY_2^\circ(\gamma) \\ Y_2(t,\gamma) &= Y_2^\circ(\gamma) \end{aligned}$$

It is clear that

$$\begin{aligned} & E\left[Y_1(t,\gamma)\right] = E\left[Y_2(t,\gamma)\right] = 0 \\ & \sigma^2 = \text{var}Y_1(t,\gamma) = E\left[Y_1^2(t,\gamma)\right] = t^2\sigma_{Y_2^\circ}^2 + \sigma_{Y_1^\circ}^2, \\ & s^2 = \text{var}Y_2(t,\gamma) = \sigma_{Y_2^\circ}^2 \\ & E\left[Y_1(t,\gamma)Y_2(t,\gamma)\right] = E\left[Y_2^{\circ 2}(\gamma)\right] t + E\left[Y_1^\circ(\gamma)Y_2^\circ(\gamma)\right] = \sigma_{Y_2^\circ}^2 t = s^2 t. \end{aligned}$$

So, the functions g_i have the form

$$y_1 = g_1(t, y_1^\circ, y_2^\circ) = y_1^\circ + t y_2^\circ$$
$$y_2 = g_2(t, y_1^\circ, y_2^\circ) = y_2^\circ$$

The inverse functions are

$$y_1^\circ = h_1(t; y_1, y_2) = y_1 - y_2 t$$
$$y_2^\circ = h_2(t; y_1, y_2) = y_2$$

The Jacobian of the inverse is

$$J = \begin{vmatrix} \dfrac{\partial h_1}{\partial y_1} & \dfrac{\partial h_1}{\partial y_2} \\ \dfrac{\partial h_2}{\partial y_1} & \dfrac{\partial h_2}{\partial y_2} \end{vmatrix} = \begin{vmatrix} 1 & -t \\ 0 & 1 \end{vmatrix} = 1$$

Therefore, the join density of the solutions is

$$f(y_1, y_2; t) = f_\circ \left[h_1(t; y_1, y_2), h_2(t; y_1, y_2)\right]$$
$$= \frac{1}{2\pi s\sigma\sqrt{1-\rho}} exp\left[-\frac{1}{2(1-\rho^2)}\left(\frac{y_1^2}{\sigma^2} - \frac{2\rho y_1 y_2}{\sigma s} + \frac{y_2^2}{s^2}\right)\right]$$

where ρ is the correlation coefficient of $Y_1(t)$ and $Y_2(t)$ at time t:

$$\rho = \frac{E\left[Y_1(t)Y_2(t)\right]}{\left\{E\left[Y_1^2(t)\right]E\left[Y_2^2(t)\right]\right\}^{1/2}} = \frac{s^2 t}{\sigma s} = \frac{s}{\sigma} t = \frac{\sigma_{Y_2^\circ} t}{\left[\sigma_{Y_2^\circ}^2 t^2 + \sigma_{Y_1^\circ}^2\right]^{1/2}}$$

It is seen that the joint distribution of $Y(t)$ and $\frac{dY(t)}{dt}$ at instant t is bivariate Gaussian. It should be noticed that although $Y(t)$ and $\frac{dY(t)}{dt}$ at $t = 0$ are independent, they are dependent for $t > 0$ and the correlation coefficient depends on time.

Other examples the reader can find in the book of Soong [132]; cf also Gaonkar [175] and Syski [247].

25.2. Liouville equation

Another way of obtaining the probability distribution of the solution of the stochastic initial-value problem (4.3) is associated with the Liouville equation in the general theory of dynamical systems (cf. [212]). In this theory one of the basic results is the Liouville theorem which asserts that each function $f(\mathbf{y}, t)$ whose integral (with respect to the spatial variable) is invariant during a motion of the system described by (4.3), satisfies the following equation

$$\frac{\partial f(\mathbf{y},t)}{\partial t}+\sum_{i=1}^{n}\frac{\partial}{\partial y_i}\left[f(\mathbf{y},t)F_i(t,\mathbf{y})\right]=0 \tag{4.14}$$

which is commonly known as the *Liouville equation.*

Let $\mathbf{y}_\circ$ be an initial state of our system; it is a particular realization of the random vector $\mathbf{Y}^\circ(\gamma)$. The probability, that $\mathbf{y}_\circ$ belongs to a region $S_\circ$ in the phase space is, of course, given by

$$P\{\mathbf{y}_\circ\in S_\circ\}=\int\underset{S_\circ}{\cdots}\int f_\circ(y_1^\circ,y_2^\circ,\ldots,y_n^\circ)dy_1^\circ dy_2^\circ\ldots dy_n^\circ \tag{4.15}$$

At time t our dynamical system evolved to the point $\mathbf{y}(t)=[y_1,y_2,\ldots,y_n]$ which is a realization (or sample point) of the random vector $\mathbf{Y}(t,\gamma)$ at time t. If $S_t=g(S_\circ,t)$ is a region whose elements at time t are the points of the phase space that at time $t_\circ$ formed the region $S_\circ$, then

$$P\{\mathbf{y}(t)\in S_t\}=\int\underset{S_t}{\cdots}\int f(y_1,y_2,\ldots,y_n;t)dy_1dy_2\ldots dy_n. \tag{4.16}$$

Since the motion of a system is governed by deterministic laws (equation (4.1) or (4.3)) the probabilities (4.15) and (4.16) are equal

$$\int_{S_\circ}f_\circ(\mathbf{y}^\circ)d\mathbf{y}^\circ=\int_{S_t}f(\mathbf{y},t)d\mathbf{y}. \tag{4.17}$$

The above equality means that the probability density $f(\mathbf{y},t)$ is an integral invariant of the system considered; therefore,the Liouville equation (4.14) holds for $f(t,\mathbf{y})$.

The Liouville equation (4.14) is a partial differential equation of the first order and can be solved, in general, by the use of Lagrange method. If the Liouville equation is written in the from

$$\frac{\partial f}{\partial t}+f\sum_{i=1}^{n}\frac{\partial F_i}{\partial y_i}+\sum_{i=1}^{n}F_i\frac{\partial f}{\partial y_i}=0 \tag{4.14'}$$

the Lagrange system is

$$\frac{dt}{1}=-\frac{df}{\sum_{i=1}^{n}\frac{\partial F_i}{\partial y_i}}=\frac{dy_1}{F_1}=\frac{dy_2}{F_2}=\cdots=\frac{dy_n}{F_n}. \tag{4.18}$$

The first equality and the initial condition give the solution

$$f(y_1, \ldots, y_n; t) =$$

$$= f_0(y_1^0, \ldots, y_n^0) exp\left\{ - \int_{t_0}^{t} \frac{\partial F_i\,(\tau, \mathbf{y} = g(t_0, \tau, \mathbf{y}^0))}{\partial g_i} d\tau \right\} \tag{4.19}$$

It can be shown (cf. [131]) that the above formula and formula (4.5′) coincide. Formula (4.19) gives also an explicit expression for the Jacobian occuring in (4.5′).

In more complicated cases the Liouville equation can be used for numerical evaluation of the probability density $f(\mathbf{y}, t)$.

Remark: The reader will easily notice that if system (4.3) is treated as a special case of the Itô stochastic differential equation (for vector process $\mathbf{Y}(t)$) in which a diffusion term vanishes the Liouville equation (4.14) is a special case of the Fokker-Planck-Kolmogorov equation.

Remark: It is worth emphasizing the fact that equations of the form (4.14) express, in general, conservation of some quantities in time. For example, in continuum mechanics, such equation is known as the continuity equation, the equation of conservation of mass or the transport equation. Here, the Liouville equation expresses the conservation of probability during a motion of the dynamical system (4.3).

26. LINEAR SYSTEMS WITH RANDOM EXCITATION

26.1. Solution and its properties

In this section we shall discuss stochastic differential equations of the form

$$\begin{aligned} &\frac{d\mathbf{Y}}{dt} = \mathbf{A}(t)\mathbf{Y}(t) + \mathbf{X}(t, \gamma), \quad t \geq t_0, \quad \gamma \in \Gamma \\ &\mathbf{Y}(t_0) = \mathbf{Y}_0 \end{aligned} \tag{4.20}$$

where $\mathbf{A}(t)$ is an $n \times n$ real matrix whose elements are continuous functions of t; $\mathbf{X}(t, \gamma)$ is a given stochastic process which is continuous (in the appropriate sense). Usually, the mean-square interpretation of equation (420) is a very suitable one and the mean-square calculus of the second order stochastic processes constitutes the basis for analysis. Here, this interpretation is adopted and the process $\mathbf{X}(t, \gamma)$ is assumed to be continuous in the mean-square sense. The initial value $\mathbf{Y}_0$ can be deterministic of random variable of a second order.

Under the above conditions there exists a unique m.s. solution of (4.20) and it has the representation (m.s. sense)

$$\mathbf{Y}(t) = \mathbf{\Phi}(t, t_0)\mathbf{Y}_0 + \int_{t_0}^{t} \mathbf{\Phi}(t, s)\mathbf{X}(s) ds \quad t \geq t_0 \tag{4.21}$$

where $\mathbf{\Phi}(t, t_o)$ is the fundamental matrix of the system considered and γ has been omitted.

Formula (4.21) constitutes a basis for evaluating the moments of the solution process. The first term in (4.21) describes the effect of the initial condition and can always be accounted for. In what follows, we focus our attention on the particular solution (assuming that $\mathbf{Y}_o = 0$),

$$\mathbf{Y}(t) = \int_{t_o}^{t} \mathbf{\Phi}(t, s)\mathbf{X}(s)ds \tag{4.22}$$

The solution process is, therefore, a linear functional of a given regular second order process $\mathbf{X}(t)$. According to the principles presented in Sec. 14 we have

$$E[\mathbf{Y}(t)] = \int_{t_o}^{t} \mathbf{\Phi}(t, s)E[\mathbf{X}(s)]\, ds \tag{4.23}$$

$$\begin{aligned} &K_Y(t_1, t_2) \\ &= E\left\{\left[\int_{t_o}^{t_1} \mathbf{\Phi}(t_1, u)\mathbf{X}(u)du\right]\left[\int_{t_o}^{t_2} \mathbf{\Phi}(t_2, v)\mathbf{X}(v)dv\right]^T\right\} = \\ &= \int_{t_o}^{t_1}\int_{t_o}^{t_2} \mathbf{\Phi}(t_1, u)K_X(u, v)\mathbf{\Phi}^T(t_2, v)dudv \end{aligned} \tag{4.24}$$

where $K_X(u, v) = E\left[\mathbf{X}(u)\mathbf{X}^T(v)\right]$ is a given correlation function matrix. The moments of higher order can be represented similarily. It is seen that evaluation of the moments is reduced to an integration.

Often, a mathematical model of a dynamical system has the form of one equation of n-th order

$$\frac{d^nY(t)}{dt^n} + a_1(t)\frac{d^{n-1}Y(t)}{dt^{n-1}} + \cdots + a_nY(t) = X(t, \gamma) \tag{4.25}$$

where $X(t, \gamma)$ is a given second order scalar stochastic process. In this case it is convenient to represent the solution in the form

$$Y(t) = \int_{t_o}^{t} p(t, s)X(s)ds \tag{4.26}$$

where $p(t, s)$ is the *impulse response function* or the weighting function of the system (in theory of differential equations it is also called the Green function). The function $p(t, s)$ characterizes the response of a system at time t to a unit impulsive input at time $s < t$.

In the case where the coefficeints in equation (4.25) are constant, $p(t,s) = p(t-s)$ and

$$Y(t) = \int_{t_0}^{t} p(t-s)X(s)ds \tag{4.27}$$

The mean and correlation function of the solution of (4.25) are, respectively

$$E[Y(t)] = \int_{t_0}^{t} p(t,s)E[X(s)]\,ds \tag{4.28}$$

$$K_Y(t_1,t_2) = \int_{t_0}^{t_1}\int_{t_0}^{t_2} p(t_1,s_1)p(t_2,s_2)K_X(s_1,s_2)ds_1ds_2 \tag{4.29}$$

These two functions completely specify a response if the excitation $X(t)$ is Gaussian. But even in the non-Gaussian case the first and second order moments of the response still provide the most important information about the process and they are useful in estimating the quality of the system; non-Gaussian responses shall be discussed in the last part of this section.

Let us consider briefly equation (4.20) where the excitation $\mathbf{X}(t) = \boldsymbol{\xi}(t)$ is a Gaussian white noise, that is

$$\frac{d\mathbf{Y}}{dt} = \mathbf{A}(t)\mathbf{Y}(t) + \boldsymbol{\xi}(t), \quad t \geq t_0 \tag{4.30}$$

In this case "equation" (4.30) can be interpreted as the stochastic Itô equation

$$d\mathbf{Y}(t) = \mathbf{A}(t)\mathbf{Y}(t) + d\mathbf{W}(t) \tag{4.31}$$

where $\mathbf{W}(t)$ is an n-dimensional Wiener process. Making use of the theory of Itô stochastic equations we can obtain the solution of (4.31) and its statistical properties. It turns out, however, that in this case (that is, in the case where the equation is linear and a white noise acts additively) the statistical properties of $\mathbf{Y}(t)$ obtained from equation (4.31) with use of the Itô calculus are the same as those calculated formally from equation (4.30), as $\boldsymbol{\xi}(t)$ would be a regular process. This coincidence has its origin in the fact that integration of the Itô equation (4.31) does not require the Itô integral; the integral with respect to the Wiener process exists in this case as a m.s. Riemann-Stieltjes integral. The fact that formal solution of (4.30) agrees with the correct solution of (4.31) indicates that an integrated white noise makes sense in performing mathematical operations.

Examples

1. Let us consider the Langevin equation

$$\frac{dY(t)}{dt} + aY(t) = \xi(t), \quad Y(0) = Y_0 \tag{4.32}$$

where $a > 0$ is constant and $\xi(t)$ is a Gaussian white noise such that

$$E\xi(t) = 0$$
$$K_\xi(t_1, t_2) = 2D\delta(t_2 - t_1)$$

If equation (4.32) is interpreted as the Itô equation then we have to solve equation

$$dY(t) = -aY(t)dt + \sqrt{2D}dW(t)$$

where $W(t)$ is standad Wiener process.

The solution $Y(t)$ is a Gaussian Markov diffusion process which can be obtained directly as

$$Y(t) = Y_0 e^{-at} + \sqrt{2D}\int_0^t e^{-a(t-s)}dW(s)$$

Mean and variance can easily be evaluated.

Another way of evaluating the solution process is based on the fact that the probability density of the solution $f(y,t)$ satisfies the Fokker-Planck-Kolmogorov equation

$$\frac{\partial f(y,t)}{\partial t} = a\frac{\partial}{\partial y}\left[yf(y,t)\right] + D\frac{\partial^2 f(y,t)}{\partial y^2}$$

with the initial condition: $f(y,0) = f_0(y)$ and "boundary" condition at infinity: $f(\pm\infty, t) = 0$. We shall assume that $f_0(y) = \delta(y - y_0)$.

The above F-P-K equation can be solved by use of one of the methods presented in Sec. 6.3. Use of the method of separation of variables leads to the result

$$f(y,t\,|\,y_0,0) = \frac{1}{\left[2\pi D(1-e^{-2at})\big/ a\right]^{1/2}} exp\left[-\frac{a(y - y_0 e^{-at})^2}{2D(1-e^{-2at})}\right]$$

It is seen that the solution has a Gaussian density with mean: $y_0 exp(-at)$ and variance: $\sigma^2\left[1 - exp(-2at)\right]$, where $\sigma^2 = \frac{D}{a}$.

For the stationary solution $t \to \infty$ we have

$$f_{st}(y) = \frac{1}{\sigma\sqrt{2\pi}} exp\left(-\frac{y^2}{2\sigma^2}\right).$$

It turns out that formal integration of equation (4.32) gives the same result. Indeed, formal solution of (4.32) is

$$Y(t) = Y_0 e^{-at} + \int_0^t e^{-a(t-s)}\xi(s)ds$$

The mean and correlation function are, respectively

$$
\begin{aligned}
E[Y(t)] &= y_0 e^{-at}, \\
K_Y(t_1,t_2) &= E\left\{\left[Y(t_1) - Y_0 e^{-at_1}\right]\left[Y(t_2) - Y_0 e^{-at_2}\right]\right\} \\
&= e^{-a(t_1,t_2)} \int_0^{t_1}\int_0^{t_2} e^{a(s_1+s_2)} E\left[\xi(s_1)\xi(s_2)\right] ds_1 ds_2 \\
&= 2De^{-a(t_1+t_2)} \int_0^{t_1}\int_0^{t_2} e^{a(s_1+s_2)} \delta(s_2 - s_2) ds_1 ds_2 \\
&= \sigma^2 e^{-a|\tau|}(1 - e^{-2a\min(t_1,t_2)}); \quad \sigma^2 = \frac{D}{a}, \quad \tau = |t_2 - t_1|
\end{aligned}
$$

Therefore,

$$
\mathrm{var} Y(t) = K_Y(t,t) = \sigma^2(1 - e^{-2at})
$$

As $t \to \infty$, that is in the stationary case

$$
K_Y(t_1,t_2) = K_Y(t_2 - t_1) = K_Y(\tau) = sigma^2 e^{-a|\tau|}
$$

2. Let us consider the equation

$$
\begin{aligned}
&\frac{dY(t)}{dt} + aY(t) = X(t), \\
&Y(0) = Y_0 \qquad (4.33)
\end{aligned}
$$

where $X(t)$ is a stationary Gaussian process such that

$$
E[X(t)] = 0, \quad K_X(\tau) = \sigma_X^2 e^{-\alpha|\tau|}, \quad \tau = t_2 - t_1
$$

The solution is

$$
Y(t) = y_0 e^{-at} + e^{-at} \int_0^t e^{as} X(s) ds
$$

The first two moments are, respectively

$$
\begin{aligned}
E[Y(t)] &= Y_0 e^{-at} \\
K_Y(t_1,t_2) &= \frac{\sigma_X^2}{a^2 - \alpha^2}\left[e^{-\alpha|\tau|} - \frac{\alpha}{a} e^{-a|\tau|}\left(1 + \frac{\alpha}{a}\right) e^{-a(t_1+t_2)}\right. \\
&\left. - e^{-at_1 - \alpha t_2} - e^{-at_2 - \alpha t_1}\right] \\
\mathrm{var} Y(t) &= \frac{\sigma_X^2}{a^2 - \alpha^2}\left[\left(1 - \frac{\alpha}{a}\right) + \left(1 + \frac{\alpha}{a}\right) e^{-2at} - 2e^{-(a+\alpha)t}\right]
\end{aligned}
$$

For $t \to \infty$ (stationary solution) we have

$$E[Y(t)] = 0,$$

$$K_Y(\tau) = \text{var}Y(t)\frac{1}{a-\alpha}\left[ae^{-\alpha|\tau|} - \alpha e^{-a|\tau|}\right],$$

$$\text{var}Y(t) = \frac{\sigma_X^2}{a(a+\alpha)}.$$

3. Let us consider a motion of a damped harmonic oscillator driven by a stationary excitation. The governing equation is

$$\frac{d^2Y(t)}{dt^2} + 2\omega_o\zeta\frac{dY(t)}{dt} + \omega_o^2 Y(t) = X(t), \quad t \geq 0 \tag{4.34}$$

where $X(t)$ is a stationary stochastic process with zero mean and correlation function $K_X(\tau)$. We assume that the initial conditions are: $Y(0) = \dot{Y}(0) = 0$. We wish to evaluate the mean-square of the response.

According to (4.27) the solution is

$$Y(t) = \int_0^t p(t-s)X(s)ds \tag{4.35}$$

where

$$p(t) = \frac{1}{\lambda_o}e^{-\zeta\omega_o t}\sin\lambda_o t, \quad \lambda_o = \omega_o(1-\zeta^2)^{1/2} \tag{4.36}$$

In accordance with (4.29) we have

$$K_Y(t_1,t_2) = \int_0^{t_1}\int_0^{t_2} p(t_1-s_1)p(t_2-s_2)K_X(s_2-s_1)ds_2ds_1 \tag{4.37}$$

In order to solve the problem we have to evaluate the above double integral for a specific form of $K_X(\tau)$. Of course, the exact analytical result can be difficult to obtain in practical situations, so the numerical integration can be helpful. Expression (4.37) can be simplified if we make use of the spectral density of the excitation process $X(t)$.

According to formula (1.71) we have

$$K_X(s_2-s_1) = \int_{-\infty}^{+\infty} g_X(\omega)e^{i\omega(s_2-s_1)}d\omega$$

Introducing the above to (4.37) gives

$$K_Y(t_1,t_2) =$$

$$= \int_{-\infty}^{+\infty}\int_0^{t_1}\int_0^{t_2} g_X(\omega)e^{i\omega(s_2-s_1)}p(t_1-s_2)p(t_2-s_2)ds_1ds_2d\omega \tag{4.38}$$

The double integral with respect to s_1 and s_2 can be evaluated using the expression (4.36) for $p(t)$. The result provided by Caughey and Stumpf [164] is

(4.39)

$$\begin{aligned}
&K_Y(t_1,t_2)\\
&= \int_{-\infty}^{+\infty} g_X(\omega)\,|H(\omega)|^2 \Big\{ exp\Big[i\omega(t_2-t_1)\Big]\\
&- exp(-\zeta\omega_o t_2)\Big[\Big(\cos\lambda_o t_2 + \frac{\zeta\omega_o}{\lambda_o}\sin\lambda_o t_2\Big)\cos\omega t_1 + \frac{\omega_o}{\lambda_o}\sin\lambda_o t_2 \sin\omega t_1\Big]\\
&- exp(-\zeta\omega_o t_1)\Big[(\cos\lambda_o t_1 + \frac{\zeta\omega_o}{\lambda_o}\sin\lambda_o t_1)\cos\omega t_2 + \frac{\omega}{\lambda_o}\sin\lambda_o t_1 \sin\omega t_2\Big]\\
&+ exp\Big[-\zeta\omega_o(t_2+t_1)\Big]\Big[\cos\lambda_o t_2\cos\lambda_o t_1 + \frac{\zeta\omega_o}{\lambda_o}\sin\lambda_o(t_2+t_1)\\
&+ \frac{\zeta^2\omega_o^2+\omega^2}{\lambda_o}\sin\lambda_o t_2 \sin\lambda_o t_1\Big]\Big\}.d\omega
\end{aligned}$$

The mean-square of the response is obtained from the above by letting $t_1 = t_2 = t$:

(4.40)

$$\begin{aligned}
E\left[Y^2(t)\right] = \int_{-\infty}^{+\infty} &|H(\omega)|^2 g_X(\omega)\Big\{1 + exp(-2\omega_o\zeta t)\Big[1\\
&+ \frac{2\omega_o\zeta}{\lambda_o}\sin\lambda_o t\cos\lambda_o t - exp(\omega_o\zeta t)\Big(2\cos\lambda_o t\\
&+ \frac{2\omega_o\zeta}{\lambda_o}\sin\lambda_o t\Big)\cos\lambda_o t - \frac{2\omega}{\lambda_o} exp(\omega_o\zeta t)\sin\lambda_o t\sin\omega t +\\
&+ \frac{1}{\lambda_o^2}(\omega_o^2\zeta^2 - \lambda_o^2 + \omega^2)\sin^2\lambda_o t\Big]\Big\}d\omega.
\end{aligned}$$

where

$$H(\omega) = \frac{1}{\omega_o^2 - \omega^2 + 2i\omega\omega_o\zeta}.$$

The above expressions provide the following information:

a) in general, the mean-square of the response depends on time; this means that the response is non-stationary though the excitation was stationary;

b) for large t ($t \to \infty$)

$$E\left[Y^2(t)\right] \longrightarrow \int_{-\infty}^{+\infty} |H(\omega)|^2 g_X(\omega)d\omega; \tag{4.41}$$

for large t_1 and t_2 $(t_1, t_2 \to \infty)$

$$K_Y(t_1,t_2) \longrightarrow K_Y(t_2-t_1) =$$

$$= \int_{-\infty}^{+\infty} |H(\omega)|^2 g_X(\omega) e^{i\omega(t_2-t_1)} d\omega. \tag{4.42}$$

Hence, the response becomes weakly stationary as $t \to \infty$.

Let us assume now that $X(t) = \xi(t)$ is a Gaussian white noise. This means that $g_X(\omega) = g_o$ =constant. In this case the integral in (4.40) can be evaluated by contour integration with the result

$$E\left[Y^2(t)\right] = \frac{\pi g_o}{2\zeta\omega_o^3}\left\{1 - \frac{1}{\lambda_o^2} exp(-2\omega_o\zeta t)\left[\lambda_o^2 \right.\right.$$

$$\left.\left. + 2(\omega_o\zeta)^2 \sin^2 \lambda_o t + \omega_o \lambda_o \zeta \sin 2\lambda_o t\right]\right\} \tag{4.43}$$

Let us assume that damping of the system tends to zero; expanding the exponential term in the above expression into a series and then taking the limit $\zeta \to 0$ gives the result

$$E\left[Y^2(t)\right] \underset{\zeta \to 0}{\longrightarrow} \frac{\pi g_o}{2\omega_o^3}\left(2\omega_o t - \sin 2\omega_o t\right). \tag{4.44}$$

This expression is unbounded as t increases.

In stationary state $(t \to \infty)$ formula (4.43) gives

$$E\left[Y^2(t)\right] \underset{t \to \infty}{\longrightarrow} \frac{\pi g_o}{2\omega_o^3 \zeta} \underset{\zeta \to 0}{\longrightarrow} \infty. \tag{4.45}$$

26.2. Stationary solutions; spectral method

It is of interest to know when the solution of equation (4.20) is a weakly stationary process. The examples given above give the answer in particular cases. The general statement is the following.

If the system considered is time-invariant (matrix $\mathbf{A}(t)$) in (4.20) is constant) and asymptotically stable (eigenvalues of matrix $\mathbf{A}$ have negative real parts) and the excitation is a weakly stationary process then the solution $\mathbf{Y}(t)$ considered for sufficiently large t (when the effect of initial conditions is negligible) is a weakly stationary process and it can be represented as

$$\mathbf{Y}(t) = \int_{-\infty}^{t} \mathbf{\Phi}(t-s)\mathbf{X}(s)ds = \int_{0}^{\infty} \mathbf{\Phi}(\tau)\mathbf{X}(t-s)d\tau \tag{4.46}$$

The mean and correlation function of the above process can be evaluated in the usual way. The mean is constant and the correlation function is a function of $\tau = t_2 - t_1$.

Now, let us consider the n-th order differential equation (4.25) with solution (4.27). Its correlation function is given by (4.29), where the impulse response function $p(t-s)$ is associated with (4.25). In the steady-state (i.e. $t \to \infty$) the correlation function (4.29) takes the form

$$K_Y(t_1,t_2) = K_Y(t_2-t_1) =$$
$$= \int_{-\infty}^{+\infty}\int_{-\infty}^{t_1}\int_{-\infty}^{t_2} g_X(\omega)e^{i\omega(s_2-s_1)}p(t_1-s_1)p(t_2-s_2)ds_2ds_1d\omega =$$
$$= \int_{-\infty}^{+\infty}\int_{-\infty}^{+\infty}\int_{-\infty}^{+\infty} g_X(\omega)e^{i\omega(s_2-s_1)}p(t_1-s_1)p(t_2-s_2)ds_1ds_2d\omega \tag{4.47}$$

since the impulse response function vanishes for negative arguments. Let

$$u = t_1 - s_1, \quad v = t_2 - s_2$$

Formula (4.47) takes the form

$$K_Y(t_2-t_1) =$$
$$= \int_{-\infty}^{+\infty}\int_{-\infty}^{+\infty}\int_{-\infty}^{+\infty} g_X(\omega)e^{i\omega(t_2-t_1-v+u)}p(u)p(v)dudvd\omega$$

Denoting

$$H(\omega) = \int_{-\infty}^{+\infty} p(t)e^{-i\omega t}dt \tag{4.48}$$

we obtain

$$K_Y(t_2-t_1) = \int_{-\infty}^{+\infty} |H(\omega)|^2 g_X(\omega)e^{i\omega(t_2-t_1)}d\omega. \tag{4.49}$$

On the other hand, however, $K_Y(t_2-t_1)$ can be represented according to (1.72′). As a result of comparison we obtain

$$g_Y(\omega) = |H(\omega)|^2 g_X(\omega). \tag{4.50}$$

The above formula is the basic relation in spectral analysis of linear systems. The function $|H(\omega)|^2$ is called the *transmittancy of the system* and it prescribes the fraction of energy to be transmitted through the system at various frequencies. For this reason a linear system is often called a *linear filter*.

Let us consider a deterministic version of equation (4.25). It can be written symbolically as

$$Q_n(p)y(t) = x(t) \tag{4.25'}$$

where $Q_n(p)$ is a polynomial of degree n with constant coefficients $a_1, a_2, \ldots, a_n$ and $p = \frac{d}{dt}$. Let us apply to both sides of (4.25′) the Fourier transform (we assume zero initial conditions). We obtain the following relationship between the transform $\tilde{x}(\omega)$ of $x(t)$ and the transform $\tilde{y}(\omega)$ of the solution $y(t)$

$$\tilde{y}(\omega) = \frac{1}{Q_n(i\omega)}\tilde{x}(\omega) \tag{4.51}$$

The function

$$H(\omega) = \frac{1}{Q_n(i\omega)} \tag{4.52}$$

is the transmittancy of the system. Expression (4.51) and the convolution theorem associated with the Fourier transform lead to the conclusion that the transmittancy is the Fourier transform of the impulse response function. This confirms (4.48).

Examples

1. Given the first order system

$$\frac{dY(t)}{dt} + aY(t) = X(t) \tag{4.53}$$

where $X(t)$ is a stationary process with zero mean and with spectral density $g_X(\omega) = \frac{\sigma^2}{\pi}\frac{\alpha}{\alpha^2+\omega^2}$. The spectral density of the stationary solution of (4.53) is

$$\begin{aligned} g_Y(\omega) &= \frac{1}{|i\omega + a|^2} \cdot \frac{\sigma^2}{\pi}\frac{\alpha}{\omega^2+\alpha^2} = \frac{\sigma^2\alpha}{\pi}\frac{1}{\omega^2+a^2}\frac{1}{\omega^2+\alpha^2} \\ &= \frac{\sigma^2\alpha}{\pi}\frac{1}{\omega^4 + (a^2+\alpha^2)\omega^2 + a^2\alpha^2} \end{aligned}$$

The variance of the solution is:

$$\begin{aligned} \mathrm{var}Y(t) &= \int_{-\infty}^{+\infty} g_Y(\omega)d\omega = \\ &= \frac{\sigma^2\alpha}{\pi}\int_{-\infty}^{+\infty}\frac{d\omega}{\omega^4 + (a^2+\alpha^2)\omega^2 + a^2\alpha^2} = \frac{\sigma^2}{a(a+\alpha)}. \end{aligned}$$

2. Given a vibratory system

$$\frac{d^2Y}{dt^2} + a_1\frac{dY}{dt} + a_2Y = X(t) \tag{4.54}$$

where $X(t)$ is a stationary process with spectral density as in the previous example.
The spectral density of the stationary solution is

$$g_Y(\omega) = \frac{1}{|a_2 + a_1 i\omega - \omega^2|^2}\frac{\alpha\sigma^2}{\pi(\omega^2+\alpha^2)}.$$

The variance of the solution is

$$\begin{aligned}\text{var}Y(t) &= \frac{\alpha\sigma^2}{\pi}\int_{-\infty}^{+\infty}\frac{d\omega}{(\omega^2+\alpha^2)\left[(a_2-\omega^2)^2+a_1^2\omega^2\right]} = \\ &= \frac{\sigma^2(a_1+\alpha)}{a_1a_2(\alpha^2+a_1\alpha+a_2)}.\end{aligned}$$

Remark: In applications of the spectral method the following integral is needed.

$$I_n = \frac{1}{2\pi}\int_{-\infty}^{+\infty}\frac{\varphi_n(i\omega)}{\Psi_n(i\omega)\Psi_n(-i\omega)}d\omega = \frac{(-1)^{n+1}N_n}{2a_oD_n} \tag{4.55}$$

where

$$\begin{aligned}\varphi(x) &= b_ox^{2n-2} + b_1x^{2n-4} + \cdots + b_{n-1}\\ \Psi_n(x) &= a_ox^n + a_1x^{n-1} + \cdots + a_n\end{aligned}$$

$$D_n = \begin{vmatrix} d_{11} & d_{12} & \dots & d_{1n}\\ d_{21} & d_{22} & \dots & d_{2n}\\ \dots & \dots & \dots & \dots\\ \dots & \dots & \dots & \dots\\ d_{n1} & d_{n2} & \dots & d_{nn}\end{vmatrix},$$

$$d_{mr} = a_{2m-r}$$

$$a_s = 0, \quad s<0 \quad s>n$$

N_n is a determinant obtained from D_n by replacing the elements of the first column by the coefficients $b_o, b_1, \ldots, b_{n-1}$. In particular:

$$I_1 = \frac{b_o}{2a_oa_1}, \quad I_2 = \frac{-b_o + \frac{a_ob_1}{a_2}}{2a_oa_1}$$

Application of the spectral method to various problems of structural dynamics is discussed in the book of Bolotin [313] and Elishakoff [320].

26.3. Nonstationary excitations; random impulses

26.3.1. Excitations reducible to stationary

In many situations the random excitation acting on a dynamical system should be modelled by a nonstationary process (for example, this is the case of the earthquake-type disturbance — see Chapter VI). This leads to differential equations with random nonstationary inhomogeneous term. However, in numerous practical cases the nonstationarity of the input process may appear not due to the inherent nonstationarity of the original excitation but because of the fact that the action of the stationary disturbance on the system varies in time. To deal with such situations it is natural to distinguish a class of nonstationary processes which can be termed as the *reducible to stationary*.

The simplest, termed as the reducible to stationary, nonstationary process can be represented as

$$Z(t) = f(t)X(t) + g(t) \tag{4.56}$$

where $X(t)$ is a stationary stochastic process, and $f(t)$ and $g(t)$ are real deterministic functions. The first moments of $Z(t)$ are given by the formulae

$$\begin{aligned} E\left[Z(t)\right] &= f(t)E\left[X(t)\right] + g(t), \\ K_Z(t_1,t_2) &= f(t_1)f(t_2)K_X(t_2-t_1), \\ \sigma_Z^2(t) &= f^2(t)K_X(0). \end{aligned}$$

The normalized correlation function is

$$\rho_z(t_1,t_2) = \frac{K_Z(t_2,t_1)}{\sigma_Z(t_1)\sigma_Z(t_2)} = \frac{K_X(t_2-t_1)}{K_X(0)} = \rho_X(t_2-t_1) \tag{4.57}$$

which is the same as the normalized covariance function of $X(t)$. Thus, an indication that a particular nonstationary stochastic process can be represented by a stationary stochastic process in the form (4.56) is the fact that its normalized correlation function depends solely on the argument difference $t_2 - t_1$.

A more general type of nonstationary reducible process has the form

$$Z(t) = \sum_{k=1}^{n} f_k(t)X_k(t) + g(t) \tag{4.56'}$$

Without going into details we wish to state that if a system considered is governed by equation (4.25) written symbolically as

$$Q_n(p)Y(t) = Z(t), \quad Z(t) = f(t)X(t) \tag{4.58}$$

where $X(t)$ is a stationary process, and $f(t)$ is a given deterministic function, then the response process $Y(t)$ can be characterized for a wide variety of functions $f(t)$ — cf. Sweshnikov [246].

If

$$f(t) = e^{\beta t}$$

then the correlation function of $Y(t)$ is given by the formula

$$K_Y(t_1, t_2) = = e^{\beta(t_1+t_2)} \int_{-\infty}^{+\infty} e^{i\omega(t_2-t_1)} \frac{g_X(\omega)}{|Q_n(\beta + i\omega)|^2} d\omega. \tag{4.59}$$

In the case of a harmonic oscillator, when

$$\frac{d^2Y}{dt^2} + 2h\frac{dY}{dt} + \omega_0^2 Y = e^{\omega_0 t} X(t)$$

and

$$g_X(\omega) = \frac{\alpha\sigma^2}{\pi(\omega^2 + \alpha^2)}$$

Formula (4.59) yields

$$K_Y(t_1, t_2) = = e^{\omega_0(t_1+t_2)} \int_{-\infty}^{+\infty} e^{i\omega(t_2-t_1)} \frac{\sigma^2 \alpha d\omega}{\pi(\omega^2 + \alpha^2)\left[\omega^4 + 4h(h + \omega_0)\omega^2 + 4\omega_0^2(h + \omega_0)^2\right]}$$

Contour integration gives the result

$$\begin{aligned} K_Y(t_1, t_2) &= \\ &= \frac{\sigma^2 e^{\omega_0(t_1+t_2)}}{\left[\alpha^2 - 2h(h + \omega_0)\right]^2 + 4\lambda_0(h + \omega_0)^2} \Big\{ e^{-\alpha|\tau|} \\ &+ \frac{\alpha}{2\lambda_\alpha\left[\lambda_0^2 + (h + \omega_0)^2\right]} \Big[(\alpha^2 - 2\lambda_0^2 - 6h(h + \omega_0) \cos \lambda_0 \tau \\ &+ 2(\lambda_0^2 - h\omega_0 - h^2 + \frac{1}{2}\alpha^2) \sin \lambda_0 |\tau| \Big] e^{-(h+\omega_0)|\tau|} \Big\} \end{aligned}$$

where

$$\lambda_o = (\omega_o^2 - h^2)^{1/2}, \quad \tau = t_2 - t_1$$

Remark: The solution indicated above can be extended to the case of a harmonic oscillator excited by the process

$$\begin{aligned} Z(t) &= f(t)X(t) \\ f(t) &= c\left[e^{-\alpha t} - e^{-\beta t}\right], \quad t > 0 \\ &= 0, \qquad\qquad\qquad t < 0 \end{aligned}$$

In a paper by Buciarelli, Kuo [157] the authors determine the mean-square of the response of a lightly damped harmonic oscillator when the excitation has the form

$$Z(t) = f(t)X(t)$$

where $f(t)$ is a slowly varying deterministic function and $K_X(\tau) = \sigma_X^2 exp(-\alpha|\tau|$ $\cos\beta\tau$.

In a paper by Shinozuka [230] the author considers a harmonic oscillator with exciation in the form

$$Z(t) = H(t)e^{-\alpha\omega_o t}X(t)$$

where $X(t)$ is a Gaussian and stationary process and $H(t)$ is a unit step function. The bounds for the probability of the first excursion of the response $Y(t)$ outside the interval $[-\lambda_1, \lambda_2]$ within time interval $(0, T)$ are obtained.

26.3.2. Evolutionary spectra

The idea of representing a nonstationary process in terms of a stationary one can seriously be enriched by introducing (instead of (4.56)–(4.56′)) an integral form representation. Such an approach is closely related to the concept of so called evolutionary spectra introduced by Priestley [216] and it stems from the effort to extend the spectral theory of stationary processes to nonstationary processes (cf. also Lyons [207]).

Without going into details of this interesting subject we wish to sketch here only the basic features of the method as it is used in practice (cf. Hammond [180], Madsen, Krenk, Lind [344]).

Let us consider systems governed by an n-th order linear stochastic equation

$$Q_n(p)Y(t) = Z(t), \quad p = \frac{d}{dt} \tag{4.25''}$$

where $Z(t)$ is a nonstationary process represented as the response of a linear time-dependent system with unit response function $h(t,\tau)$ to a stationary random excitation $X(t)$, i.e.

$$Z(t) = \int_0^\infty h(t,\tau)X(t-\tau)d\tau \tag{4.60}$$

It is seen that at arbitrary time t the process $Z(t)$ is a weighted average of the stationary process $X(t)$. Of course,

(4.61) $$E[Z(t)] = \int_0^\infty h(t,\tau)E[X(t-\tau)]\,d\tau = m_X \int_0^\infty h(t,\tau)d\tau$$

$$K_Z(t_1,t_2) =$$
$$= \int_0^\infty \int_0^\infty h(t_1,\tau_1)h(t_2,\tau_2)K_X(t_1-\tau_1.t_2-\tau_2)d\tau_1 d\tau_2.$$

Since $X(t)$ is a weakly stationary process, its correlation function depends on the difference of its arguments, i.e.

$$K_X(t_1-\tau_1,t_2-\tau_2) = K_X\left(t_2-\tau_2-(t_1-\tau_1)\right) = K_X(t_2-t_1-\tau_2+\tau_1)$$

If we express the above correlation function in terms of the corresponding spectral density $g_X(\omega)$ we get

$$K_Z(t_1,t_2) =$$
$$= \int_{-\infty}^{+\infty} \int_0^\infty \int_0^\infty h(t_1,\tau_1)h(t_2,\tau_2)g_X(\omega)e^{i\omega(t_2-t_1-\tau_2+\tau_1)}d\tau_1 d\tau_2 d\omega =$$
$$= \int_{-\infty}^{+\infty} \left\{\int_0^\infty h(t_1,\tau_1)e^{i\omega\tau_1}d\tau_1\right\}\left\{\int_0^\infty h(t_2,\tau_2)e^{-i\omega\tau_2}d\tau_2\right\} g_X(\omega)e^{i\omega(t_2-t_1)}d\omega$$

If we introduce the function

(4.62) $$A(\omega,t) = \int_0^\infty h(t,\tau)e^{-i\omega\tau}d\tau$$

we obtain

(4.63) $$K_Z(t_1,t_2) = \int_{-\infty}^{+\infty} A(\omega,t_2)\overline{A(\omega,t_1)}g_X(\omega)e^{i\omega(t_2-t_1)}d\omega$$

This is a generalization of (4.49) for stationary response of a linear system.

For $t_1 = t_2 = t$ formula (4.63) can be written as

$$K_Z(t) = \int_{-\infty}^{+\infty} g_z(\omega,t)d\omega$$

where

(4.64) $$g_Z(\omega,t) = |A(\omega,t|^2\, g_X(\omega)$$

defines the so called *evolutionary power spectrum* of the nonstationary process $Z(t)$.

Coming back to equation (4.25″) driven by the nonstationary process (4.60) we have

$$Y(t) = \int_{-\infty}^{t} p(t-\tau)Z(\tau)d\tau,$$

$$E\left[Y(t)\right] = \int_{-\infty}^{t} p(t-\tau)E\left[Z(\tau)\right]d\tau = m_X \int_{-\infty}^{t} A(0,\tau)p(t-\tau)d\tau$$

The correlation function of the solution is

$$K_Y(t_1,t_2) = \int_{-\infty}^{t_1}\int_{-\infty}^{t_2} p(t_1-\tau_1)p(t_2-\tau_2)K_Z(\tau_1,\tau_2)d\tau_1 d\tau_2$$

When formula (4.63) is used the result contains convolution integral of $A(\omega,\tau)$ and $p(\tau)$, i.e.

$$\begin{aligned} K_Y(t_1,t_2) &= \\ = \int_{-\infty}^{+\infty} &\left\{\int_{-\infty}^{t_1} \overline{A(\omega,\tau_1)}p(t_1-\tau_1)e^{-i\omega\tau_1}d\tau_1\right\} \times \\ \times &\left\{\int_{-\infty}^{t_2} A(\omega,\tau_2)p(t_2-\tau_2)e^{i\omega\tau_2}\right\} g_X(\omega)d\omega \end{aligned}$$

If we introduce the time-dependent transmittancy (or transfer function)

$$H(\omega,t) = \int_{0}^{\infty} A(\omega,t-\tau)p(\tau)e^{-i\omega\tau}d\tau,$$

the average and correlation function of the solution are, respectively

$$E\left\{Y(t)\right\} = m_X H(0,t),$$

$$K_Y(t_1,t_2) \int_{-\infty}^{+\infty} H(\omega,t_2)\overline{H(\omega,t_1)}g_X(\omega)e^{i\omega(t_2-t_1)}d\omega \tag{4.65}$$

An application of the described method to vibratory system is straightforward. The computational effort necessary to evaluate (4.65) is reduced if $H(\omega,t)$ can be represented in a closed form.

26.3.3. Generalized spectral density

Let a nonstationary process $X(t)$ have correlation function $K_X(t_1,t_2)$. Let $g_X(\omega_1,\omega_2)$ be a Fourier transform of $K_X(t_1,t_2)$, that is

$$g_X(\omega_1,\omega_2) = \frac{1}{(2\pi)^2}\int_{-\infty}^{+\infty}\int_{-\infty}^{+\infty} K_X(t_1,t_2)exp\left[-i(\omega_1 t_1 - \omega_2 t_2)\right] dt_1 dt_2 \tag{4.66}$$

The function $g_X(\omega_1,\omega_2)$ is called the *generalized spectral density* of the nonstationary process $X(t)$; it exists if $K_X(t_1,t_2)$ is absolutely integrable.

Making use of the general formula for the correlation function of the solution (4.29) which implies that $X(t)=0$ for $t<t_o$ and that the impulse response function vanishes for negative arguments we have

$$\begin{aligned} g_Y(\omega_1,\omega_2) &= \\ &= \frac{1}{(2\pi)^2}\int_{-\infty}^{+\infty}\int_{-\infty}^{+\infty} K_Y(t_1,t_2)exp\left[-i(\omega_1 t_1 - \omega_2 t_2)\right] dt_1 dt_2 = \\ &= \frac{1}{(2\pi)^2}\int_{-\infty}^{+\infty}\int_{-\infty}^{+\infty}\int_{-\infty}^{+\infty}\int_{-\infty}^{+\infty} K_X(s_1,s_2)p(t_1-s_1)p(t_2-s_2) \\ &\qquad exp\left[-i(\omega_1 t_1 - \omega_2 t_2)\right] ds_1 ds_2 dt_1 dt_2 \end{aligned}$$

This means that

$$g_Y(\omega_1,\omega_2) = H(\omega_1)H^*(\omega_2)g_X(\omega_1,\omega_2). \tag{4.67}$$

This relation is an analogue to formula (4.50) for the stationary case and shows how the generalized spectral density of the nonstationary response can be evaluated.

Application of formula (4.67) in an analysis of a harmonic oscillator driven by the so called evolutionary white noise (i.e. by process $X(t)$ with correlation function: $K_X(t_1,t_2) = I(t_1)\delta(t_2-t_1)$) has been presented by Roberts [220].

26.3.4. Stochastic integral equation formulation

As we have mentioned in Sec. 21.3 a stochastic differential equation (3.18), that is the equation

$$\frac{d^nY}{dt^n} + a_1(t)\frac{d^{n-1}Y}{dt^{n-1}} + \cdots + a_{n-1}\frac{dY}{dt} + a_n(t)Y = X(t,\gamma) \tag{4.68}$$

with initial conditions

$$\begin{aligned} &Y(t_o) = Y_o, \\ &\frac{dY}{dt}(t_o) = Y_1, \ldots, \frac{d^{n-1}Y}{dt^{n-1}}(t_o) = Y_{n-1} \end{aligned} \tag{4.69}$$

can be converted into stochastic integral equation of the Volterra type. The integral equation formulation of the problem (4.68)–(4.69) can be especially useful in the case of time-varying coefficients.

Let us assume that the coefficeints $a_k(t)$, $k = 1, \dots, n$ are defined and continuous on $[a, b]$ and $t_o \in [a, b]$.

The reduction of the initial value problem (4.68)–(4.69) to the Volterra integral equation is accomplished by introducing a new unknown process $\varphi(t)$ as follows:

$$\frac{d^n Y(t)}{dt^n} = \varphi(t) \tag{4.70}$$

Making use of the initial conditions we have

$$\frac{d^{n-1}Y}{dt^{n-1}} = \int_{t_o}^{t} \varphi(s) + Y_{n-1},$$

$$\frac{d^{n-2}Y}{dt^{n-2}} = \int_{t_o}^{t} \frac{d^{n-1}Y}{dt^{n-1}}(s_n)ds_n =$$

$$= \int_{t_o}^{t} \int_{t_o}^{s_n} \varphi(s_{n-1})ds_{n-1}ds_n + (t - t_o)Y_{n-1} + Y_{n-2},$$

$$\vdots$$

$$\frac{dY}{dt}$$

$$= \int_{t_o}^{t} \int_{t_o}^{s_n} \cdots \int_{t_o}^{s_3} \varphi(s_2)ds_2 \dots ds_{n-1}ds_n$$

$$+ \frac{Y_{n-1}}{(n-2)!}(t - t_o)^{n-2} + \cdots + Y_2(t - t_o) + Y_1,$$

$$Y(t) =$$

$$= \int_{t_o}^{t} \int_{t_o}^{s_n} \cdots \int_{t_o}^{s_3} \int_{t_o}^{s_2} \varphi(s_1)ds_1 ds_2 \dots ds_{n-1}ds_n$$

$$+ \frac{Y_{n-1}}{(n-1)!}(t - t_o)^{n-1} + \frac{Y_{n-2}}{(n-2)!}(t - t_o)^{n-2} + \cdots + Y_1(t - t_o) + Y_o \tag{4.71}$$

Making use of the Cauchy formula (see Appendix)

$$\int_a^s \int_a^{s_n} \cdots \int_a^{s_3} \int_a^{s_2} F(s_1)ds_1 ds_2 \dots ds_{n-1}ds_n =$$

$$= \frac{1}{(n-1)!} \int_a^s (s - \tau)^{n-1} F(\tau)d\tau$$

the multiple integrals occuring in (4.71) are expressed by the single integrals. Now, if we multiply relations (4.70) and (4.71) subsequently by 1, $a_1(t)$, $a_2(t), \dots, a_n(t)$ and add, we find that the initial-value problem (4.68)–(4.69) is reduced to the Volterra integral equation of the second kind

$$\varphi(t,\gamma) = \Psi(t) + \int_{t_o}^{t} K(t,\tau)\varphi(\tau,\gamma)d\tau \tag{4.72}$$

where

$$K(t,\tau) = -\sum_{k=1}^{n} a_k(t)\frac{(t-\tau)^{k-1}}{(k-1)!} \tag{4.73}$$

$$\begin{aligned} &\Psi(t,\gamma) \\ &= X(t,\gamma) - Y_{n-1}a_1(t) - [(t-t_o)Y_{n-2} + Y_{n-2}]\,a_2(t) \\ &- \cdots - \left[\frac{(t-t_o)^{n-1}}{(n-1)!}Y_{n-1} + \cdots + (t-t_o)Y_1 + Y_o\right]a_n(t). \end{aligned} \tag{4.74}$$

Conversely, if $\varphi(t,\gamma)$ satisfies integral equation (4.72), then process $Y(t)$ defined by the last equation of the system (4.71), that is

$$\begin{aligned} &Y(t,\gamma) \\ &= \int_{t_o}^{t} \frac{(t-\tau)^{n-1}}{(n-1)!}\varphi(\tau,\gamma)d\tau + \frac{(t-t_o)^{n-1}}{(n-1)!}Y_{n-1} + \cdots + Y_1(t-t_o) + Y_o \end{aligned} \tag{4.75}$$

satisfies differential equation (4.68) and initial conditions (4.69).

Equation (4.72) is a linear Volterra integral equation with random inhomogeneous term. The solution $\varphi(t,\gamma)$ is a new stochastic process, the probabilistic properties of which depend on the probabilistic process $\Psi(t,\gamma)$, which — according to (4.74) — is defined by process $X(t,\gamma)$, the coefficients of the original equation (4.68) and the initial conditions. The solution of equation (4.72) is

$$\varphi(t,\gamma) = \Psi(t,\gamma) + \int_{t_o}^{t} \Lambda(t,\gamma)\Psi(\tau,\gamma)d\tau \tag{4.76}$$

where, like in deterministic case, the resolvent kernel $\Lambda(t,\tau)$ is given by the Neumann series

$$\Lambda(t,\tau) = \sum_{k=1}^{\infty} R^{(k)}(t,\tau) \tag{4.77}$$

and the iterated kernels $R^{(k)}(t,\tau)$ are defined by

$$\begin{aligned} &R^{(1)}(t,\tau) = K(t,\tau), \\ &R^{(k)}(t,\tau) = \int_{t_o}^{t} K(t,\xi)R^{(k-1)}(\xi,\tau)d\xi \end{aligned} \tag{4.78}$$

for $k = 2, 3, \ldots$.
Therefore, the determination of the solution of stochastic integral equation (4.72) requires in specific problems that the convergence of the Neumann series (4.77) be established and the resolvant kernel $\Lambda(t, \tau)$ is calculated. Calculation of the mean and the correlation function of solution (4.76) is straightforward.

An application of the above method to the analysis of vibratory system with variable inertia subjected to random loading is discussed by Szopa in his paper [248].

26.3.5. Application of orthogonal expansions

Another possibility in studying nonstationary processes governed by linear differential equations with random inhomogeneous term lies in application of the Karhunen-Loéve orthogonal expansion (cf. Sec. 2.5).

If the random excitation $X(t)$ is represented by its orthogonal expansion (2.43), i.e.

$$X(t) = \sum_k Z_k(\gamma)\varphi_k(t) \tag{4.79}$$

then the mean-square continuous second order solution of equation (4.25) is given by the expansion

$$Y(t) = \sum_k Z_k(\gamma)\Psi_k(t) \tag{4.80}$$

where

$$\Psi_k(t) = \int_0^t p(t, s)\varphi_k(s)ds \tag{4.81}$$

If functions $\Psi_k(t)$ are determined then the correlation function and the variance of the solution are, respectively

$$K_Y(t_1, t_2) = \sum_k \sigma_k^2 \Psi_k(t_1)\overline{\Psi_k(t_2)}, \quad \text{var}Y(t) = K_Y(t, t) = \sum_k \sigma_k^2 \, |\Psi_k(t)|^2 \tag{4.82}$$

26.3.6. Systems subjected to random impulses

In many problems of structural dynamics an excitation acting on a specific structure, consists of a series of impulses occuring at random times and with random magnitude. For example, among the many stochastic processes that have been suggested for modelling vehicular traffic flow and the corresponding loading of highway bridges a process of randomly arriving impulses has been successfully used (cf. Tung [249]). Another example is an excitation due to earthquake; a number of authors have simulated earthquake process as a series of impulses (cf. [203], [252]). Lin [339] indicated the conditions under which there may be valid reasons for modelling an earthquake excitation by uncorrelated and correlated random impulses, whereas Verne-Jones [252] using earthquake data from New Zealand, focussed his attention on modelling an earthquake excitation by a general point process (stream of correlated random events).

One of the simplest random nonstationary impulsive excitation is a *nonstationary shot noise*, that is a process (cf. formula (1.183) where $w(t,\tau_k) = \delta(t-\tau_k)$).

$$X(t) = \sum_{k=1}^{N(t)} F_k(\gamma)\delta\left(t - \tau_k(\gamma)\right) \tag{4.83}$$

where $N(t)$ is a Poisson process with time dependent intensity $N(t)$, and the $F_k(\gamma)$ are independent and identically distributed random variables; usually they are additionally assumed to be independent of the random times $\tau_k(\gamma)$. If the Poisson process $N(t)$ is homogeneous in time, that is $\nu(t) = \nu_0$, where ν_0 is constant then process (4.83) is often called a *stationary shot noise*. Response of dynamical systems to the excitation of form (4.83) has been studied in many papers. A systematic analysis with application to a vibratory system can be found in the papers by Roberts [221], [222].

The model (4.83) assumes that the number of impulses occuring in any finite collection of non-overlapping time intervals form a set of independent random variables or — in other words — that the points on the time axis corresponding to the times of occurrence of impulses are distributed in accordance to the Poisson law with constant (stationary stream of events) or with variable (nonstationary stream of events) arrival rate (cf. [203], [221], [222]).

As the examples indicated above (connected with vehicular loading of bridges and with earthquake) show, the assumption about independence may not always be adequate. There is therefore a need for more general models of impulsive random excitations. One such model is obtained by the following superposition of impulses

$$X(t) = \sum_{k=1}^{N(t)} Z_k(\gamma)\omega(t,\tau_k) \tag{4.84}$$

where $\omega(t,\tau_k)$ describes the shape of a pulse at time $\tau_k(\gamma)$, $Z_k(\gamma)$ is a random amplitude of the pulse and $N(t)$ is a stochastic counting process. Of course, model (4.84) includes shot noise as a special case. The random variables $Z_k(\gamma)$ are assumed to be independent of the random times $\tau_k(\gamma)$.

When the excitation is represented in the form (4.84) the response of a linear system described by the equation (4.25) can be written as

$$Y(t) = \sum_{k=1}^{N(t)} Z_k(\gamma) W(t, \tau_k) \tag{4.85}$$

where

$$W(t, \tau_k) = \int_0^t p(t-\tau)\omega(\tau, \tau_k)d\tau \tag{4.86}$$

and where we have assumed that the coefficients of (4.25) are constant (in this case $p(t,\tau) = p(t-\tau)$).

The solution (4.85) can be written in an alternative form, as a integral with respect to $N(t)$

$$Y(t) = \int_0^t W(t,\tau)Z(\tau)dN(\tau) \tag{4.85}$$

where $N(t)$ is a counting process describing the number of pulses in the time interval $[0, t]$; it is a process whose realizations are step functions (with changes of values at each time when impulse occurs).

In the case when the excitation is the train of Dirac delta impulses, i.e. $X(t)$ is given by (4.83) then

$$Y(t) = \sum_{k=1}^{N(t)} F_k(\gamma)p(t-\tau_k) = \int_0^t p(t-\tau)F(\tau)dN(\tau) \tag{4.87}$$

The moments of the solution can be evaluated in a straightforward way. The mean and variance are (cf. [188])

$$\begin{aligned} E[Y(t)] &= \int_0^t W(t,\tau)E[Z(\tau)]\,E[dN(\tau)] \\ &= \int_0^t W(t,\tau)E[Z(\tau)]\,f_1(\tau)d\tau, \end{aligned} \tag{4.88}$$

$$\begin{aligned} &E\left[Y^2(t)\right] \\ &= \int_0^t\int_0^t W(t,\tau_1)W(t,\tau_2)E\left[Z(\tau_1)Z(\tau_2)\right]E\left[dN(\tau_1)dN(\tau_2)\right] \\ &= \int_0^t W^2(t,\tau)E\left[Z^2(\tau)\right]f_1(\tau)d\tau \\ &+ \int_0^t\int_0^t W(t,\tau_1)W(t,\tau_2)E\left[Z(\tau_1)Z(\tau_2)\right]f_2(\tau_1,\tau_2)d\tau_1 d\tau_2 \end{aligned} \tag{4.89}$$

where $f_1(\tau)$ and $f_2(\tau_1, \tau_2)$ are the first-order and second-order product densities of the counting process $N(t)$ — cf. Sec. 8. Function $f_1(\tau)$ characterizes the expected arrival rate of impulses and $f_2(\tau_1, \tau_2)$ describes the correlation between the arrival times.

The variance of the solution is

$$\begin{aligned}
\mathrm{var} Y(t) &= E\left[Y^2(t)\right] - \{E\left[Y(t)\right]\}^2 \\
&= \int_0^t W^2(t,\tau) E\left[Z^2(\tau)\right] f_1(\tau) d\tau \\
&+ \int_0^t \int_0^t W(t,\tau_1) W(t\tau_2) \Big\{ E\left[Z(\tau_1) Z(\tau_2)\right] f_2(\tau_1 - \tau_2) \\
&- E\left[Z(\tau_1)\right] E\left[Z(\tau_2)\right] f_1(\tau_1) f_2(\tau_2) \Big\} d\tau_1 d\tau_2.
\end{aligned} \tag{4.90}$$

In the case when the arrival times are still correlated but the amplitudes are mutually independent and identically distributed we have

$$\begin{aligned}
\mathrm{var} Y(t) &= E\left[Z^2\right] \int_0^t W^2(t,\tau) g_1(\tau) d\tau \\
&+ E\left[Z^2\right] \int_0^t \int_0^t W(t,\tau_1) W(t,\tau_2) g_2(\tau_1, \tau_2) d\tau_1 d\tau_2
\end{aligned} \tag{4.91}$$

where $g_1(\tau) = f_1(\tau)$ and $g_2(\tau_1, \tau_2) = f_2(\tau_1, \tau_2) - f_1(\tau_1) f_2(\tau_2)$. In the case when $N(t)$ is a Poisson process $g_2(\tau_1, \tau_2) = 0$ and in formula (4.91) only the first term remains. If, additionally the shape of the pulses are characterized by the Dirac delta, that is the excitation is a shot noise the formula (4.91) reduces to

$$\mathrm{var} Y(t) = E[Z^2] \int_0^t p^2(t-\tau) \nu(\tau) d\tau \tag{4.92}$$

where $\nu(\tau) = g_1(\tau)$.

The application of the above formulae to specific equations arising in practice is, therefore, reduced to integration (when characteristics of counting process $N(t)$ and statistics of random variables $Z_k(\gamma)$ are given). As linear systems are concerned, a detailed analysis (including numerical calculations for specific data) of the correlation function and variance have been presented by a number of authors (cf. Roberts [221],[222], Lin [203], [339], Srinivasan et al. [245], Iwankiewicz and Śniady [187], Iwankiewicz and Sobczyk [188]).

Of course, the response of linear systems to random impulses in the form (4.83) or (4.84) is, in general, non-Gaussian. Hence, the determination of the probability distribution of process $Y(t)$ is concerned with essential difficulties and usually only some approximations are possible (we discuss this problem in the next sub-section).

26.4. Linear systems and normality

26.4.1. General remarks

So far, we concentrated our attention on the methods which allow us to determine the first order moments of the solution process; in particular, the mean and the correlation function. Such characterization of the solution is complete if the random excitation (acting on a linear system) is Gaussian. The assumption of the normality of the excitation is a very convenient one, since it simplifies the analysis. However, in many real situations it turns out to be very difficult to justify correctly the normality of the process in question. For instance, in engineering theory of random signals it is usually believed that, if a stationary stochastic process is the input of a linear system, then the response is "approximately normally distributed" as the bandwidth of the system tends to zero. Obviously, such a statement can not be true in general and in cases when it could be justified one should be able to sketch out a range within which such an assertion would be valid. In spite of this we need not only the asymptotic result (if some conditions are satisfied) but, primarily, the true probability distribution of the response in any practical situation or, at least, the estimation of the departure of the response from normality.

In what follows we shall present briefly the possible approaches to the problem with a special attention to the departure of the solution (response of a system) from Gaussian distribution (cf. Kotulski, Sobczyk [191], also [210],[215],[233].

26.4.2. Measures of departure from normality

The definition and properties of Gaussian stochastic processes imply that measures of departure from normality should be expressed in terms of the joint probability distributions or in terms of the moments of order greater than two. In general, departure from normality can be characterized in the following way.

Let $X(t)$ be a non-Gaussian second order stochastic process and $X_G(t)$ a Gaussian process associated with $X(t)$. Let $\mathbf{I}$ be any n-dimensional interval in R_n. Let us denote by $\mathbf{X}$ the n-dimensional random vector $[X(t_1),\ldots,X(t_n)]$, obtained from the process $\mathbf{X}(t)$ for an arbitrary set $\{t_1,\ldots,t_n\}$ of the values of parameter t. Let $\mathbf{X}_G$ be the n-dimensional Gaussian vector associated with $\mathbf{X}$ (having the same first and second moments). The quantity

$$\Delta = P(\mathbf{X}\in I) - P(\mathbf{X}_G \in I) \tag{4.93}$$

where P denotes the probability of event (.), characterizes fully the departure of the process $\mathbf{X}_G(t)$ from normality. Unfortunately, the quantity (4.93) — including all joint probability distributions — is too complicated to serve as a tool in measuring non-normality of real processes.

Historically, first measures of departure from normality have been introduced in connection with central limit theorems and some problems of mathematical statistics. Simpler measures are mainly based on one-dimensional distributions. As far as random processes are concerned, the statements obtained by use of one-dimensional distributions provide sufficient information when the departure from normality is significant. Small departure from normality indicated by a one-dimensional distribution does not mean, of course, that the process as a whole is nearly Gaussian. The most popular measures of departure from normality can be defined as follows:

$$\begin{aligned} \varepsilon_X &= |F(x) - \Phi(x)|\,, \\ \tilde{\varepsilon}_X &= \sup_x |F(x) - \Phi(x)| \end{aligned} \tag{4.94}$$

where $F(x)$ is the distribution function of a non-Gaussian variable and $\Phi(x)$ is the Gaussian distribution with the same two first moments as $F(x)$.

If the stochastic process under consideration has finite moments of order $p \geq 1$, then a convenient measure of departure from normality can be defined as a distance in the metric space L_p, i.e.

$$\eta_p(F,\Phi) = \left\{ \int_{-\infty}^{+\infty} |F(x) - \Phi(x)|^p \, dx \right\}^{1/p} \tag{4.95}$$

A systematic and uniform approach to measuring of the distance from normality with application to discrete linear filters is presented in paper by Mallows [210]. The starting point in the analysis given in [210] is the following measure of the departure from normality (in the one-dimensional case)

$$\delta(F,\Phi) = \int_{-\infty}^{+\infty} [F(x) - \Phi(x)]^2 \, dx \tag{4.96}$$

It can be shown easily that $\delta(F,\Phi) = 0$ iff $F(x)$ is Gaussian. Using this measure Mallows defines the measure of departure for the finite- and infinite-dimensional cases (for random process as a whole).

A simple (parametric) measure of departure from normality which turns out to be very useful in applied probability and statistics is the asymmetry and excess coefficient, defined respectively as

$$\zeta_1 = \frac{\mu_3}{\sigma^3}, \qquad \zeta_2 = \frac{\mu_4}{\sigma^4} - 3 \tag{4.97}$$

where σ is the standard deviation of the considered process and μ_3 and μ_4 are the third and fourth order central moments, respectively. For the Gaussian distribution holds $\zeta_1 = \zeta_2 = 0$. The value ζ_1 and ζ_2 different from zero characterize the departure of a one-dimensional distribution from normality.

Concluding these remarks it is worth noting that some extensions of Chebyshev's inequality lead to bounds of measure (4.94). Also, quantity (4.94) can be bounded in terms of (4.96). The following theorem is valid (cf. [210]).

If

$$\frac{dF}{dx} \le A, \qquad \frac{d\Phi}{dx} \le B \quad \text{for all} \quad x,$$

then

$$|F(x) - \Phi(x)| \le \left[\frac{3AB}{A+B}\delta(F,\Phi)\right]^{1/3} \tag{4.98}$$

26.4.3. Linear systems subjected to non-Gaussian excitation

A. Asymptotic result

Let the input to a linear system be denoted by $X(t)$. We assume that $X(t)$ is given stationary non-Gaussian process with $t \in T \subset [0,\infty)$. The resulting response is represented as

$$Y(t) = \int_{t_0}^{t} p(t,\tau)X(\tau)d\tau \tag{4.99}$$

where $p(t,\tau)$ is a unit response function. Integral (4.99) can be approximated by a finite sum of random variables:

$$Y(t) \approx \sum_{i=1}^{n} \xi_i,$$
$$\xi_i = \int_{(i-1)\Delta}^{j\Delta} p(t,\tau)X(\tau)d\tau \tag{4.100}$$

where Δ denotes the length of an elementary interval in the division of the interval $[0,t]$ into n parts. If the variables ξ_i were independent, then — by virtue of the central limit theorem — the distribution of $Y(t)$ could be considered approximately as Gaussian as $n \to \infty$. But the variables ξ_i are evidently dependent since they are defined by the values of the same stochastic process.

A number of central limit theorems has recently been proven for dependent variables. This more general form of the central limit theorem imposes, however, specific conditions on the random variables in question. Let us take, for illustration, the limit theorem proposed by Rosenblat [225]. He showed that, subject to appropriate conditions, weighted integrals of stationary random processes of the form (4.99) become asymptotically normal. The conditions are:

1) strong mixing: $X(t)$ is said to be strongly mixing, if there exists a non-negative, monotone decreasing function $\varphi(t)$, $\varphi(t) \to 0$ for $t \to \infty$, such that, if A is any event dependent only on values of $X(t)$ for $t \leq t_1$ and B any event dependent only on values of $X(t)$ for $t \geq t_2$, $t_2 > t_1$, then

$$|P(AB) - P(A)P(B)| < \varphi(t_2 - t_1);$$

2) the correlation function and fourth order moments of $X(t)$ are absolutely integrable;
3) the spectral density is non-zero at any frequency;
4) the weighting function $P(t) = \int_0^t p^2(t,\tau)d\tau$ increases to infinity as $t \to \infty$; the filter must be progressively narrower.

The assumptions stated above are rather restrictive. Especially the problem of testing the strong mixing condition 1) in real physical situations is not easy. Strong mixing is used to show that integrals over adjecent segments of $X(t)$ become statistically asymptotically independent. Only for Gaussian stationary processes this condition can be replaced by a simpler one, but we are interested in the non-Gaussian case. In spite of this it was shown in [191] that the conditions of Rosenblat's theorem concerning the properties of the filter are not satisfied for the most popular practical systems described by the differential equations of the first and second orders (for real parameter values). The band of these systems is not infinitely narrow and the response departs from normality significantly. In [191] it was also shown, how the departure from normality is affected by the correlation time of the excitation and the parameters of the systems.

B. Departure from normality

Let us consider a first order differential equation

$$\frac{dY(t)}{dt} + aY(t) = X(t), \quad Y(t_o) = 0 \tag{4.101}$$

where $X(t) = Z^2(t)$, and $Z(t)$ is a stationary Gaussian process

$$\begin{aligned} &E[Z(t)] = 0, \\ &K_Z(\tau) = \sigma_Z^2 e^{-\alpha|\tau|}, \quad \alpha > 0. \end{aligned} \tag{4.102}$$

In this case, the one-dimensional probability density of process $X(t)$ has the form

$$f_{X(t)}(x) = \begin{cases} \dfrac{1}{\sigma_Z\sqrt{2\pi}}\dfrac{1}{\sqrt{x}}e^{-\frac{x}{2\sigma_Z^2}} & , x > 0 \\ 0 & , x \leq 0 \end{cases} \tag{4.103}$$

and $\zeta_1^X = 3.9$, $\zeta_2^X = 15.0$. Calculations of the third and fourth order moments of the stationary solution (by use of the two-dimensional Itô equation — cf.[191]) yields the following expressions for the asymmetry and excess coefficients

$$\zeta_1^Y = \frac{2}{a+\alpha}\sqrt{2a(a+\alpha)},$$
$$\zeta_2^Y = 3\left[\frac{15a^2+25a\alpha+2\alpha^2}{(a+\alpha)(3a+2\alpha)} - 1\right]. \tag{4.104}$$

Analysis of these expressions indicates that even for large values of time (for stationary solution) the coefficients characterizing the departure from normality take significant values. For example, for $\frac{a}{\alpha} = 0.5$, $\zeta_1^Y \approx 1.5$, $\zeta_2 \approx 7.4$; for $\frac{a}{\alpha} = 5.0$, $\zeta_1^Y \approx 2.0$, $\zeta_2^Y \approx 11,8$.

Evaluation of the coefficients ζ_1^y and ζ_2^Y for the harmonic oscillator

$$\frac{d^2Y}{dt^2} + 2h\frac{dY}{dt} + \omega_0^2 Y = X(t) \tag{4.105}$$

with the same process $X(t) = Z^2(t)$ gives the dependence of ζ_1^Y and ζ_2^Y on α, ω_0 and h. The departure from normality increases when the damping coefficient increases. The similar observations have been obtained also by Grigoriu and Ariaratnam [179] for the equation (4.105) in which the excitation process $X(t)$ is a polynomial function of $Z(t)$, i.e.

$$X(t) = \sum_{k=1}^{l} a_k Z^k(t). \tag{4.106}$$

Let us assume now that a linear system is subjected to the series of random impulses, i.e.

$$X(t) = \sum_{i=1}^{N(t)} \delta(t - t_i) \tag{4.107}$$

where $N(t)$ is a Poisson process with intensity $\nu(t)$. In this case

$$Y(t) = \sum_{i=1}^{N(t)} p(t - t_i) \tag{4.108}$$

and

$$\zeta_1^Y = \left[\frac{I_3^2(t)}{I_2^3(t)}\right]^{1/2}, \quad \zeta_2^Y = \frac{I_4(t)}{I_2^2(t)} \tag{4.109}$$

where

$$I_n(t) = \int_{-\infty}^{+\infty} \nu(t)\, |p(t-\tau)|^n \, d\tau$$

If $\nu(t) = \nu =$ const. and the system is governed by equation (4.105) then

$$\zeta_1^Y = \frac{3}{8}\left(\frac{4h^3\omega_o^2}{\nu(8h^2+\omega_o^2)^2}\right)^{1/2},$$

$$\zeta_2^Y = \frac{6h\omega_o^2}{\nu(12h^2+\omega_o^2)} \tag{4.110}$$

If $\nu \to \infty$ (and ω_o, h are fixed) ζ_1^Y and ζ_2^Y converge to zero; so, the solution can be considered as Gaussian (with respect to one-dimensional distribution).

C. Probability distributions

Using the Hilbert space formulation (cf. Sec. 23.3) the complete probabilistic characterization of the process described by equations (4.105), (4.107) has been obtained in the form of the characteristic functional:

$$\Phi_Y[\lambda] = exp\left\{\nu \int_o^T \left[exp\{\frac{i}{\omega_h}\int_o^T H(t-t')e^{-h(t-t')}\times \right.\right.$$

$$\left.\left. \times \sin\omega_h(t-t')\lambda(t)dt\} - 1\right]dt'\right\} \tag{4.111}$$

where $H(t)$ is a Heaviside function, $t \in [0,T]$ and $\omega_h = \omega_o^2 - h^2$. The functional differentiation of (4.111) and simple integration yield formulae (4.110).

Although functional (4.111) characterizes the solution process $Y(t)$ completely, it is not easy to obtain from it the probability distribution, especially — the one-dimensional probability density. This is an analogous situation to that when we have the characteristic function of the response to the shot noise; the problem of inverting it to obtain the required probability density is not solvable analytically. To obtain the density function Racicot and Moses [217] have used a numerical technique, whereas Roberts [223] has adopted the so called, "saddle point" approximation; the approximate formulae for the probability density of the solution of equation (4.105) with $X(t)$ in form (4.107) obtained in [223] turned out to be in close agreement with simulation results even when the response depart significantly from the Gaussian distribution.

It seems that the most popular way of approximate representation of the probability density of non-Gaussian process is the use of the orthogonal expansions in the form of the Gram-Charlier or the Edgworth series. If $\varphi_t(y)$ is the one-dimensional density of Gaussian distribution $N(0,t)$ then any standardized probability density $f_t(y)$ can be represented as

$$f_t(y) = c_o\varphi_t(y) + \frac{c_1}{1!}\varphi_t'(y) + \frac{c_2}{2!}\varphi_t''(y) + \dots \tag{4.112}$$

where c_ν are constants and $\varphi_t'(y), \varphi_t''(y), \ldots$ are the derivatives of $\varphi_t(y)$. Using the fact of orthogonality of the Hermite polynomials $H_\nu(y)$ with respect to $\varphi_t(y)$ the constants $c_o, c_1, \ldots$ are simply expressible in terms of moments, that is

$$
\begin{aligned}
c_o &= 1, \\
c_1 &= c_2 = 0, \\
c_3 &= -\frac{\mu_3}{\sigma^3} = -\zeta_1, \\
c_4 &= \frac{\mu_4}{\sigma^4} - 3 = \zeta_2, \quad \text{etc.}
\end{aligned}
\tag{4.113}
$$

So, the Gram-Charlier expansion has the form

$$
f_t(y) \approx \varphi_t(y) - \frac{\zeta_1}{3!}\varphi'''(y) + \frac{\zeta_2}{5!}\varphi^{\text{iv}}(y) + \ldots \tag{4.114}
$$

As it has been shown by Cramer, series (4.114) is convergent (for every y) to the probability density $f_t(y)$ if $f_t(y)$ is of bounded variation in $(-\infty, +\infty)$ and

$$
\int_{-\infty}^{+\infty} e^{\frac{y^2}{4}} f_t(y) dy < \infty \tag{4.115}
$$

It is clear that there exist many distributions which do not satisfy condition (4.115). In spite of this, three other things should be kept in mind:

(i) the sum of infinte number of terms in (4.114) may give negative values of $f_t(y)$, particularly near the tails;

(ii) the series (4.114) may behave irregularly in the sense that the sum of k terms may give a worse fit than the sum of $(k-1)$ terms;

(iii) the probability density curves $f_t(y)$ may be multimodal.

Therefore, the question which requires the answer is: what is the region in $(c_1, c_2, \ldots, c_n)$-space, or more practically, in (ζ_1, ζ_2)-plane for which $f_t(y)$ is positive and unimodal? The analysis of this problem indicates that for (ζ_1, ζ_2) which are different from the Gaussian values the curve $f_t(y)$ has negative values for some part of the range of y; if $\zeta_1 < 0.6$ and $\zeta_2 < 1.5$ then the distribution $f_t(y)$ represented by (4.114) can be expected to be unimodal. Therefore, as it has also been confirmed by the examples in stochastic dynamics, expansion (4.114) is useful only when the probability distribution is very nearly Gaussian.

Another possibility of the approximate characterization of non-Gaussian distributions lies in the maximum entropy principle (cf. sec. 27.3).

27. NONLINEAR SYSTEMS WITH RANDOM EXCITATION

27.1. White noise excitation

In this section we shall consider dynamical systems governed by stochastic differential equations of the type

$$\begin{aligned} &\frac{d\mathbf{Y}(t)}{dt} = \mathbf{F}[\mathbf{Y}(t), t] + \mathbf{X}(t), \quad t \in [t_o, \infty) \\ &\mathbf{Y}(t_o) = \mathbf{Y}_o \end{aligned} \tag{4.116}$$

where $\mathbf{Y}(t)$ is an unknown vector stochastic process; $\mathbf{X}(t)$ is a given vector stochastic process (called the excitation) and $\mathbf{Y}_o$ is a vector (deterministic or random) characterizing the initial state of the system.

This subsection is devoted to the study of systems (4.116) in the case where the excitation process $X(t)$ is a vector white noise, i.e. $\mathbf{X}(t) = \boldsymbol{\xi}(t) = [\xi_1(t), \xi_2(t), \ldots, \xi_m(t)]$ where $\xi_i(t)$, $i = 1, 2, \ldots, m$ are independent white noises. In this case we shall represent the system by the more general Langevin-type vector "equation"

$$\begin{aligned} &\frac{d\mathbf{Y}(t)}{dt} = \mathbf{F}[\mathbf{Y}(t), t] + \boldsymbol{\sigma}[\mathbf{Y}(t), t]\boldsymbol{\xi}(t) \\ &\mathbf{Y}(t_o) = \mathbf{Y}_o \end{aligned} \tag{4.117}$$

or, in explicit form

$$\begin{aligned} \frac{dY_i}{dt} = &F_i[Y_1(t), \ldots, Y_n(t), t] \\ &+ \sum_{j=1}^{m} \sigma_{ij}[Y_1(t), \ldots, Y_n(t), t]\, \xi_j(t) \end{aligned} \tag{4.118}$$

$i = 1.2. \ldots, n.$

As we know from the previous chapter the above system of "equations" can be interpreted as the Itô vector equation (3.60) or as the Stratonovich equation (3.73) which is equivalent to the Itô equation with modified drift vector according to formula (3.74) or in the scalar case to (3.71).

If the functions F_i and σ_{ij} satsify the conditions of the theorems given in Section 22.1, then there exists a solution of (4.117)–(4.118) which is a diffusion Markov process and its transition probability density $p(\mathbf{y}, t; \mathbf{y}_o, t_o)$ satisfies the Fokker-Planck-Kolmogorov (F-P-K) equation (1.191), i.e.

$$\frac{\partial p}{\partial t} + \sum_{i=1}^{n} \frac{\partial}{\partial y_i}[a_i(\mathbf{y}, t)p] - \frac{1}{2}\sum_{i,j=1}^{n} \frac{\partial^2}{\partial y_i \partial y_j}[b_{ij}(\mathbf{y}, t)p] = 0 \tag{4.119}$$

with initial condition

$$p(y, t_o; y_o, t_o) = \prod_{j=i}^{n} \delta(y_j - y_{oj})$$

The coefficients of the F-P-K equation, according to formulae (3.61)–(3.62) are

$$\begin{aligned} a_i(\mathbf{y}, t) &= F_i(\mathbf{y}, t), \quad i = 1, 2, \ldots, n \\ b_{ij}(\mathbf{y}, t) &= \sum_{r=1}^{n} \sigma_{ir}(\mathbf{y}, t)\sigma_{jr}(\mathbf{y}, t), \quad j = 1, 2, \ldots, m \end{aligned} \tag{4.120}$$

27.1.1. First order equations

In order to illustrate the method of analysis we start from the first order equations driven by white noise. Such stochastic equations can serve as models of various phenomena in physics, biology as well as in mechanics where they have recently been used for modelling of random fatigue crack growth in materials.

Examples

1. Let us consider the following "equation"

$$\begin{aligned} \frac{dY}{dt} + f\,(Y(t)) &= D\xi(t), \\ Y(0) &= Y_o \end{aligned} \tag{4.121}$$

where D is a positive constant. The Itô stochastic equation corresponding to (4.121) is

$$dY(t) = -f\,(Y(t))\,dt + D dW(t) \tag{4.121'}$$

The F-P-K equation for the transition probability density is given by

$$\begin{aligned} \frac{\partial p}{\partial t} &= \frac{\partial}{\partial y}\left[f(y)p\right] + \frac{1}{2}D^2\frac{\partial^2 p}{\partial y^2}, \\ p(y, 0; y_o) &= \delta(y - y_o) \end{aligned} \tag{4.122}$$

If $f(y)$ is such that the condition (1.116) for the existence of a stationary solution is satisfied then a steady-state probability density $p_{st}(y)$ is a solution to the equation

$$\frac{\partial}{\partial y}\left[f(y)\right] + \frac{1}{2}D^2\frac{\partial^2 p_{st}}{\partial y^2} = 0$$

or

$$\frac{\partial}{\partial y}\left\{f(y)p_{st} + \frac{1}{2}D^2\frac{\partial p_{st}}{\partial y}\right\} = 0$$

Direct integration yields (cf. formula (1.120))

$$p_{st}(y) = \frac{1}{C}exp\left[-\frac{2}{D^2}\int_0^y f(z)dz\right] \tag{4.123}$$

where

$$C = \int_{-\infty}^{+\infty} exp\left[-\frac{2}{D^2}\int_0^x f(z)dz\right]dx. \tag{4.124}$$

2. Given the following "equation"

$$\begin{gathered}\frac{dY}{dt} + aY - \frac{b}{Y} = D\xi(t), \quad Y(t) > 0\\ Y(t_\circ) = Y_\circ\end{gathered} \tag{4.125}$$

the corresponding Itô equation takes the form

$$dY(t) = \left[-aY(t) + \frac{b}{Y}\right]dt + DdW(t) \tag{4.125'}$$

The drift and diffusion coefficients of the solution are

$$a(y) = -ay + \frac{b}{y}, \quad b(y) = D^2$$

Let $p(y, y_\circ, t)$ be a probability density of the transition from $y_\circ = y(t_\circ)$ to $y = y(t)$ within time $\tau = t - t_\circ > 0$. The F-P-K equation is

$$\frac{\partial p}{\partial t} = \frac{\partial}{\partial y}\left[\left(ay - \frac{b}{y}\right)p\right] + \frac{1}{2}D^2\frac{\partial^2 p}{\partial y^2}$$

The equation for a steady-state density p_{st} is

$$\frac{\partial}{\partial y}\left\{\left(ay - \frac{b}{y}\right)p_{st} + \frac{1}{2}D^2\frac{\partial p_{st}}{\partial y}\right\} = 0$$

Making use of formula (1.120) and assuming the absorbing boundaries at $y = 0$ and $y = \infty$, we obtain

$$\begin{aligned}p_{st}(y) &= Cexp\left[-\frac{2}{D^2}\left(\frac{1}{2}ay^2 - blny\right)\right]\\ &= Cy^{\frac{b}{a\sigma^2}}exp\left(-\frac{y^2}{2\sigma^2}\right)\end{aligned} \tag{4.126}$$

where

$$\sigma^2 = \frac{D^2}{2a}, \quad C = \frac{2}{(2\sigma^2)^{\frac{1}{2}+\frac{b}{D^2}}\Gamma\left(\frac{1}{2}+\frac{b}{D^2}\right)}. \tag{4.127}$$

Taking into account values of the gamma function we come to the conclusion that the above formula for the stationary probability density reduces for $b = 0$ to one-sided Gaussian density and for $b = \frac{D^2}{2}$, to the density of the Raleigh distribution.

The time-dependent (nonstationary) transition probability density of the solution of equation (4.125) can be found by use of the method of separation of variables presented in Sec. 6.3. Indeed, if we assume that $p(t,y) = \Lambda(y)T(y)$ then we obtain the equations

$$(*) \qquad \frac{dT(t)}{dt} + \lambda^2 T(t) = 0, \quad \sigma^2\frac{d^2\Lambda(y)}{dy^2} + \frac{d}{dy}(y\Lambda) - \frac{b}{a}\frac{d}{dy}\left(\frac{\Lambda}{y}\right) + \frac{\lambda^2}{a}\Lambda = 0$$

According to (1.136) the solution is

$$p(y, y_o, t - t_o) = \sum_{n=0}^{\infty} \frac{\Lambda_n(y_o)\Lambda_n(y)}{f_{st}(y_o)} e^{-\lambda_n^2(t-t_o)}$$

Let us introduce in equation (*) the new variables

$$U = z^{-\mu}\Lambda, \quad z = \frac{y^2}{2\sigma^2} = \frac{a}{D^2}y^2, \quad \mu = \frac{4b-1}{8D^2}$$

Equation (*) takes the form of the hypergeometric equation

$$\frac{d^2U}{dz^2} + \frac{dU}{dz} + \left[\frac{\frac{1}{2}+\mu+\frac{\lambda_n^2}{2a}}{z} + \frac{\frac{1}{4}-\mu}{z^2}\right]U = 0$$

If we assume the eigenvalues $\lambda_n^2 = 2na$ then

$$U_n(z) = z^{\mu+\frac{1}{2}}e^{-z}L_n^{2\mu}(z),$$

where

$$L_n^{\alpha}(z) = \frac{1}{n!}e^z z^{-\alpha}\frac{d^n}{dz^n}(e^{-z}z^{n+\alpha}).$$

are the Laguerre polynomials. Coming back to original variables and making use of the orthogonality conditions (1.137) we obtain

$$\Lambda_n(y) = \frac{\sqrt{2}}{\sigma}\frac{z^{2\mu+\frac{1}{2}}e^{-z}L_n^{2\mu}(z)}{[n!\Gamma(n+2\mu+1)\Gamma(2\mu+1)]^{1/2}}, \; z = \frac{y^2}{2\sigma^2}$$

Therefore, the final expression for $p(y, y_o, t - t_o)$ is

(4.128)
$$p(y, y_o, t - t_o) = = \frac{\sqrt{2}}{\sigma z^{2\mu+\frac{1}{2}} e^{-z}} \sum_{n=0}^{\infty} \frac{L_n^{2\mu}(z) L_n^{2\mu}(z_o)}{n!\Gamma(n + 2\mu + 1)} e^{-2na(t-t_o)}$$

where $z_o = \frac{y_o^2}{2\sigma^2}$. Making use of the properties of polynomials $L_n(z)$ and passing to the limit as $t - t_o \to \infty$ the stationary density (4.126) can be obtained.

3. Given the following "equation"

(4.129)
$$\frac{dY}{dt} + Y ln \frac{Y}{\beta} = aY\xi(t), Y(t) > 0, \quad \beta > 0$$

Let us interpret this "equation" in the sense of Stratonovich; then the corresponding Itô stochastic equation (according to (3.71))

(4.129′)
$$dY(t) = \left[-aY(t) ln \frac{Y(t)}{\beta} + \frac{1}{2} a^2 Y \right] dt + aY(t) dW(t)$$

The drift and diffusion coefficients of the solution are

$$a(y) = -ayln\frac{y}{\beta} + \frac{1}{2}a^2 y, \qquad b(y) = a^2 y^2$$

The F-P-K equation takes the form

$$\frac{\partial p}{\partial t} = a \frac{\partial}{\partial y} y \left[ln \frac{y}{\beta} - \frac{1}{4} a \right] p + \frac{1}{2} a^2 \frac{\partial^2}{\partial y^2} (y^2 p)$$

Application of formula (1.120) gives the following expression for the steady-state density

(4.130)
$$p_{st}(y) = \frac{1}{\sigma_y \sqrt{2\pi}} exp \left[-\frac{(lny - ln\beta)^2}{2\sigma^2} \right], \quad \sigma^2 = \frac{a}{8}$$

It is the well known log-normal distribution.

4. Given the following "equation"

(4.131)
$$\frac{dY(t)}{dt} = \Theta Y(t) [1 - Y(t)] + \sigma Y^2(t) \xi(t)$$

If we interpret this "equation" as the Stratonovich stochastic differential equation we have

$$(4.131') \qquad (S) \quad dY(t) = \Theta Y(t)\,[1 - Y(t)]\,dt + \sigma Y^2(t) dW(t)$$

The above equation is equivalent to the following Itô equation

$$(4.131'') \qquad (I) \quad dY(t) = \left[\Theta Y(t)\,(1 - Y(t)) + \sigma^2 Y^3(t)\right] dt + \sigma Y^2(t) dW(t).$$

Let us take the following state transformation

$$Z_t = \frac{1}{Y_t}$$

Use of the Itô formula gives

$$\begin{aligned} dZ_t &= -\frac{1}{Y_t^2}\left[\Theta Y_t(1 - Y_t) + \sigma^2 Y_t^3\right] dt + \\ &+ \frac{1}{Y_t^3}\sigma^2 Y_t^4 dt - \sigma dW_t \\ &= (-\Theta Z_t + \Theta)dt - \sigma dW_t \end{aligned}$$

that is, a linear stochastic equation for process $Z(t)$. Taking

$$Z_1(t) = 1 - Z(t)$$

we get

$$dZ_1(t) = -\Theta Z_1(t)dt + \sigma dW(t)$$

It is seen that for $\Theta > 0$, the process $Z_1(t)$ is the Ornstein-Uhlenbeck process. The process $Y(t)$ has the representation

$$(4.132) \qquad Y(t) = \frac{1}{1 - \sigma \int_0^t exp\,[-\Theta(t - s)]\,dW(s)}, \quad \Theta > 0$$

Since the denominator has a Gaussian distribution, the distribution of $Y(t)$ can be easily determined.

5. In studying of fatigue crack growth the following stochastic differential model has been introduced (cf. Sobczyk [236],[367])

$$(4.133) \qquad \begin{aligned} \frac{dY}{dt} &= f_1(t) Y^p(t) + f_2(t) Y^p(t) \xi(t) \\ Y(t_0) &= Y_0 \end{aligned}$$

where $\xi(t)$ is a Gaussian white noise with intensity $2D$, p is a positive constant occuring in experimental equations of fatigue crack growth and the unknown stochastic process $Y(t)$ characterizes a crack length at time t. The functions $f_1(t)$ and $f_2(t)$ describe the external loading (acting on the structural element considered) and include also the constant material parameters. The multiplicative noise $\xi(t)$ is introduced to account for randomness of other unpredictable factors provoking fatigue and contributing to the scatter of fatigue data.

If we adopt the Stratonovich interpretation of the above "equation" we obtain the following stochastic Itô differential equation

$$\begin{aligned} dY(t) &= m(Y,t)dt + \sigma(Y,t)dW(t) \\ Y(t_\circ) &= Y_\circ \end{aligned} \tag{4.134}$$

where,

$$\begin{aligned} m(y,t) &= f_1(t)y^p + pD^2 f_2^2(t)y^{2p-1} \\ \sigma(y,t) &= 2Df_2(t)y^p \end{aligned} \tag{4.135}$$

According to the theory presented in Sec. 22.3 in the case $p = 1$ (linear equation) we have the solution

$$Y(t) = Y_\circ exp\left\{\int_{t_\circ}^{t} \alpha(s)ds + \int_{t_\circ}^{t} \beta(s)dW(s)\right\} \tag{4.136}$$

where $\alpha(t) = f_1(t) + D^2 f_2^2(t)$, $\beta(t) = 2Df_2(t)$. The moments are

$$E\left[Y^r(t)\right] = Y_\circ^r exp\left\{r\int_{t_\circ}^{t} \alpha(s)ds + \frac{r^2}{2}\int_{t_\circ}^{t} \beta(s)ds\right\} \tag{4.137}$$

If $p > 1$ equation (4.134) defines the solution process only up to a random explosion time $\tau(\gamma)$. To solve the equation in the interval $[t_\circ, \tau(\gamma)]$ we make use of the state transformation:

$$Z_t = \varphi(Y_t) = Y_t^{1-p}, \quad Z_\circ = Y_0^{1-p} \tag{4.138}$$

After application of the Itô formula we obtain a simpler linear equation with the solution

$$Z_t = Z_\circ - (p-1)\int_{t_\circ}^{t} f_1(s)ds - (p-1)\int_{t_\circ}^{t} f_2(s)dW(s) \tag{4.139}$$

If $f_1(t) = f_1^\circ$ =constant, $f_2(t) = f_2^\circ$ =constant, we have

$$Z_t = Z_\circ - (p-1)f_1^\circ(t - t_\circ) - (p-1)f_2^\circ\left[W(t) - W(t_\circ)\right] \tag{4.139'}$$

It is seen that $\Lambda_t = Z_\circ - Z_t = \frac{1}{Y_\circ^{p-1}} - \frac{1}{Y_t^{p-1}}$ has Gaussian distribution with the mean and variance, respectively:

$$E[\Lambda_t] = (p-1)\int_{t_\circ}^{t} f_1(s)ds,$$

$$\text{var}\Lambda_t = (p-1)^2 \text{var}\int_{t_\circ}^{t} f_2(s)dW(s) \tag{4.140}$$

or, when (4.139′) is used we have, respevtively

$$\begin{gathered}(p-1)f_1^\circ(t-t_\circ),\\ 2D(p-1)^2 f_2^\circ(t-t_\circ)\end{gathered} \tag{4.140′}$$

Therefore, the density of the transition $p(y,t;y_\circ,t_\circ)$ is obtained from (4.139), (4.140). This probability density is given by the formula

$$\begin{aligned}p(y,t) &= p(z,t)\frac{dz}{dy}\\ &= \frac{p-1}{y_t^p}\frac{1}{[2\pi\text{var}\Lambda_t]^{1/2}}exp\left\{-\frac{\left[y_\circ^{1-p} - y_t^{1-p} - E[\Lambda_t]\right]^2}{2\text{var}\Lambda_t}\right\}\end{aligned} \tag{4.141}$$

In the case considered here, that is when $p > 1$, it is of interest to characterize the random explosion time $\tau(\gamma)$ in which the process escapes to infinity. This time is defined as

$$\tau(\gamma) = \inf\left\{t : (p-1)\int_{t_\circ}^{t} f_1(s)ds + (p-1)\int_{t_\circ}^{t} f_2(s)dW(s) \geq \frac{1}{Y_\circ^{p-1}}\right\} \tag{4.142}$$

The above first passage time can be solved exactly in the case when $f_1(t)$ and $f_2(t)$ are constant. It has been found in the paper [236] that the random variable $\tau(\gamma)$ has the inverse Gaussian distribution.

27.1.2. Second order equation

The main approach to the analysis of non-linear systems excited by white noise and governed by second order differential equations (e.g. vibratory mechanical systems) is based on the representation of the equation in the form of a system (4.117) of first order equations and solving the corresponding Fokker-Planck-Kolmogorov equation (1.119) with appropriate conditions.

The method of solving F-P-K equations have been briefly summarized in Sec. 6.3. Exhaustive presentation of problems and results associated with the F-P-K equation is given in the recent book of Risken [29].

In general, the solution process is characterized by the time-dependent (non-stationary) solution of the appropriate F-P-K. However, usually such nonstationary solutions can not be obtained in exact and explicit form. Exact solutions of the F-P-K equation associated with second order non-linear equation driven by white noise have been found only for the steady-state probability density.

To illustrate the procedure we shall consider one particular equation (describing a vibratory motion of non-linear oscillator)

$$\frac{d^2Y}{dt^2} + \beta\frac{dY}{dt} + f(Y) = \xi(t) \tag{4.143}$$

where $\xi(t)$ is a white noise with zero mean and the intensity equal to $2D$.

The above equation takes the form $(Y(t) = Y_1(t))$

$$\begin{aligned} \frac{dY_1}{dt} &= Y_2(t) \\ \frac{dY_2}{dt} &= -\beta Y_2(t) - f\left(Y_1(t)\right) + \xi(t) \end{aligned} \tag{4.144}$$

The coefficients of the F-P-K equation are

$$\begin{aligned} &a_1 = F_1 = y_2, \quad a_2 = F_2 = -\beta y_2 - f(y_1) \\ &b_{11} = b_{12} = b_{21} = 0, \quad b_{22} = 2D \end{aligned}$$

The F-P-K equation takes the form

$$\frac{\partial(y_1, y_2; t)}{\partial t} = -\frac{\partial}{\partial y_1}(y_2 p) + \frac{\partial}{\partial y_2}\left\{[\beta y_2 + f(y_1)]\, p\right\} + D\frac{\partial^2 p}{\partial y_2^2}$$

The stationary solution $p_{st}(y_1, y_2)$ satisfies the equation

$$D\frac{\partial^2 p_{st}}{\partial y_2^2} - \frac{\partial}{\partial y_1}(y_2 p_{st}) + \frac{\partial}{\partial y_2}\left\{[\beta y_2 + f(y_1)]\, p_{st}\right\} = 0$$

After rearranging the terms in the equation above, we find that it can be written in the form

$$\begin{aligned} &\frac{\partial}{\partial y_2}\left[f(y_1)p_{st} + \frac{D}{\beta}\frac{\partial p_{st}}{\partial y_1}\right] + \\ &+ \left(\beta\frac{\partial}{\partial y_2} - \frac{\partial}{\partial y_1}\right)\left[y_2 p_{st} + \frac{D}{\beta}\frac{\partial p_{st}}{\partial y_2}\right] = 0 \end{aligned}$$

This equation is satisfied if

$$f(y_1)p_{st} + \frac{D}{\beta}\frac{\partial p_{st}}{\partial y_1} = 0,$$

$$y_2 p_{st} + \frac{D}{\beta}\frac{\partial p_{st}}{\partial y_2} = 0$$

The above equations are easily integrated. The result is

$$p_{st}(y_1, y_2) = p(y, \dot{y}) =$$

$$= Cexp\left\{-\frac{\beta}{D}\left[\frac{y_2^2}{2} + \int_0^{y_1} f(x)dx\right]\right\} \tag{4.145}$$

where the constant C is determined by the normalization condition. For the well known Duffing equation

$$f(y) = \omega_0^2(y + \varepsilon y^3)$$

the normalization constant C is

$$C = \frac{\left(\frac{\varepsilon\beta}{D}\right)^{1/2} exp\left(-\frac{\beta\omega_0^2}{8\varepsilon D}\right)}{K_{\frac{1}{4}}\left(\frac{\beta\omega_0^2}{8\varepsilon D}\right)}$$

where K_ν is the Bessel function of the second order with imaginary argument.

It is seen that the displacement $Y(t)$ and the velocity $\dot{Y}(t)$ are statistically independent, i.e. $p_{st}(y, \dot{y}) = p_1(y)p_2(\dot{y})$. The velocity process $Y_2(t)$ *at an arbitrary* t is a Gaussian random variable, whereas the displacement process $Y_1(t)$ is not. However, $Y_2(t)$ can not be a Gaussian process because its integral $Y_1(t)$ is not.

It is worth noting that the term $\frac{y_2^2}{2}$ and $\int_0^{y_1} f(x)dx$ are the kinetic energy and the strain energy of the system. If we write

$$\mathcal{E} = \frac{y_2^2}{2} + \int_0^{y_1} f(x)dx \tag{4.146}$$

and solution (4.145) as

$$p_{st} = Cexp(-A\mathcal{E}) \tag{4.147}$$

OBwe see that in the stationary state the energy as a random process is exponentially distributed. Probability distribution (1.147) is known as the Maxwell-Boltzmann distribution and it describes the kinetic energy of the molecules of gas.

Since, in general, it is difficult to obtain exact solutions of the F-P-K equations associated with the second order non-linear systems under random excitation, a number of techniques have been developed to obtain the approximate solutions (cf. Risken [29], Caughey [162], Wen [255]).

27.1.3. Multidimensional systems

In general situation of multidimensional systems (4.117) the process in question takes on an essentially more complex behaviour than is possible in the case of one variable. This complexity manifests itself especially when we consider a system in the bounded spatial region. Boundaries are no longer simple end points of a line but curves and surfaces, and the nature of the boundary can be different at different segments.

As we know (cf. Sec. 6.4) the F-P-K equation (4.119) associated with system (4.117) can be represented in form of conservation equation (1.152) where the probability flux $\mathbf{\Phi}$ is defined by (1.153). Although the full variety of possible boundary conditions for an arbitrary multidimensional F-P-K equation does not seem to have been specified, usually such conditions are formulated in terms of the probability flux $\mathbf{\Phi}$. Most often we deal with reflecting boundary, what means that (at boundary)

$$\mathbf{n}\mathbf{\Phi} = 0 \tag{4.148}$$

where $\mathbf{n}$ is a normal vector to the surface, and with absorbing boundary, what means that

$$f(\mathbf{x}, t) = 0 \tag{4.149}$$

In certain situations, some part of the surface may be reflecting and another absorbing.

Exact solution for the vectorial non-stationary F-P-K is not knwon. Here, we confine our attention on equations characterizing the stationary states of process $\mathbf{Y}(t)$, i.e.

$$\sum_{i=1}^{n} \frac{\partial}{\partial y_i} [a_i(\mathbf{y}) p_{st}] - \frac{1}{2} \sum_{i,j=1}^{n} \frac{\partial^2}{\partial y_i \partial y_j} [b_{ij}(\mathbf{y}) p_{st}] = 0 \tag{4.150}$$

and on the boundary S at which the probability flux vanishes, namely

$$\Phi_i = 0 \quad \text{on} \quad S \tag{4.151}$$

where Φ_i is the component of vector $\mathbf{\Phi}$ in y_i-direction.

For one-dimensional diffusion processes, the probability flux is, of course, also one-dimensional and it must vanish everywhere if it does so at the boundaries. In multi-dimensional case the situation is more complex. The vanishing of the flux at boundary does not guarantee that the probability flux vanishes everywhere, since circulatory flows are possible.

Let us consider the situation when circulatory flows are absent and the probability flux vanishes for all y in the considered domain. It means that

$$\Phi_i = a_i(\mathbf{y})p_{st}(\mathbf{y}) - \frac{1}{2}\sum_j \frac{\partial}{\partial y_j}[b_{ij}(\mathbf{y})p_{st}(\mathbf{y})] = 0 \tag{4.152}$$

The above equation can be rearranged to the form

$$\frac{1}{2}\sum_j b_{ij}(\mathbf{y})\frac{\partial p_{st}(\mathbf{y})}{\partial y_j} = p_{st}(\mathbf{y})\left[a_i(\mathbf{y}) - \frac{1}{2}\sum_j \frac{\partial}{\partial y_j}b_{ij}(\mathbf{y})\right] \tag{4.153}$$

Let us assume that matrix $\mathbf{B} = \{b_{ij}\}$ is *non-singular* for all $\mathbf{y}$ and its inverse is $\mathbf{B}^{-1}(\mathbf{y})$. Then equation (4.153) is

$$\frac{\partial}{\partial y_i} ln\,[p_{st}(\mathbf{y})] = \sum_k \{B^{-1}(\mathbf{y})\}_{ik}\left[2a_k(\mathbf{y}) - \sum_j \frac{\partial}{\partial y_j}b_{kj}(\mathbf{y})\right] \equiv$$

$$\equiv \Psi_i(\mathbf{A}, \mathbf{B}, \mathbf{y}). \tag{4.154}$$

This equation can not be solved for arbitrary $b_{ij}(\mathbf{y})$ and $a_i(\mathbf{y})$ since the left hand side is a gradient. So, Ψ_i must also have the form of a gradient, and a necessary and sufficient condition for that is the vanishing of curl, i.e.

$$\frac{\partial \Psi_i}{\partial y_j} = \frac{\partial \Psi_j}{\partial y_i} \tag{4.155}$$

If this condition is satisfied, the stationary solution of equation (4.150) with conditions (4.151) is obtained by integration of (4.154):

$$p_{st}(\mathbf{y}) = exp\left[\int^{\mathbf{y}} \Psi(\mathbf{A}, \mathbf{B}, d\mathbf{y}')d\mathbf{y}'\right] = exp(-\mathbf{U}(\mathbf{y})) \tag{4.156}$$

where

$$U(\mathbf{y}) = \int^{\mathbf{y}} \Psi(\mathbf{A}, \mathbf{B}, \mathbf{y}')d\mathbf{y}' \tag{4.157}$$

Conditions (4.155) are known as the *potential conditions.*

In order to avoid an explicit use of the assumption about non-singularity of matrix $\mathbf{B}(\mathbf{y})$ we can argue as follows. Let us look for the solution in the form

$$p_{st}(\mathbf{y}) = Cexp(-U(\mathbf{y})) \tag{4.158}$$

By substituting (4.158) into (4.152) we obtain

$$2a_i(\mathbf{y}) - \sum_j \frac{\partial}{\partial y_j} b_{ij}(\mathbf{y}) + \sum_j b_{ij}(\mathbf{y}) \frac{\partial U}{\partial y_j} = 0 \tag{4.159}$$

for $i = 1, 2, \ldots, n$. Since the coefficients $a_i(\mathbf{y})$ and $b_{ij}(\mathbf{y})$ are given (in each specific problem) an appropriate function (the potential) $U(\mathbf{y})$ should be determined from (4.159). Solvability of system (4.159) depends on the form of $a_i(\mathbf{y})$ and $b_{ij}(\mathbf{y})$. If there exists such function $U(\mathbf{y})$ which satisfies all n equations (4.159) then we say that the problem belongs to the class of *stationary potential.*

In general, circularity probability flows exist in multidimensional phase space (when the appropriate $U(\mathbf{y})$ is not obtainable from (4.159)). To treat such situations, it has been proposed (cf. Graham, Haken [178], Lin [204]) to split the drift coefficients into two parts:

$$a_i(\mathbf{y}) = a_i^{(1)} + a_i^{(2)} \tag{4.160}$$

where $a_i^{(1)}(\mathbf{y})$ should satisfy equations (4.159), that is

$$2a_i^{(1)}(\mathbf{y}) - \sum_j \frac{\partial}{\partial y_j} b_{ij}(\mathbf{y}) + \sum_j b_{ij}(\mathbf{y}) \frac{\partial U}{\partial y_j} = 0 \tag{4.161}$$

and $a_i^{(2)}(\mathbf{y})$ is responsible for the circularity flows.

By substituting (4.158), (4.159) and (4.161) into equation (4.150), we obtain

$$\sum_i \frac{\partial a_i^{(2)}(\mathbf{y})}{\partial y_i} - \sum_i a_i^{(2)}(\mathbf{y}) \frac{\partial U(\mathbf{y})}{\partial y_i} = 0 \tag{4.162}$$

If function $U(\mathbf{y})$ can be obtained from (4.161), (4.162) then we say that the problem belongs to the class of *generalized stationary potentials* (cf. [205]). It is worth noting that conditions stated above have been used (Lin,Cai [205]) for constructing the *equivalent stochastic systems* in a sense that they have different F-P-K equations but they share an identical form of the stationary probability density.

The fact that stationary solutions of certain F-P-K equations correspond to a vanishing probability flux is a manifestation of more general phenomenon called *detailed balance* (cf. [29]). A given system of stochastic equations having a stationary solution, may or may not satisfy the conditions for detailed balance. However, if the conditions for detailed balance are satisfied, then they can be used to obtain the stationary solution, that is the solution of the reduced F-P-K equation (4.150). For detailed discussion of this idea with its illustrations by several examples the reader is referred to the papers: Yong, Lin [258] and Cai, Lin [159], Lin, Cai [205].

27.1.4. Equations for moments

Because of great difficulties in obtaining nonstationary solutions of F-P-K equations it has always been interesting to determine the time-dependent moments of the solution process.

The differential equations for moments can be obtained by two ways. One way is to use the F-P-K equation. Namely, if we denote:

$$h(\mathbf{Y}) = Y_1^{k_1}(t)Y_2^{k_2}(t)\dots Y_n^{k_n}(t) \tag{4.163}$$

then

$$\begin{aligned} m_{k_1k_2\dots k_n}(t) &= E\left[Y_1^{k_1}(t)Y_2^{k_2}\dots Y_n^{k_n}(t)\right] = \\ &= \int_{-\infty}^{+\infty}\dots\int_{-\infty}^{+\infty} h(\mathbf{y})p(\mathbf{y},t;\mathbf{y}_o,t_o)f(\mathbf{y}_o,t_o)d\mathbf{y}d\mathbf{y}_o \end{aligned} \tag{4.164}$$

The time derivative of (4.164) gives

$$\begin{aligned} \frac{dm_{k_1k_2\dots k_n}}{dt} &= \\ &= \int_{-\infty}^{+\infty}\dots\int_{-\infty}^{+\infty} h(\mathbf{y})\frac{\partial p(\mathbf{y},t;\mathbf{y}_o,t_o)}{\partial t}f(\mathbf{y}_o,t_o)d\mathbf{y}d\mathbf{y}_o \end{aligned} \tag{4.165}$$

If now $\frac{\partial p}{\partial t}$ in (4.165) is taken from the F-P-K equation (4.119) and the integration is performed by parts over entire state space ($-\infty < y_i, y_{oj} < +\infty$), the resulting equation is the first order ordinary differential equation containing the moments of the solution process; for different values of $k_1, k_2, \dots, k_n$, a system of first order differential equations is obtained.

Another — most natural way of deriving equations for moments is based on the usage of the Itô formula to the function $h(\mathbf{y})$ and taking the average. Namely, according to (2.106)

$$\begin{aligned} \frac{d}{dt}E\left[h(\mathbf{Y})\right] &= \sum_j E\left[a_i(\mathbf{Y})\frac{\partial h(\mathbf{Y})}{\partial Y_i}\right] \\ &+ \frac{1}{2}\sum_l\sum_{i,j} E\left[b_{il}(\mathbf{Y})b_{jl}(\mathbf{Y})\frac{\partial^2 h(\mathbf{Y})}{\partial y_i \partial y_j}\right] \end{aligned} \tag{4.166}$$

Equations for moments are obtained directly by letting $h(\mathbf{Y})$ be as in (4.163).

Let us consider an illustrative example ($n = 1$)

$$\frac{dY(t)}{dt} + Y(t) + \mu Y^3(t) = \xi(t) \tag{4.167}$$

or, in the Itô interpretation

$$dY(t) = -\left[Y(t) + \mu Y^3(t)\right] dt + dW(t)$$

Let $h(Y) = Y^k$. We have $k_1 = k,\ k_2 = k_3 = \cdots = k_n = 0$

$$a_1 = -(Y + \mu y^3), \quad b = 1,$$
$$\frac{\partial h}{\partial Y} = kY^{k-1}, \quad \frac{\partial^2 h}{\partial Y^2} = k(k-1)Y^{k-2}$$

Substitution of these quantities into equation (4.166) yields

$$\frac{dm_k(t)}{dt} = -km_k(t) - k\mu m_{k+2} + \frac{1}{2}k(k-1)m_{k-2} \tag{4.168}$$

$k = 1, 2, \ldots$. For example

$$\begin{aligned} \frac{dm_1(t)}{dt} &= -m_1(t) - \mu m_3(t) \\ \frac{dm_2(t)}{dt} &= -2m_2(t) - 2\mu m_4(t) + 1 \end{aligned} \tag{4.169}$$

Let us return now to the second order equation (4.143). Application of equation (4.166) to this system results in ($n = 2,\ m_1(t) = E\left[Y_1(t)\right],\ m_{i,k}(t) = E\left[Y_i Y_k\right]$)

$$\begin{aligned} \frac{dm_1(t)}{dt} &= m_2(t) \\ \frac{dm_2(t)}{dt} &= -\beta m_2(t) - E\left[f_1(Y_1)\right] \\ \frac{dm_{1,1}(t)}{dt} &= 2m_{1,2}(t) \\ \frac{dm_{1,2}(t)}{dt} &= -\beta m_{1,2}(t) - E\left[Y_1 f(Y_1)\right] + m_{2,2}(t) \\ \frac{dm_{2,2}(t)}{dt} &= -2\beta m_{2,2}(t) - 2E\left[Y_2 f(Y_1)\right] + 2D \\ &\vdots \end{aligned} \tag{4.170}$$

It is seen from (4.168) that the equation for $m_k(t)$ contains moments of higher order than k. Similarly, since $f(Y_1)$ is non-linear, system (4.170) is not closed in the sense that equation for a lower moment may contain terms of higher order moments. In general, the system of non-linear Itô stochastic equations generates an *infinite hierarchy of moment equations.*

In order to obtain a closed form of moment equations some *closure approximations* have to be introduced to truncate the hierarchy (usually we are interested only in some lower order moments); commonly we wish to express the higher order moments by the moments of lower orders. Since many years such closure procedures have been widely used in turbulence theory (cf. [202]). More recently they attracted an interest of researchers in control and vibratory systems (cf. [149],[170],[189],[257]).

One of the simplest closure approximations is the *Gaussian closure* scheme according to which higher order moments are expressed in terms of the first and second moments as if the random process involved were normally distributed. Although it has been used in numerous problems, it was found to be unsuitable in some important cases (e.g. it can lead to unacceptable errors in the second order moments when the non-linearity is not small (cf. [149]).

A better scheme is to neglect the third and fourth order central moments. Since

$$\begin{aligned} \mu_3 &= m_3 - 3m_1 m_2 - 2m_1^3 \\ \mu_4 &= m_4 - 4m_1 m_3 + 6m_1^2 m_2 - 3m_1^4 \end{aligned} \tag{4.171}$$

then assuming $\mu_3 = \mu_4 = 0$ we can determine the moments m_3 and m_4 as functions of m_1 and m_2.

Another generalization of the Gaussian closure is focussed on the properties of cumulants or semi-invariants. As it is known (cf. formula (1.11)) they can be expressed in terms of moments. Since the third and higher order cumulants (including joint cumulants for multi-dimensional random variables) of Gaussian distribution are zero, the Gaussian closure scheme is, therefore, equivalent to neglecting the cumulants above the second order.

Thus, remaining in the moment equations all those which correspond to non-zero higher order cumulants (third, fourth, etc.) we obtain a group of closure schemes which are called the *cumulant-neglect* hypotheses. A systematic and detailed discussion of this procedure along with interesting examples showing versatility of the scheme has been presented by Wu and Lin [257].

The moments and cumulants of random variables are related to each other as follows (cf. [35],[257]).

$$
\begin{aligned}
&E[X] = m_1 = \kappa_1(X),\\
&E[X_jX_k] = m_{j,k} = \kappa_2(X_j,X_k) + \kappa_1(X_j)\kappa_2(X_k),\\
&E[X_jX_kX_l] = m_{j,k,l}\\
&= \kappa_3(X_j,X_k,X_l)\\
&+ 3\{\kappa_1(X_j)\kappa_2(X_k,X_l)\}_S\\
&+ \kappa_1(X_j)\kappa_1(X_k)\kappa_1(X_l),\\
&E[X_jX_kX_lX_m] = m_{j,k,l,m}\\
&= \kappa_4(X_j,X_k,X_l,X_m)\\
&+ 3\{\kappa_2(X_j,X_k)\kappa_2(X_l,X_m)\}_S\\
&+ 4\{\kappa_1(X_j)\kappa_3(X_k,X_l,X_m)\}_S\\
&+ 6\{\kappa_1(X_j)\kappa_1(X_k)\kappa_2(X_l,X_m)\}_S\\
&+ \kappa_1(X_j)\kappa_1(X_k)\kappa_1(X_l)\kappa_1(X_m)
\end{aligned}
\tag{4.172}
$$

where $\{\ \}_S$ indicates a symmetrizing operation with respect to all its arguments; namely, taking the arithmetic mean of different permuted terms similar to the one within the braces. For example,

$$
\begin{aligned}
\{\kappa_1(X_j)\kappa_2(X_k,X_l)\}_S = \frac{1}{3}\Big\{&\kappa_1(X_l)\kappa_2(X_j,X_k)+\\
&+\kappa_1(X_j)\kappa_2(X_k,X_l)+\\
&+\kappa_1(X_k)\kappa_2(X_j,X_l)\Big\}.
\end{aligned}
\tag{4.173}
$$

It is seen that if the first order cumulants κ_1 are zero, then formulae (4.172) become much simpler.

One of the examples elaborated by Wu and Lin [257] concerns the second order equation (4.143). Since in this case we are able to make comparison with solution (for the stationary moments) it will be instructive to report here the results.

Let in equation (4.143) $\beta = 1$, $D = 1$ and $f(y) = y + \varepsilon y^3$ where ε is the parameter representing the degree of non-linearity. So, we have the equation

$$
\frac{d^2Y}{dt^2} + \frac{dY}{dt} + (Y + \varepsilon Y^3) = \xi(t)
\tag{4.143'}
$$

Making use of the general formula (4.145) for the solution we obtain

$$
p_{st}(y_1,y_2) = C exp\left[-\frac{y_2^2}{2} - \left(\frac{y_1^2}{2} + \varepsilon\frac{y_1^4}{4}\right)\right]
\tag{4.145'}
$$

The mean-square of the stationary solution $Y(t) = Y_1(t)$ is [257]:

$$E(Y^2) = C\int_{-\infty}^{+\infty} y_1^2 exp\left[-\left(\frac{y_1^2}{2}+\varepsilon\frac{y_1^4}{4}\right)\right] dy_1 =$$

$$= C_1\left(\frac{\varepsilon}{2}\right)^{-\frac{3}{4}} exp\left(\frac{1}{8\varepsilon}\right) D_{-\frac{3}{2}}\left(\frac{1}{\sqrt{2\varepsilon}}\right) \tag{4.174}$$

where $D_{-\frac{3}{2}}$ is a parabolic cylinder function [144].

For Gaussian closure (neglecting the cumulants of third and higher order) only the group of equations (4.170) is required (in which $f(Y_1) = Y_1 + \varepsilon Y_1^3$). For stationary solution we let the time derivative be zero and we obtain a system of algebraic equations. Letting $\kappa_k = 0$ for $k \geq 3$ and truncate the hierarchy consistently ($E(Y_1^3) = 0$, $E(Y_1^4) = 3\left[E(Y_1^2)\right]^2$, $E(Y_1^3 Y_2) = 3E(Y_1^2)E(Y_1 Y_2)$) one obtains the following Gaussian closure result for the mean-square

$$E(Y^2) = \frac{-1+\sqrt{1+12\varepsilon}}{6\varepsilon} \tag{4.175}$$

Additionally, we obtain the exact values: $E(Y_2^2) = 1$, $E(Y_1 Y_2) = 0$.

It has been shown [257] that the fourth order cumulant neglect approximation ($\kappa_k = 0$ $k \geq 4$) gives the following algebraic equation for $\zeta = E(Y^2)$:

$$30\zeta^4 + 15\varepsilon\zeta^2 + (1-12\varepsilon)\zeta - 1 = 0 \tag{4.176}$$

whereas the sixth order cumulant neglect approximation ($\kappa_k = 0$, $k \geq 6$) yields

$$714\varepsilon^3\zeta^4 + 420\varepsilon^2\zeta^3 + (63-336\varepsilon)\varepsilon\zeta^2 + (1-90\varepsilon)\zeta - (1-30\varepsilon) = 0 \tag{4.177}$$

Of course, it has been of interest to compare the approximate results (4.175), (4.176), (4.177) with exact expression (4.174). For a *small* ε one obtains:
exact solution:

$$E(Y^2) \approx 1 - 3\varepsilon + 24\varepsilon^2 - 297\varepsilon^3 + 4896\varepsilon^4 - 100278\varepsilon^5 + \ldots,$$

Gaussian closure:

$$E(Y^2) \approx 1 - 3\varepsilon + 18\varepsilon^2 - \ldots,$$

fourth order cumulant neglect:

$$E(Y^2) \approx 1 - 3\varepsilon + 24\varepsilon^2 - 297\varepsilon^3 + 4536\varepsilon^4 - \ldots,$$

sixth order cumulant neglect:

$$E(Y^2) \approx 1 - 3\varepsilon + 24\varepsilon^2 - 297\varepsilon^3 + 4896\varepsilon^4 - 100278\varepsilon^5 + \ldots$$

For a *large* ε the asymptotic result is:
exact solution: $E(Y^2) \approx 0.6760\varepsilon^{-\frac{1}{2}}$
Gaussian closure: $E(Y^2) \approx 0.5774\varepsilon^{-\frac{1}{2}}$
sixth order cumulant neglect: $E(Y^2) \approx 0.6480\varepsilon^{-\frac{1}{2}}$

Other non-Gaussian closure approximations the reader will find in papers by Crandall [169][170] and Hampl and Schuëller [181].

When the nonstationary solution is required, the time derivatives in moment equations have to be retained. Application of a cumulant-neglect closure to the right-hand sides results in non-linear terms and, in general, the problem needs numerical integration to obtain the solution for moments.

It is clear that all the closure approximations described above have not sound mathematical basis. Only a careful physical justification in specific problems can make them credible. A kind of a necessary condition for applicability a specific closure approximation is preservation of the basic moment properties (e.g. the variance of the solution should be positive), cf. [152].

27.2. Real random excitation

27.2.1. Extension of state space

As it has been indicated in Sec. 22.5 the methods based on the Itô stochastic equations (or in other words, the methods of diffusion Markov process theory) can be extended to the important case where the excitation process is not white noise but it can be obtained as a response of a dynamical system to a white noise excitation.

To illustrate this approach let us consider a second order equation (everywhere in sequel, $\frac{dY}{dt} = \dot{Y}$)

$$\frac{d^2Y}{dt^2} = F\left[Y(t), \dot{Y}(t), t\right] + X(t) \tag{4.178}$$

where $X(t)$ is a real (non-white) stochastic process. However, if $X(t)$ is a diffusion Markov process it can be represented as a solution of the following "equation"

$$\frac{dX}{dt} = m\left[X(t), t\right] + \sigma\left[X(t), t\right]\xi(t) \tag{4.179}$$

It is clear that the vector process $[Y_1(t), Y_2(t), Y_3(t)] = \left[Y(t), \dot{Y}(t), X(t)\right]$ satisfies the system of Itô equations

$$\begin{aligned} dY_1(t) &= Y_2(t)dt \\ dY_2(t) &= F\left[Y_1(t), Y_2(t), t\right]dt + Y_3(t)dt \\ dY_3(t) &= m\left[Y_3(t), t\right]dt + \sigma\left[Y_3(t), t\right]dW(t) \end{aligned} \tag{4.180}$$

There is an important special case when the process $X(t)$ is a solution of a linear equation with white noise excitation. It can easily be shown (cf. [313],[339]) that an arbitrary Gaussian and stationary random process with rational spectral density can be obtained as an output of a time-invariant linear system (filter), with white noise as the input. In other words, an arbitrary Gaussian stationary process with rational spectral density is a component of a multi-dimensional diffusion Markov process governed by a set of linear stochastic equations.

Examples.

1. Given a Gaussian stationary process $X(t)$ with zero mean and with the correlation function

$$K_X(\tau) = \sigma^2 e^{-\alpha|\tau|}\left(\cos\beta\tau + \frac{\alpha}{\beta}\sin\beta|\tau|\right)$$

It can be verified that this process has the following spectral density

$$g_X(\omega) = \frac{2\alpha\sigma^2(\alpha^2+\beta^2)}{\pi}\,\frac{1}{|-\omega^2+2i\alpha\omega+\alpha^2+\beta^2|^2}$$

It is clear (from the spectral method) that the process $X(t)$ is a stationary solution of the equation

$$\frac{d^2X}{dt^2} + 2\alpha\frac{dX}{dt} + (\alpha^2+\beta^2)X = 2\sigma\sqrt{\alpha(\alpha^2+\beta^2)}\,\xi(t)$$

where $\xi(t)$ is a white noise with intensity $\frac{1}{2\pi}$. The equation for the process $X(t)$ is of second order, so the process $X(t)$ itself is not a Markov process (its future behaviour does not depend solely on the value $X(t_0)$ but also on the initial value of the derivative $\dot{X}(t_0)$). A two-dimensional process $[X_1(t), X_2(t)]$ where $X_2(t) = \dot{X}(t)$ is a Markov process and it is governed by the equations

$$\frac{dX_1}{dt} = X_2(t)$$
$$\frac{dX_2}{dt} = -(\alpha^2+\beta^2)X_1 - 2\alpha X_2 + 2\sigma\sqrt{\alpha(\alpha^2+\beta^2)}\,\xi(t)$$

where $\xi(t)$ is a white noise. The drift and diffusion coefficients of the diffusion process $[X_1(t), X_2(t)]$ are

$$a_1(x_1,x_2,t) = x_2, \quad a_2(x_1,x_2,t) = -(\alpha^2+\beta^2)x_1 - 2\alpha x_2$$
$$b_{11} = b_{12} = b_{21} = 0, \quad b_{22} = \frac{2}{\pi}\sigma^2\alpha(\alpha^2+\beta^2)$$

-The F-P-K equation for the probability density $f(x_1,x_2;t)$ the of process $[X_1(t),$ $X_2(t)]$ is

$$\frac{\partial f}{\partial t} + \frac{\partial}{\partial x_1}(x_2 f) - \frac{\partial}{\partial x_2}\left\{\left[(\alpha^2+\beta^2)x_1 + 2\alpha x_2\right]f\right\} - \frac{1}{\pi}\sigma^2\alpha(\alpha^2+\beta^2)\frac{\partial^2 f}{\partial x_2^2} = 0.$$

2. Given stochastic equation

$$\frac{dY}{dt} + aY^2(t) = kX^2(t)$$

where $X(t)$ is a Gaussian stationary process such that

$$E[X(t)] = 0,$$
$$K_X(\tau) = \sigma_X^2 e^{-\alpha|\tau|}$$

The spectral density of $X(t)$ is

$$g_X(\omega) = \frac{\sigma_X^2 \alpha}{\pi(\omega^2 + \alpha^2)}$$

therefore, the process $X(t)$ is governed by the "equation"

$$\frac{dX}{dt} + \alpha X(t) = \sigma_X \sqrt{2\alpha}\xi(t)$$

where $K_\xi(\tau) = \delta(\tau)$. The process $[Y(t), X(t)]$ is governed by the system of Itô equations

$$dY(t) = \left[-aY^2(t) + kX^2(t)\right] dt$$
$$dX(t) = -\alpha X(t)dt + \sigma_X \sqrt{2\alpha} dW$$

The drift and diffusion coefficients of $[Y(t), X(t)]$ are:

$$a_1(y, x; t) = -ay^2 + kx^2, \quad a_2(y, x; t) = -\alpha x,$$
$$b_{11} = b_{12} = b_{21} = 0, \quad b_{22} = 2\alpha\sigma_X^2.$$

27.2.2. Stochastic averaging method

In sec. 21.4. we have presented the mathematical results associated with the averaging principle and Khasminskii limit theorem for stochastic non-linear systems. These results constitute a ground for an effective method of analysis of a wide class of stochastic equations occuring in practice; this method is known as the *stochastic averaging method*. In order to indicate the essence of stochastic averaging we shall consider two examples. To illustrate the averaging principle described in Sec. 21.4.2. let us consider a first order system of the form (3.32), i.e.

$$\frac{d\mathbf{Y}^\epsilon(t)}{dt} = \mathbf{F}\left[\mathbf{Y}^\epsilon(t), \mathbf{X}\left(\frac{t}{\varepsilon}\right)\right],$$
$$\mathbf{Y}^\epsilon(0) = \mathbf{Y}_0 \tag{4.181}$$

in which the right-hand side *does not depend on the process* $\mathbf{Y}(t)$, and $\mathbf{X}(t)$ is a stationary process; ε is a small parameter. In this particular case process $\mathbf{Y}^\varepsilon(t)$ can be represented as

$$\mathbf{Y}^\varepsilon(t) = \mathbf{Y}_0 + \int_0^t \mathbf{F}\left[\mathbf{X}\left(\frac{s}{\varepsilon}\right)\right] ds = \\ = \mathbf{Y}_0 + t\left(\frac{t}{\varepsilon}\right)^{-1} \int_0^{\frac{t}{\varepsilon}} \mathbf{F}(\mathbf{X}(s))\, ds$$

Let us denote

$$\mathbf{m} = E\left[\mathbf{F}(\mathbf{X}(s))\right] \\ K_{ij}(\tau) = E\left\{[F_i(\mathbf{X}(s+\tau)) - m_i]\,[F_j(\mathbf{X}(s)) - m_j]\right\}$$

and assume that

$$\sum_{i=1}^{n} K_{ii}(\tau) \longrightarrow 0 \quad \text{as } \tau \to \infty$$

By virtue of the Tchebyshev inequality, for each $\delta > 0$ we have (cf. theorem 3.7 and condition (3.35))

$$P\left\{\left|\frac{1}{T}\int_t^{t+T} \mathbf{F}(\mathbf{X}(s))\, ds - \mathbf{m}\right| > \delta\right\} \leq \\ \leq \frac{1}{T^2\delta^2}\int_t^{t+T}\int_t^{t+T}\sum_{i=1}^{n} E\left\{[F_i(\mathbf{X}(s)) - m_i]\,[F_j(\mathbf{X}(u)) - m_j]\right\} ds du = \\ = \frac{1}{T^2\delta^2}\int_t^{t+T}\int_t^{t+T}\sum_{i=1}^{n} K_{ij}(u-s) du ds \to 0, \text{ as } T \to \infty$$

Therefore, according to theorem 3.7

$$\lim_{\varepsilon\to 0} P\left\{\sup_{0\leq t\leq T} |\mathbf{Y}^\varepsilon(t) - \mathbf{Y}_1(t)| > \delta\right\} = 0$$

where $\mathbf{Y}_1(t)$ is defined as

$$\frac{d\mathbf{Y}_1(t)}{dt} = \mathbf{m}, \\ \mathbf{Y}_1(t) = \mathbf{Y}_0 + \mathbf{m}t$$

If the process $\mathbf{X}(t)$ satisfies the strong mixing condition and $\mathbf{F}(\mathbf{x})$ is bounded, then the family of processes

$$\zeta^\varepsilon(t) = \frac{1}{\sqrt{\varepsilon}}\left[\mathbf{Y}^\varepsilon(t) - \mathbf{m}t - \mathbf{Y}_0\right] = \sqrt{\varepsilon}\int_0^{\frac{t}{\varepsilon}} \left[\mathbf{F}(\mathbf{X}(s)) - \mathbf{m}\right] ds$$

converges weakly (as $\varepsilon \to 0$) to Gaussian process $\zeta^{\circ}(t)$, cf. [136].

In order to illustrate the application of Khasminskii theorem (cf. eqs (3.39)–(3.40) and (3.41)–(3.42) let us consider the equation

$$\frac{d^2Y}{dt^2} + \omega_o^2 Y + \varepsilon F\left(Y, \dot{Y}, X(t,\gamma)\right) = 0 \tag{4.182}$$

where $F(y, \dot{y}, x)$ is sufficiently smooth, bounded function and the parameter ε is used to quantify the strength of the non-linear term. This equation can be written in the form of a system of equations for the displacement Y and velocity $\dot{Y}$. However, these variables are usually rapidly flucuating in time and we cannot apply directly the Khasminskii theorem. The appropriate transformation of variables has to be introduced to bring the equation in question to the form required by the theorem. The transformation which is commonly adopted is as follows

$$Y(t) = A(t)\cos\left[\omega_o t + \Psi(t)\right] \tag{4.183}$$

$$\dot{Y}(t) = -A(t)\omega_o \sin\left[\omega_o t + \Psi(t)\right] \tag{4.184}$$

Here the amplitude envelope process $A(t)$ and the phase process $\Psi(t)$ are slowly varying with respect to time when ε is small.

Differentiation of expression (4.183) with respect to time yields

$$\dot{Y}(t) = \dot{A}(t)\cos\Phi(t) - A(t)\left\{\omega_o + \dot{\Psi}(t)\right\}\sin\Phi(t) \tag{4.185}$$

where

$$\Phi = \omega_o t + \Psi \tag{4.186}$$

Equating (4.184) and (4.185) gives

$$\dot{A}(t)\cos\Phi - A(t)\dot{\Psi}(t)\sin\Phi(t) = 0 \tag{4.187}$$

Differentiation (4.184) with respect to t, yields

$$\begin{aligned} \ddot{Y}(t) &= -\omega_o^2 A(t)\cos\Phi \\ &- \omega_o\dot{\Psi}(t)A(t)\cos\Phi(t) - \omega_o\dot{A}(t)\sin\Phi(t) \end{aligned} \tag{4.188}$$

Substituting (4.188) into equation (4.182) and using (4.183), (4.184) leads to the result

$$\begin{aligned} &\omega_o\dot{A}\sin\Phi + \omega_o\dot{\Psi}A\cos\Phi = \\ &= \varepsilon F\left[A\cos\Phi, -\omega_o A\sin\Phi, X(t,\gamma)\right] \end{aligned} \tag{4.189}$$

Solving (4.187) and (4.189) for $\dot{A}, \dot{\Psi}$ results in the system of (exact) equations

$$\text{(4.190)} \qquad \begin{aligned} \frac{dA}{dt} &= \frac{\varepsilon}{\omega_o} F\left[A\cos\Phi, -A\omega_o \sin\Phi, X(t,\gamma)\right] \sin\Phi(t) \\ \frac{d\Psi}{dt} &= \frac{\varepsilon}{\omega_o A} F\left[A\cos\Phi, -A\omega_o \sin\Phi, X(t,\gamma)\right] \cos\Phi(t) \end{aligned}$$

The above pair of equations is often referred to as the "standard form" and it is equivalent to the equation (4.182). It is seen that both $\dot{A}(t)$ and $\dot{\Psi}(t)$ are of order ε.

To make further analysis more expedient, let us consider a specific form of equation (4.182), namely (cf. [224])

$$\text{(4.182')} \qquad \frac{d^2Y}{dt^2} + \varepsilon^2 h(Y, \dot{Y}) + \omega_o^2 Y = \varepsilon X(t,\gamma)$$

In this case, equations (4.190) are

$$\text{(4.190')} \qquad \begin{aligned} \frac{dA}{dt} &= \frac{\varepsilon^2}{\omega_o^2} h\left[A\cos\Phi, -A\omega_o \sin\Phi\right] \sin\Phi - \frac{\varepsilon}{A\omega_o} X(t,\gamma) \sin\Phi \\ \frac{d\Psi}{dt} &= \frac{\varepsilon^2}{A\omega_o} h\left[A\cos\Phi, -A\omega_o \sin\Phi\right] \cos\Phi - \frac{\varepsilon}{A\omega_o} X(t,\gamma) \cos\Phi \end{aligned}$$

The above equations have the form corresponding to general equation (3.39), (3.40) in the formulation of the Khasminskii theorem. Assuming that function $h(y, \dot{y})$ and process $X(t,\gamma)$ satisfy all the conditions of the theorem we assert that process $[A(t), \Psi(t)]$ converges weakly, as $\varepsilon \to 0$, to a Markov diffusion process $[A^{\circ}(t), \Psi^{\circ}(t)]$ with drift vector and diffusion matrix determined (as time averages) according to formulae (3.41), (3.42).

Let us assume that stochastic process $X(t,\gamma)$ is a broad band stationary stochastic process with zero mean and with spectral density $g(\omega)$. Due to the periodic nature the term involved, one obtains the following expressions for components of the drift and diffusion of the limiting process:

$$\text{(4.191)} \qquad \begin{aligned} &a_A = \frac{-1}{\omega_o} f_1(a) + \frac{\pi g_o(\omega_o)}{2a\omega_o^2}, \quad a_\Psi = \frac{-1}{a\omega_o} f_2(a), \\ &b_{A,A} = \frac{\pi g(\omega_o)}{\omega_o^2}, \quad b_{A,\Psi} = b_{\Psi,A} = 0, \quad b_{\Psi,\Psi} = \frac{\pi g(\omega_o)}{\omega_o^2 a^2} \end{aligned}$$

where

$$\text{(4.192)} \qquad \begin{aligned} f_1(a) &= -\frac{1}{2\pi}\int_0^{2\pi} h\left[a\cos\Phi, -a\omega_o \sin\Phi\right] \sin\Phi d\Phi \\ f_2(a) &= -\frac{1}{2\pi}\int_0^{2\pi} h\left[a\cos\Phi, -a\omega_o \sin\Phi\right] \cos\Phi d\Phi \end{aligned}$$

Therefore, the transition density function, $p(a, \psi, t)$ of the limiting Markov process is governed by the F-P-K equation

$$\frac{\partial p}{\partial t} = \frac{\partial}{\partial a}\left[\left(\frac{f_1(a)}{\omega_o} - \frac{\pi g(\omega_o)}{2a\omega_o^2}\right)p\right] + \frac{f_2(a)}{a\omega_o}\frac{\partial p}{\partial \psi} + \frac{\pi g(\omega_o)}{2\omega_o^2}\frac{\partial^2 p}{\partial a^2} + \frac{\pi g(\omega_o)}{2\omega_o^2 a^2}\frac{\partial^2 p}{\partial \psi^2} \tag{4.193}$$

An inspection of formulae (4.191) indicates that the limiting process A^o is uncoupled from the phase process Ψ^o (coefficients a_A and $b_{A,A}$ do not depend on ψ). Therefore, the transition density function of A^o is governed by the following F-P-K equation

$$\frac{\partial \overline{p}}{\partial t} = \frac{\partial}{\partial a}\left[\left(\frac{f_1(a)}{\omega_o} - \frac{\pi g(\omega_o)}{2a\omega_o^2}\right)\overline{p}\right] + \frac{\pi g(\omega_o)}{2\omega_o^2}\frac{\partial^2 \overline{p}}{\partial a^2} \tag{4.194}$$

Equations (4.193) and (4.194) should be solved subject to the following initial conditions:

$$p(a, \psi, t|a_o, \psi_o, t_o) = \delta(a - a_o)\delta(\psi - \psi_o), \quad \text{for} \quad t = t_o, \tag{4.195}$$

$$\overline{p}(a, t|a_o, t_o) = \delta(a - a_o) \quad \text{for} \quad t = t_o. \tag{4.196}$$

It is worth noting that if $h(y, \dot{y})$ is a function of y only, then, from equations (4.192) it follows that $f_1(a) = 0$.
Similarly, if $h(y, \dot{y}) = h(\dot{y})$, then $f_2(a) = 0$.

The stationary solution of equation (4.194) (when $t \to \infty$ and $\frac{\partial \overline{p}}{\partial t} \to 0$) is found as a solution of the ordinary differential equation and it has the form

$$\overline{p}_{st}(a) = Ca\, exp\left\{-\frac{2\omega_o}{\pi g(\omega_o)}\int_o^a f_1(\xi)d\xi\right\} \tag{4.197}$$

where C is a normalisation constant. In the *linear case* ($h(y, \dot{y}) = 2\zeta\omega_o\dot{y}$) formula (4.197) gives

$$\overline{p}_{st}(a) = \frac{a}{\sigma^2} exp\left\{-\frac{a^2}{2\sigma^2}\right\}, \quad \sigma^2 = \frac{\pi g(\omega_o)}{2\zeta\omega_o^3}. \tag{4.198}$$

In the special case of white noise excitation, where $g(\omega) = g_o$ =constant, the above formula agrees with known exact result.

In the linear case one also can find a solution of complete equation (4.194) — cf. Spanos, Solomos [242]. The result is

$$\overline{p}(a, t|a_o, t_o) = \frac{a}{c(t_o, t)} exp\left\{-\frac{a^2 + a_o^2 e^{-2\zeta\omega_o t}}{2c(t_o, t)}\right\} I_o\left\{\frac{a a_o e^{-\zeta\omega_o t}}{c(t_o, t)}\right\} \tag{4.199}$$

where

$$c(t_o, t) =$$

$$= \frac{\pi}{\omega_o^2} exp(-2\zeta\omega_o t) \int_{t_o}^{t} exp(2\zeta\omega_o \tau')g(\tau')d\tau' \tag{4.200}$$

and I_o denotes the Bessel function of zero order, $\tau = t - t_o$.

It is clear that in more complicated situations (higher-dimensional systems of equations) the analytical solutions of the F-P-K equations for the limiting process are usually not obtainable. Hence, the approximate methods have to be used even for obtaining the stationary solutions. The moment equations which can be derived from the F-P-K equations for amplitude and phase processes are, however, not amenable to such approximations as closure procedures because they include non-polynomial terms. To overcome this difficulty the polar coordinate transformation (4.183), (4.184) can be replaced by an equivalent transformation to complex random processes (cf. [156]). Namely, in the case considered above this transformation is as follows

$$\begin{aligned} Y(t) &= Z(t)e^{i\omega t} + \overline{Z}(t)e^{-i\omega t} \\ \dot{Y}(t) &= i\omega\left[Z(t)e^{i\omega t} - \overline{Z}(t)e^{-i\omega t}\right] \end{aligned} \tag{4.201}$$

where $Z(t)$ is a complex-valued random process and $\overline{Z}(t)$ is its complex conjugate. It is seen that $Z(t)$ and $\overline{Z}(t)$ are two different linear combinations of $Y(t)$ and $\dot{Y}(t)$. Processes $Z(t)$ and $\overline{Z}(t)$ vary much more slowly in time than $Y(t)$, $\dot{Y}(t)$.

$$\begin{aligned} |Z(t)|^2 &= \frac{1}{4}A^2(t) \\ ReZ(t) &= \frac{1}{2}A(t)\cos\Psi(t), \\ ImZ(t) &= \frac{1}{2}A(t)\sin\Psi(t) \end{aligned} \tag{4.202}$$

27.2.3. Statistical linearization technique

A.

In various engineering applications some approximate techniques have been elaborated with desire to study dynamics of non-linear systems with random inputs. One of such procedures which has gained an exceptional popularity in practice is the *statistical* or stochastic linerization technique.

The essence of statistical linerization is that a given non-linear system with random excitation is replaced by an "equivalent" linear system which should describe approximately (in some sense and to a certain extend) the behaviour of the non-linear system under consideration. This idea was developed first by Booton (1954) and Kazakov (1956) in their study of non-linear control systems with random inputs. The procedure has further been developed due to problems in non-linear random vibrations of mechanical and structural systems (cf. Caughey [161], Wen [256], Spanos [241], Casciati and Faravelli [160]). Now, the statistical linearization technique is well described in the literature, including many books. An expertized review of the mathematical and applicatory problems associated with statistical linearization has been provided by Spanos [241]. Recently, Kozin [198] presented a renewed look at the procedure and its relationship to parameter identification of linear systems.

In order to illustrate the basic idea of the technique let us consider a special system described by the differential equation (cf. [132], [339]):

$$\ddot{Y}(t) + f\left[Y(t), \dot{Y}(t)\right] = X(t) \tag{4.203}$$

where $X(t)$ is a given stationary stochastic process. The above equation is approximated by the linear ("equivalent") equation

$$\ddot{Y}(t) + \beta_e \dot{Y}(t) + k_e Y(t) = X(t) \tag{4.204}$$

where the parameters β_e and k_e should be selected so that the above linear equation approximates best the original non-linear equation (4.203). When the parameters β_e and k_e are determined (what constitutes the main problem) the properties of the solution process $Y(t)$ are obtained from (4.204) by use of the methods of linear theory.

The quantity which can be minimized in order to obtain a good approximation of (4.203) by (4.204) is the error

$$\Delta = \beta_e \dot{Y}(t) + k_e Y(t) - f\left[Y(t), \dot{Y}(t)\right] \tag{4.205}$$

Since $\Delta = \Delta(t)$ is a stochastic process, a common criterion is to minimize the mean-square of the error $\Delta(t)$. So, the parameters β_e and k_e should be determined from the condition

$$E[\Delta^2] = E\left[\left\{\beta_e \dot{Y}(t) + k_e Y(t) - f\left[Y(t), \dot{Y}(t)\right]\right\}^2\right] = \min \tag{4.206}$$

Calculation of the first and second order derivatives of $E\left[\Delta^2(t)\right]$ shows that the conditions

$$\frac{\partial}{\partial \beta_e} E[\Delta^2] = 0, \qquad \frac{\partial}{\partial k_e} E[\Delta^2] = 0 \tag{4.207}$$

lead to the minimization of $E[\Delta^2(t)]$. The parameters β_e and k_e are the solutions of the algebraic equations

$$\begin{aligned}
&\beta_e E\left[\dot{Y}^2(t)\right] + k_e E\left[Y(t)\dot{Y}(t)\right] - E\left[\dot{Y} f[Y,\dot{Y}]\right] = 0, \\
&k_e E\left[Y^2(t)\right] + \beta_e E\left[Y(t)\dot{Y}(t)\right] - E\left[Y f[Y,\dot{Y}]\right] = 0
\end{aligned} \tag{4.208}$$

Solving the above equations, we obtain

$$\beta_e = \frac{E[Y^2]E\left[\dot{Y} f[Y,\dot{Y}]\right] - E[Y\dot{Y}]E\left[Y f[Y,\dot{Y}]\right]}{E[Y^2]E[\dot{Y}^2] - \left\{E[Y\dot{Y}]\right\}^2} \tag{4.209}$$

$$k_e = \frac{E[\dot{Y}^2]E\left[Y f[Y,\dot{Y}]\right] - E[Y\dot{Y}]E\left[\dot{Y} f[Y,\dot{Y}]\right]}{E[Y^2]E[\dot{Y}^2] - \left\{E[Y\dot{Y}]\right\}^2} \tag{4.210}$$

It is seen that the parameters β_e and k_e are expressed in terms of averages of the unknown process $Y(t)$. This means that in order to evaluate β_e and k_e we should know the probability distribution of the unknown process! This is the main methodical deficiency of the statistical linearization procedure.

In order to make the procedure effective, the appropriate "approximations" have to be introduced. One possibility lies in replacing the joint probability density $p(y,\dot{y};t)$ needed for calculation of the averages in (4.209), (4.210) by the stationary density $p_{st}(y,\dot{y})$ determined from the original non-linear equation (4.203); in some cases it can be obtained from the F-P-K equation. Another reasoning can be adopted in the situations where the general analytical form of $p(y,\dot{y};t)$ — with some unknown parameters — can be deduced from the knowledge of the physics of the process considered. If unknown parameters of this probability density function are related to the moments of the solution one can determine the parameters β_e and k_e. These parameters will depend on the moments introduced and the solution of the linearized equation (4.204) will provide a relation between unknown moments of the solution; for the stationary solution this relationship and appropriate approximations lead to an algebraic or transcendental equation. Of course, the accuracy of the results will depend on the correctness of our "guess" about an unknown distribution.

In applications most often a *Gaussian approximation* of unknown process is adopted. If the non-linear terms in the equation are small and excitation process is Gaussian, such an approximation is usually regarded as acceptable (since a solution of a "linear part" of the orginal equation is, in this case, Gaussian). If the excitation $X(t)$ is a Gaussian and stationary process then, of course, the processes $Y(t)$ and $\dot{Y}(t)$ obtained from the linearized equation (4.204) are also Gaussian and they tend to a stationary process as $t \to \infty$ (if the dynamical system is stable). If we are interested in the stationary (steady-state) response we can take $Y(t)$ and $\dot{Y}(t)$ to be jointly stationary, and we have $[Y\dot{Y}] = 0$. Formulae (4.209) and (4.210) take the form

$$\beta_e = \frac{E\left[\dot{Y} f[Y, \dot{Y}]\right]}{E[\dot{Y}^2]}, \tag{4.209'}$$

$$k_e = \frac{E\left[Y f[Y, \dot{Y}]\right]}{E[Y^2]}. \tag{4.210'}$$

If the non-linearity is of a more specific form:

$$f[Y, \dot{Y}] = f_1(Y) + f_2(\dot{Y}) \tag{4.211}$$

then

$$\beta_e = \frac{E\left[\dot{Y} f_2(\dot{Y})\right]}{E[\dot{Y}^2]} \tag{4.209''}$$

$$k_e = \frac{E\left[Y f_1(Y)\right]}{E[Y^2]} \tag{4.210''}$$

The application of the procedure sketched above to equation

$$\frac{d^2Y}{dt^2} + 2h\frac{dY}{dt} + f_1(Y) = X(t) \tag{4.212}$$

where

$$f_1(y) = \omega_0^2 y + \varepsilon y^3$$

gives

$$k_e = \frac{\omega_0^2 E[Y^2] + \varepsilon E[Y^4]}{E[Y^2]} \tag{4.123}$$

Unfortunately, the expression for k_e includes the moments of the second and fourth order. If the Gaussian closure is adopted (i.e. $E(Y^4) = 3\left[E(Y^2)\right]^2$) then

$$k_e = \omega_0^2 + 3\varepsilon E[Y^2] \tag{4.214}$$

Now, by virtue of the spectral method applied to linear equation (4.204) we get the following expression for the spectral density of a stationary solution (of course, $\beta_e = 2h$)

$$g_Y(\omega) = \frac{g_X(\omega)}{|-\omega^2 + \beta_e i\omega + k_e|^2} \tag{4.215}$$

Assuming that $E[X(t)] = 0$, we obtain

$$\sigma_Y^2 = \int_{-\infty}^{+\infty} \frac{g_X(\omega)}{(k_e - \omega^2)^2 + 4h^2\omega^2} d\omega$$

which gives (when the integral is evaluated) the equation for σ_Y^2. For example, if $g_X(\omega) = g_o$ =const., then

$$\sigma_Y^2 = \frac{\pi g_o}{2hk_e} = \frac{\pi g_o}{2h(\omega_o^2 + 3\varepsilon\sigma_Y^2)}$$

and

$$\sigma_Y^2 = \frac{\sqrt{\omega_o^4 + \frac{6\pi g_o \varepsilon}{h}} - \omega_o^2}{6\varepsilon} \tag{4.216}$$

In the last years the statistical linearization procedure has been extended to more complex situations (cf. [150] for multi-degree of freedom systems, [185] — for nonstationary excitations, [256] — for hysteretic systems).

B.

The statistical linearization technique as it was worked out in variety of applications (shortly presented above) has widely been accepted as a handy tool for the analysis of non-linear systems subjected to random excitation. Nevertheless or just therefore, we would like to summarize here the *essential deficiencies* of this procedure.

1. There is no general proof of the existence of an equivalent linear version of a non-linear differential equation with random inhomogeneous term either of the method of the error estimation; the error of statistical linearization has only been found for particular and simple examples. Though many of these examples indicate that the accuracy of the procedure is quite satisfactory for practical purposes, nevertheless one also can show examples where the error can be significant or where the results are qualitatively distorted.
2. As we have already underlined, in order to evaluate rigorously the equivalent coefficients a knwoledge of the probability distribution of the solution is required. This unknown distribution is most often "approximated" by the Gaussian distribution (with unknown moments) whereas the solution of non-linear stochastic equation can never be truly Gaussian. If such an "approximation" could be accepted in the case of small non-linearity, it is evidently not appropriate in strongly non-linear case.
   Let us add that it can be shown that the Gaussian hypothesis used often in the statistical linearization procedure leads to the same results as the Gaussian closure of the hierarchy equations for moments of the original non-linear equation. Incidentally, such coincidence of results could be expected, since — as it is known — the Gaussian approximation of non-linear transformation of

a normal random variable assures the best linear prediction according to the least mean-square criterion. Therefore, the hypothesis that the solution process (of non-linear stochastic equation) is Gaussian leads automatically to the linearization of the relationship between the solution and the excitation processes. Hence, if we want to gain true insight into the specific features of the behaviour of non-linear systems under random excitation we should try to introduce a more appropriate hypotheses.

3. In the existing literature the statistical linearization procedure is applied not only to analytical non-linearities but also to strongly non-linear non-analytical expressions (such as for example $f(y, \dot{y}) = c_1 \text{sign}\, \dot{y}$). This can rise a concern for the results, especially in cases where they cannot be verified with other findings.

In order to illustrate the remarks stated above let us consider two simple examples.

a) Given a harmonic oscillator subjected to a random process which is a square of a stationary Gaussian process $X(t)$, i.e.

$$\frac{d^2Y}{dt^2} + 2h\frac{dY}{dt} + k^2Y = k^2X^2(t),$$
$$h > 0,$$
$$\omega_0 = \sqrt{k^2 - h^2} \tag{4.217}$$

where process $X(t)$ has the correlation function

$$K_X(\tau) = \sigma_X^2 e^{-0,5h|\tau|}\left(\cos\frac{\omega_0\tau}{2} + \frac{h}{\omega_0}\sin\frac{\omega_0|\tau|}{2}\right) \tag{4.218}$$

Let us replace the non-linear term $X^2(t)$ by a linear one according to the statistical linearization, that is

$$F[X(t)] = X^2(t) = a[X(t) - m_X] + b$$

where the linearization coefficient b is determined from the condition of equality of mean values, that is

$$b = E\left[X^2(t)\right] = \sigma_X^2 + m_X^2$$

The coefficient a is evaluated from the condition

$$E\left\{F[X(t)] - a[X(t) - m_x] - b\right\}^2 = \min$$

with the result

$$a = 2m_X$$

So, the linearized problem is

$$\frac{d^2Y}{dt^2} + 2h\frac{dY}{dt} + k^2Y = \\ = k^2\left\{2m_X\left[X(t) - m_X\right] + \sigma_X^2 + m_X^2\right\}$$

By virtue of the spectral method (cf Sec. 26.2) the spectral density of the stationary solution is

$$g_Y(\omega) = \frac{4k^4 m_X^2 g_X(\omega)}{(\omega^2 - k^2)^2 + 2h^2\omega^2} =$$

$$(4.219) \qquad = \frac{\sigma_X^2 k^2 h m_X^2}{\pi\left[\left(\omega^2 - \frac{1}{4}k^2\right)^2 + h^2\omega^2\right]\left[\left(\omega^2 - k^2\right)^2 + 4h^2\omega^2\right]}.$$

On the other hand, the spectral density $g_Y(\omega)$ can be determined exactly. To do that we first evaluate the spectral density of $X^2(t)$ as it was shown in Exercise 8. — Ch. I. with the difference that here $m_X \neq 0$. Since $X^2(t)$ can be represented as

$$(*) \qquad X^2(t) = \left[X(t) - m_X\right]^2 + 2m_X\left[X(t) - m_X\right] + m_X^2$$

where process $X(t) - m_X$ has zero mean and hence $[X(t) - m_X]^2$ has the spectral density given in the Exercises. Since for the Gaussian process $X(t) - m_X$ and $\left[X(t) - m_X\right]^2$ are uncorrelated the spectral density of $X^2(t)$ is equal to the sum of the spectral densities of the first two components in (*), that is

$$g_{X^2}(\omega) = 2\int_{-\infty}^{+\infty} g_X(\omega - \omega_1)d\omega_1 + 4m_X^2 g_X(\omega).$$

Direct application of formula (4.50) gives

$$g_Y(\omega) =$$

$$= \frac{\sigma_X^2 k^2 h m_X^2}{\pi\left[\left(\omega^2 - \frac{1}{4}k^2\right)^2 + h^2\omega^2\right]\left[(\omega^2 - k^2)^2 + 4h^2\omega^2\right]} +$$

$$(4.220) \qquad + \frac{2k^4}{(\omega^2 - k^2)^2 + 4h^2\omega^2}\int_{-\infty}^{+\infty} g_X(\omega_1)(g_X(\omega - \omega_1)d\omega_1$$

It is seen that the approximate solution (4.129) obtained by statistical linearization gives only the first term in the exact solution (4.220).
Depending on the numerical values of the system parameters and the form of

$g_X(\omega)$ the second term in (4.220) can give a siginificant contribution to the spectral density of the solution. In spite of this the second term in (4.220) differs from (4.219) qualitatively; the maximum occurs for different values of ω.

b) Let us consider the equation

$$\frac{d^2Y}{dt^2} + 2h\frac{dY}{dt} + f(Y) = \xi(t), \tag{4.221}$$

where $\xi(t)$ is a Gaussian white noise with the intensity D and

$$f(y) = \omega_0^2(y + \mu \text{ sign } y)$$

The statistical linearization coefficients, according to (4.209), (4.210′) (for Gaussian approximation of the response) are

$$\beta_e = 2h, \quad k_e = \omega_0^2\left[1 + \frac{\mu}{\sigma}\sqrt{\frac{2}{\pi}}\right]$$

The variance of the approximate solution is (cf. [171])

$$\sigma_Y^2 = \frac{\sigma_0^2}{1 + \frac{\mu}{\sigma}\sqrt{\frac{2}{\pi}}}, \quad \sigma_0^2 = \frac{D}{4h\omega_0^2}$$

The equation(4.221) can be analyzed with the use of the F-P-K equation for which a stationary probability density can be obtained exactly. The exact formula for the variance is (cf. [171])

$$\sigma_Y^2 = \sigma_0^2\left\{1 + \frac{\mu^2}{\sigma_0^2} - \frac{\mu}{\sigma_0}\sqrt{\frac{2}{\pi}} exp\left(-\frac{\mu^2}{2\sigma_0^2}\right)\left[1 - \Phi\left(\frac{\mu}{\sigma_0\sqrt{2}}\right)\right]^{-1}\right\}$$

where Φ is the Laplace integral.

A comparisom of the approximate solution and the exact one shows that for $\frac{\mu}{\sigma_0} \to 0$ these results coincide. However, the difference between the approximate and exact solution increases with increase of $\frac{\mu}{\sigma_0}$.

27.2.4. Pertubation method

Another approximate technique developed in the analysis of engineering nonlinear dynamical systems subjected to random excitation is a pertubation method. Its idea comes from the classical theory of ordinary differential equations and deals with the situations when the non-linear term in the equation is concidered to be small. In such cases a small parameter $\varepsilon << 1$ is usually introduced to the equation and a solution is looked for in the form of an expansion with respect to the powers of ε.

To illustrate the idea let us consider a system governed by the second order equation

$$L_\circ Y(t) + \varepsilon f(Y, \dot{Y}) = X(t)$$
$$L_\circ = \frac{d^2}{dt^2} + a_0 \frac{d}{dt} + a_1 \tag{4.222}$$

where ε is a small parameter and $X(t)$ is a stationary Gaussian random process.

The pertubation technique is based on the assumption that the solution process $Y(t)$ can be expanded into powers of ε, that is

$$Y(t) = Y_\circ(t) + \varepsilon Y_1(t) + \varepsilon^2 Y_2(t) + \ldots \tag{4.223}$$

Substituting (4.223) into equation (4.222) and equating terms of the same power of ε gives

$$L_\circ Y_\circ(t) = X(t)$$
$$L_\circ Y_1(t) = -f\left[Y_\circ(t), \dot{Y}_\circ(t)\right]$$
$$L_\circ Y_2(t) = -f'_{Y_\circ}\left[Y_\circ(t), \dot{Y}_\circ(t)\right] Y_1(t) - f'_{\dot{Y}_\circ}\left[Y_\circ(t), \dot{Y}_\circ(t)\right] \dot{Y}_1(t) \tag{4.224}$$
$$\vdots$$

where $f'_{Y_\circ}$ denotes the derivative of $f(Y, \dot{Y})$ with respect to Y evaluated at $Y(t) = Y_\circ(t)$.

It is seen that the non-linear problem is reduced to one of solving a sequence of linear equations with random inhomogeneous term.

The steady-state solutions of the above equations are

$$Y_\circ(t) = \int_0^\infty p(\tau)X(t-\tau)d\tau \tag{4.225}$$
$$Y_1(t) = -\int_0^\infty p(\tau)f\left[Y_\circ(t-\tau), \dot{Y}(t-\tau)\right]d\tau$$
$$\vdots$$

where $p(\tau)$ is the impulse reponse of the linear operator $L_\circ$. Expressions (4.225) can be used to compute various statistics of the response. For example, the average of the solution process is given by

$$E\left[Y(t)\right] = E\left[Y_\circ(t)\right] + \varepsilon E\left[Y_1(t)\right] + \varepsilon^2\left[Y_2(t)\right] + \ldots \tag{4.226}$$

Using (4.225) in (4.226) yields

(4.227)
$$E[Y(t)] = \int_0^\infty p(\tau)E[X(t-\tau)]\,d\tau + \\ -\varepsilon\int_0^\infty p(\tau)E\left\{f\left[Y_o(t-\tau),\dot{Y}_o(t-\tau)\right]\right\}d\tau + \dots$$

To evaluate such terms as the average occuring in the second integral we usually make use of the fact that $X(t)$ is Gaussian. So, also $Y_o(t)$ is Gaussian and

$$E\left\{f\left[Y_o,\dot{Y}_o\right]\right\} =$$

(4.228)
$$= \int_{-\infty}^{+\infty}\int_{-\infty}^{+\infty} f(y_o,\dot{y}_o)p_{st}(y_o,\dot{y}_o)dy_o\,d\dot{y}_o$$

where $p_{st}(y_o,\dot{y}_o)$ is a steady-state density of the process $\left[Y_o(t),\dot{Y}_o(t)\right]$. Higher order moments can be calculated in a similar way, but the computational difficulties increase rapidly.

The pertubation technique has been worked out by Crandall (in 1961) in his study of random vibrations of non-linear mechanical systems. The problem considered by Crandall was connected with the Duffing oscillator, i.e.

(4.229)
$$\ddot{Y}(t) + 2h\dot{Y}(t) + \omega_o^2[Y(t) + \varepsilon f(Y)] = X(t)$$

In the particular case where

(4.230)
$$f(Y) = Y^3$$

it has been found that if we restrict ourselves to the first order term (in ε), i.e.

(4.231)
$$Y(t) = Y_o(t) + \varepsilon Y_1(y)$$

then (cf. [331],[339]) after implementation of the procedure presented above we have the following result for the mean-square of $Y(t)$

(4.232)
$$E\left[Y^2(t)\right] = K_{Y_o}(0)\left[1 - 6\varepsilon\omega_o^2\int_0^\infty p(\tau)K_{Y_o}(\tau)d\tau\right]$$

where

(4.233)
$$K_{Y_o}(\tau) = \int_0^\infty\int_0^\infty p(\tau_1)p(\tau_2)K_X(\tau - \tau_1 + \tau_2)d\tau_1 d\tau_2$$

A discussion of interesting questions associated with stochastic differential equations with small parameter can be found in the book by Ventzel and Freidlin [136].

27.3. Use of maximum entropy principle

27.3.1. Methodical introduction

As we underlined in Sec. 2.4., an important quantity which characterizes an overall randomness of a random variable is its *entropy*. The notion of entropy is also one of the fundamental concepts of statistical physics where it was first introduced as a certain macroscopic characteristics of the termodynamic system. Later on (Gibbs, Boltzmann) the entropy had been interpreted statistically and it is now widely used as a measure of uncertainty of the microscopic states of a system. If $p(\mathbf{y}, t)$ is the phase probability density characterizing the states of the system of material particles at time t, then the entropy is defined, according to formula (1.37) as

$$H = -\int p(\mathbf{y}, t) ln p(\mathbf{y}, t) d\mathbf{y} \quad \mathbf{y} \in R_n \tag{4.234}$$

It seems, however, that significance of entropy in physics remarkably increased when Jaynes [190] used the entropy functional for calculating probability density function and, in this way, provided a firm basis for the procedures of statistical mechanics. Before, it was a common practice to hypothesize a certain probability structure for the microscopic states of a system in particular situations, and then calculate from it the average values of various quantities of interest. Jaynes proposed to choose as the underlying probability distribution of the microscopic states of the system that distribution which had maximum entropy, among all those distributions which were in accord with the observed macroscopic values (being the values of various ensamble average quantities). This is just the *maximum entropy principle* (cf. also Ingarden and Urbanik [184]). If one interprets this principle in its narrow sense of statistical physics then various $p(\mathbf{r}, t)$ occuring in (4.234) are the probability densities of non-equilibrium states; the maximum value of H is reached when the system approaches to its thermodynamical equilibrium.

But, there is no need to restrict the idea of maximum entropy principle to statistical physics. Indeed, for example, Burg in his celebrated paper [158] used the philosophy of the maximum entropy in spectrum estimation of stationary random processes. An important problem which occurs in signal processing is concerned with estimation of the spectrum of random signal (or, spectral density of a stationary random process) on the basis of measurements of the covariance function. Of course, if we knew the entire covariance function (i.e. $K(\tau)$ for all values of the argument) we could obtain the spectrum by means of the Fourier transform (of the discrete type since the data are taken in discrete times). However, most often we can reliably measure the covariance only for a certain finite number of values of τ, say for τ_n, $n = 0, 1, 2, \ldots, p$. Since $K(\tau) = K(-\tau)$, we know the values of $K(\tau)$ for τ_n , $n = 0, 1, 2, \ldots, p$ and we do not know $K(\tau)$ for τ_n when $|n| > p$. The important question is: *how should we estimate the spectrum from only this partial knowlegde?*

The conventional approach to estimating the spectrum from $K(\tau_n)$, $|n| \leq p$ is to assume that $K(\tau_n) = 0$ for $n > p$ and to take the Fourier transform of

$\varphi(\tau_n)K(\tau_n)$, $|n| \leq p$ where φ is a certain weighting or, window function. In order to resolve the problem in a more satisfactory way Burg made use of the maximum entropy principle, which in this context can be verbalized as: choose that spectrum which corresponds to the most random or the most unpredictable process whose covariance function coincides with the given set of values.

As a matter of fact the maximum entropy principle should be applicable to any problem of interference on the basis of incomplete data. Therefore, this principle can be formulated in a more general way as the following "extended maximum entropy principle":
when we make predictions based on incomplete information, we should draw them from probability distribution which has the maximum entropy permitted by the information we have.

It should be emphasized that the above "extended" principle is entirely in the spirit of the mathematical statistics where the reasoning based on the information theory has been accepted for a long time in formulation and analysis of a number of problems (cf. [199]). It also seems to be fruitful to use the principle of maximum entropy to the prediction of a "most rational" hypothetical distribution on the basis of empirical information. It is very likely that the possible probability distributions of real phenomena might be concentrated strongly near the one of maximum entropy, that is, that distributions with much lower entropy than the maximum permitted by the data are non-typical, at least in stationary phenomena.

Let us assume now that we are interested in stationary solutions of a given (non-linear) differential equation. Most often our available information about the solution process can be given in terms of moments. If the number of moments (or, the number of moment equations) is finite, then the associated probability distributions constitute a certain class $\{F\}$ of admissible distributions. Adopting the idea of the extended maximum entropy principle it is natural to assume that the most appropriate (approximate) probability distribution of the solution process satisfying the moment conditions will be that one for which the entropy functional H assumes its maximum (on the class $\{F\}$).

It is clear that when the number of moments (or moment conditions) increases, the number of admissible distributions decreases. In the limtit, an infinite system of moment conditions determines the distribution $p(\mathbf{y})$ uniquely (if, the known Carleman conditions — cf. [27] — are satisfied).

In the sequel we shall show that the idea described above can be successfully used in approximate characterization of the probability distribution of a solution of non-linear stochastic differential equation.

27.3.2. Some general statements

A formal framework of the maximum entropy distribution is as follows. The distribution in question is represented in the form

$$f(y) = exp\left(-\sum_{l=0}^{n} \lambda_l y^l\right) \tag{4.235}$$

where λ_l, $l = 0, 1, \ldots, n$ are the unknown Lagrange multipliers. They should be determined from the moment conditions

$$\int y^k exp\left(-\sum_{l=0}^{n} \lambda_l y^l\right) dy = m_k, \quad k = 0, 1, \ldots, n \tag{4.236}$$

with $m_o = 1$.

If equations (4.236) have a unique solution $[\lambda_o, \lambda_1, \ldots, \lambda_n]$ then we say that the maximum entropy distribution exists. It can be shown (cf. [174]) that in the case when only the first two moments m_1 and m_2 are given the following assertions are valid:

(i) for finite interval (finite range of variability of y) a distribution maximizing the entropy exists and is unique;

(ii) for a semi-infinite interval $[0, \infty)$ the maximum entropy distribution exists if and only if $m_2 \leq m_1^2$, that is $m_1^2 \leq m_2 \leq 2m_1^2$.

In both cases, the maximum entropy is attained by a density function which on appropriate interval takes the form:

$$f(y) = exp\left\{-\left(\lambda_o + \lambda_1 y + \lambda_2 y^2\right)\right\}. \tag{4.237}$$

The nature of this distribution depends on the sign of λ_2. It is a truncated normal distribution if $\lambda_2 > 0$, an exponential if $\lambda_2 = 0$, and so called truncated U-distribution if $\lambda_2 < 0$.

27.3.3. Simple examples

A.

Let us consider equation (4.121)

$$\frac{dY}{dt} + f\left(Y(t)\right) = \xi(t) \tag{4.238}$$

where $\xi(t)$ is the Gaussian white noise with intensity D. The hierarchy equations for moments of a stationary solution satisfy an infinite system of equations

$$E\left[Y^k f(Y)\right] - \frac{D^2}{2} k E[Y^{k-1}] = 0, \quad k = 0, 1, \ldots \tag{4.239}$$

We shall look for the probability density of the stationary solution of equation (4.238) for which the functional

$$H = -\int_{-\infty}^{+\infty} p(y) ln p(y) dy \tag{4.240}$$

reaches its maximum under the conditions (4.239) and the normalization condition

$$\int_{-\infty}^{+\infty} p(y) dy = 1. \tag{4.241}$$

According to the Lagrange multipliers method we construct an extended functional (accounting for conditions (4.239), (4.241)):

$$\begin{aligned} \tilde{H} = &- \int_{-\infty}^{+\infty} p(y) ln p(y) dy + \\ &- (\lambda_o - 1) \int_{-\infty}^{+\infty} p(y) dy + \\ &- \int_{-\infty}^{+\infty} p(y) \sum_k \lambda_k \left[y^k f(y) - \frac{D^2}{2} k y^{k-1} \right] dy \end{aligned} \tag{4.242}$$

where $\lambda_o - 1$ and λ_k are the unknown Lagrange multipliers.

The condition for the existence of the maximum of functional $\tilde{H}$ is $\frac{\delta h}{\delta p} = 0$, where $\frac{\delta}{\delta p}$ denotes the functional derivative, and

$$h = -p(y) \left[ln p(y) + \lambda_o - 1 + \sum_k \lambda_k \left(y^k f(y) - \frac{D^2}{2} y^{k-1} \right) \right] \tag{4.243}$$

The condition $\frac{\partial h}{\partial p} = 0$ yields

$$p(y) = C exp \left\{ - \sum_k \lambda_k \left(y^k f(y) - \frac{D^2}{2} k y^{k-1} \right) \right\} \tag{4.244}$$

where $C = e^{-\lambda_o}$. In order to determine the Lagrange multipliers we substitute (4.244) into the condition equations

$$\int_{-\infty}^{+\infty} \left(y^k f(y) - \frac{D^2}{2} k y^{k-1} \right) p(y) dy = 0, \ k = 0, 1, \ldots$$

After integration by parts in the second term we have

$$\begin{aligned} \int_{-\infty}^{+\infty} \Bigg\{ f(y) - \frac{D^2}{2} \sum_k \lambda_k \Big[k y^{k-1} f(y) \\ + y^k f'(y) - \frac{D^2}{2} k(k-1) y^{k-2} \Big] \Bigg\} y^l p(y) dy = 0 \end{aligned}$$

($l = 1, 2, \ldots$). The above equation shall be satisfied if (for each l)

$$f(y) - \frac{D^2}{2}\sum_k \lambda_k \Big[ky^{k-1}f(y) + \tag{4.245}$$
$$+ y^k f'(y) - \frac{D^2}{2}k(k-1)y^{k-1}\Big] = 0$$

Let us assume that $f(y)$ is the odd function; in such situation, $k = 1, 3, \ldots$. In order to perform the calculation of λ_k, it is useful to represent function $f(y)$ in the form of its Taylor series in the neighbourhood of $y = 0$, i.e.

$$f(y) = f_o' y + \frac{1}{3!}f_o''' y^3 + \cdots + \frac{1}{m!}f_o^{(m)} y^m + \ldots \tag{4.246}$$

where

$$f_o^{(m)} = \frac{d^m f(y)}{dy^m}(0)$$

The derivative of $f(y)$ and its integral are represented as

$$f'(y) = f_o' + \frac{3}{3!}f_o''' y^2 + \ldots + \frac{m}{m!}f_o^{(m)} y^{m-1} + \ldots \tag{4.247}$$
$$\int f(y)dy = f_o' \frac{y^2}{2} + \frac{1}{3!}f_o''' \frac{y^4}{4} + \ldots + \frac{1}{m!}f_o^{(m)} \frac{y^{m+1}}{m+1} + \ldots$$

Now, we sustitute the above series into equation (4.245) and equate to zero the coefficients of the same powers of y. As a result the infinite system of linear algebraic equations for λ_k is obtained. The first equations have the form

$$\lambda_1 \frac{f_o'}{1!} - \frac{f_o'}{1!D^2} - \frac{D^2}{2}3\lambda_3 = 0$$
$$\lambda_1 \frac{f_o'''}{3!} + \lambda_3 \frac{f_o'}{1!} - \frac{f_o'''}{3!2D^2} - \frac{D^2}{2}5\lambda_5 = 0 \tag{4.248}$$
$$\lambda_1 \frac{f_o^{(v)}}{5!} + \lambda_3 \frac{f_o'''}{3!} + \lambda_5 \frac{f_o'}{1!} - \frac{f_o^{(v)}}{5!3D^2} - \frac{D^2}{2}7\lambda_7 = 0$$

It is seen that the special structure of the above system makes it possible to express all multipliers λ_k, for $k \geq 3$ in terms of $\lambda_1 = \lambda^*$. Namely,

$$\lambda_3 = -\frac{4}{3D^4}\left[\frac{f_o'}{1!2}(1 - \lambda^* D^2)\right], \tag{4.249}$$
$$\lambda_5 = -\frac{4}{5D^4}\left[\left(\frac{f_o'}{1!}\right)^2 \frac{1}{3D^2}(1 - \lambda^* D^2) + \frac{f_o'''}{3!4}(1 - 2\lambda^* D^2)\right],$$

Substituting the above multipliers into formula (4.244) for $p(y)$ and performing some transformations one obtains

$$p(y) = Cexp\left\{-\frac{2}{D^2}\left(\frac{1}{1!}f'_o\frac{y^2}{2}+\frac{1}{3!}f'''_o\frac{y^4}{4}+\dots\right)\right\} =$$

$$= Cexp\left(-\frac{2}{D^2}\int f(y)dy\right) \tag{4.250}$$

which coincides with the exact result (4.123) found by use of the F-P-K equation. The above example indicates that the principle of maximum entropy constitutes an effective method of determining the probability density when the infinite system of moments is given (assuming that they satisfy the known Carleman conditions, cf. [17]). Of course, when only few first order moments are given the principle of maximum entropy provides a certain approximation of the true probability distribution. However, it is remarkable that the principle of maximum entropy (applied to the equation considered above) with only one moment $k = 1$ in (4.236)) gives for the variance of stationary solution the result which is much better than that obtained by the statistical linearization.

B.

Let us take now a little more complicated situation:

$$\frac{dY}{dt} + f(Y) = X(t,\gamma) \tag{4.251}$$

where $X(t,\gamma)$ is a weakly stationary process with spectral density

$$g_X(\omega) = \frac{\sigma^2}{\pi}\frac{\alpha}{\omega^2+\alpha^2} \tag{4.252}$$

Performing the extention of the state space (cf. Sec. 27.2.1) we obtain the following system of two Itô equations for $Y_1(t) = Y(t)$, $Y_2(t) = X(t)$:

$$\begin{aligned} dY_1(t) &= -f(Y_1) + Y_2(t) \\ dY_2(t) &= -\alpha Y_2 + dW(t) \end{aligned} \tag{4.253}$$

where $W(t)$ is a Wiener process associated with white noise $\xi(t)$ with intensity $D = \frac{\sigma^2\alpha}{\pi}$. The joint probability density $p(y_1, y_2)$ of the stationary solution satisfies the reduced F-P-K equation

$$\frac{\partial}{\partial y_1}\{[f(y_1) - y_2]\,p\} + \alpha\frac{\partial}{\partial y_2}(y_2 p) + \frac{D}{2}\frac{\partial^2 p}{\partial y_2^2} = 0 \tag{4.254}$$

The exact solution of this equation is not known. It is, however, easy to obtain the equations for moments. Multiplying the left-hand side of (4.254) subsequently by

y_1^2, y_1, y_2, y_2^2 and integrating over entire space we obtain the first and second order moment equations

$$\int_{-\infty}^{+\infty}\int_{-\infty}^{+\infty} [y_1 f(y_1) - y_1 y_2] p(y_1, y_2) dy_1 dy_2 = 0$$

$$\int_{-\infty}^{+\infty}\int_{-\infty}^{+\infty} \left[y_2 f(y_1) + \alpha y_1 y_2 - y_2^2\right] p(y_1, y_2) dy_1 dy_2 = 0$$

$$\int_{-\infty}^{+\infty}\int_{-\infty}^{+\infty} \left(\frac{D}{2} - \alpha y_2^2\right) p(y_1, y_2) dy_1 dy_2 = 0 \tag{4.255}$$

Assuming the normalization condition and conditions (4.255) as given information about a solution we shall look for the approximate stationary probability distribution as a solution of the variational problem for the entropy functional. The extended functional of this variational problem is

$$\begin{aligned} \tilde{H} = &- \int_{-\infty}^{+\infty}\int_{-\infty}^{+\infty} p \ln p \, dy_1 dy_2 \\ &- \int_{-\infty}^{+\infty}\int_{-\infty}^{+\infty} \Big\{ (\lambda_o - 1) + \lambda_1 [y_1 f(y_1) - y_1 y_2] \\ &+ \lambda_2 \left[y_2 f(y_1) + \alpha y_1 y_2 - y_2^2\right] \\ &+ \lambda_3 \left(\frac{D}{2} - \alpha y_2^2\right) \Big\} p(y_1, y_2) dy_1 dy_2 \end{aligned} \tag{4.256}$$

The probability density which realizes the maximum entropy has the form:

$$\begin{aligned} p(y_1, y_2) = C exp \Big\{ &- \lambda_1 [y_1 f(y_1) - y_1 y_2] \\ &- \lambda_2 \left[y_2 f(y_1) - \alpha y_1 y_2 - y_2^2\right] - \lambda_3 \left(\frac{D}{2} - \alpha y_2^2 s\right) \Big\} \end{aligned} \tag{4.257}$$

where $C = e^{-\lambda_o}$. Substituting (4.257) into (4.255) we obtain three transcendental equations for calculating $\lambda_1, \lambda_2, \lambda_3$. Unfortunately, these equations can not be solved analytically. However, when $\lambda_1, \lambda_2, \lambda_3$ are calculated numerically and then substituted into (4.257) we have analytical (approximate) result for the stationary probability density of the solution. Of course,

$$p(y) = p(y_1) = \int_{-\infty}^{+\infty} p(y_1, y_2) dy_2.$$

Other examples of application of the maximum entropy principle are presented in the paper by Sobczyk and Trebicki [237].

28. STOCHASTIC SYSTEMS

28.1. General Remarks

Very often the parameters of dynamical systems undergo random temporal flutuations. Such systems are usually called stochastic systems and their motion is governed by differential equations with random coefficients. Analysis of such equations includes a number of specific questions; above all the equations of this type are directly connected with stochastic (differential) operators (cf. Sec. 12) and the problem of prime importance is that of stochastic stability.

Analysis of differential equations with time-dependent random coefficients create, in the general case, serious difficulties. The reason of this is the fact that even when an equation is linear the solution depends in a non-linear way on its random coefficients. We are not going to comprise here all the problems and methods associated with stochastic systems. This section should be treated just as a brief introduction to the subject.

In general, stochastic systems are governed by equations of the form

$$\frac{d\mathbf{Y}(t)}{dt} = \mathbf{F}\left[\mathbf{Y}(t), \mathbf{X}(t,\gamma), t\right]$$

$$\mathbf{Y}(t_o) = \mathbf{Y}_o \tag{4.258}$$

where $\mathbf{X}(t)$ is a given stochastic process occuring multiplicatively with $\mathbf{Y}(t)$.

There are two important special classes of stochastic systems. The first class includes systems which are linear with respect to $\mathbf{X}(t)$. In this case the governing equations have the form:

$$\frac{d\mathbf{Y}(t)}{dt} = \mathbf{m}\left[\mathbf{Y}(t), t\right] + \boldsymbol{\sigma}\left[\mathbf{Y}(t), t\right]\mathbf{X}(t,\gamma) \tag{4.259}$$

The second class comprises systems which are linear with respect to unknown process $\mathbf{Y}(t)$. In this case we have a linear stochastic system and the governing equations can be written as

$$\frac{d\mathbf{Y}(t)}{dt} = \mathbf{A}(t,\gamma)\mathbf{Y}(t) + \mathbf{B}(t,\gamma)$$

$$\mathbf{Y}(t_o) = \mathbf{Y}_o \tag{4.260}$$

where $\mathbf{A}(t,\gamma)$ and $\mathbf{B}(t,\gamma)$ are a given stochastic matrix and a stochastic vector process, respectively. The initial condition $\mathbf{Y}_o$ can be deterministic or random.

Basic theoretical problems associated with non-linear stochastic systems (4.258), (4.259) were discussed in Chapter III; in the case when $\mathbf{X}(t,\gamma) = \boldsymbol{\xi}(t,\gamma)$ where $\boldsymbol{\xi}(t,\gamma)$ is a white noise, the analysis is based on the theory of stochastic Itô equations. In the previous section we presented the solution methods for the systems (4.259) in which σ does not depend on $\mathbf{Y}$; i.e. the systems are subjected to additive (or, external) noises. In this section we shall concentrate on systems (4.260), which are linear ones with parametric (or multiplicative) noises. As a matter of fact, we shall restrict our attention to the systems governed by homogeneous equations of the form

$$\frac{d\mathbf{Y}(t)}{dt} = \mathbf{A}(t,\gamma)\mathbf{Y}(t), \qquad \mathbf{Y}(t_\circ) = \mathbf{Y}_\circ \tag{4.261}$$

or

$$\frac{d\mathbf{Y}(t)}{dt} = \left[\mathbf{A}_\circ(t) + \mathbf{A}_1(t,\gamma)\right]\mathbf{Y}(t), \qquad \mathbf{Y}(t_\circ) = \mathbf{Y}_\circ \tag{4.261$'$}$$

where $\mathbf{A}_\circ(t)$ is a deterministic stable matrix and $\mathbf{A}_1(t,\gamma)$ is a matrix whose elements are stochastic processes.

Of course, system (4.261′) can be written as

$$\frac{dY_i}{dt} = \sum_{j=1}^{n}\left[a_{ij}^{\circ}(t) + a_{ij}^{1}(t,\gamma)\right]Y_j(t), \qquad Y_i(t_\circ) = Y_i^\circ, \quad i = 1,2,\ldots,n. \tag{4.261$''$}$$

In order to obtain effective results concerning the solution process of the above systems we have to introduce some assumptions related to random processes occuring as coefficients.

28.2. Systems with parametric uncertainty

In many problems of applicatory interest there is a need for analysis of systems governed by differential equations whose coefficients are random variables, i.e. by equations of the form

$$\frac{d\mathbf{Y}(t)}{dt} = \mathbf{F}\left[\mathbf{Y}(t);\mathbf{V}(\gamma),t\right], \quad t \geq t_\circ \tag{4.262}$$

where the random (constant) vector $\mathbf{V}(\gamma)$ enters the equation through the coefficients. We assume that the joint probability density of $\mathbf{V}(\gamma)$ and the initial condition $\mathbf{Y}_\circ(\gamma)$ is given.

System (4.262) can be written down as

$$\frac{dY_i(t)}{dt} = F_i\left[Y_1(t),\ldots,Y_n(t);V_1(\gamma),\ldots,V_m(\gamma),t\right], \qquad Y_i(t_\circ) = Y_i^\circ(\gamma) \tag{4.262$'$}$$

If the joint distribution $f_\circ(\mathbf{y}^\circ, \mathbf{v}) = f_\circ(y_1^\circ, \ldots, y_n^\circ; v_1, \ldots, v_m)$ is given one should determine the probability distribution $f(\mathbf{y}, \mathbf{v}; t) = f(y_1, \ldots, y_n; v_1, \ldots, v_m, t)$ of the solution of (4.262)–(4.262′). It seems that the first systematic treatment of stochastic equations considered here was performed by Dostupov and Pugachev [173] and Kozin [194],[195].

A.

The situation represented by the initial-value problem (4.262) has much in common with the equations with random initial conditions considered in Sec. 25.. More explicitly, problem (4.262) can be equivalently reduced to the equations with random initial conditions only. Let us introduce the vector stochastic process

$$\tilde{\mathbf{Y}}(t) = \begin{bmatrix} \mathbf{Y}(t) \\ \mathbf{V}(\gamma) \end{bmatrix} \tag{4.263}$$

Then we can represent problem (4.262) in the form

$$\begin{aligned} &\frac{\tilde{\mathbf{Y}}(t)}{dt} = \tilde{\mathbf{F}}\left[\tilde{\mathbf{Y}}(t), t\right], \quad \tilde{\mathbf{F}} = \begin{bmatrix} \mathbf{F} \\ 0 \end{bmatrix} \\ &\tilde{\mathbf{Y}}(t_\circ) = \tilde{\mathbf{Y}}_\circ, \qquad \tilde{\mathbf{Y}}_\circ = \begin{bmatrix} \mathbf{Y}_\circ \\ \mathbf{V}(\gamma) \end{bmatrix} \end{aligned} \tag{4.264}$$

The above implies that the probability distribution of the solution of (4.262) can be determined by use of the same reasoning as that presented in Sec. 25.

For example, in the case of the first order equation

$$\begin{aligned} &\frac{dY(t)}{dt} = F\left[Y(t), V_1(\gamma), \ldots, V_m(\gamma), t\right] \\ &Y(t_\circ) = Y_\circ \end{aligned} \tag{4.265}$$

with joint probability density $f_\circ(y_\circ, v_1, \ldots, v_m)$, system (4.264) forms a system of $(m+1)$ equations. It is clear that if

$$Y(t) = g\left[t, t_\circ, Y_\circ, V_1, \ldots, V_m\right]$$

is a solution of (4.265) for which there exists, for each t, the inverse transformation

$$\begin{aligned} &Y_\circ = h[t, t_\circ, Y, V_1, \ldots, V_m] \\ &V_1 = V_1, \ldots, V_m = V_m \end{aligned}$$

then the probability density of the solution (of system (4.264)) is

$$f(y, v_1, \ldots, v_m) = f_\circ\left[h(t, t_\circ, y, v_1, \ldots, v_m)\right] |J| \tag{4.266}$$

where

$$J = \begin{vmatrix} \frac{\partial h}{\partial y} & \frac{\partial h}{\partial v_1} & \cdots & \frac{\partial h}{\partial v_m} \\ 0 & 1 & \ldots & 1 \\ \ldots & \ldots & \ldots & \ldots \\ 0 & 0 & \ldots & 1 \end{vmatrix} \tag{4.267}$$

Let us consider two illustrative examples.

1. Let

$$\begin{aligned} &\frac{dY(t)}{dt} + V(\gamma)Y = 0, \\ &Y(0) = y_o \end{aligned} \tag{4.268}$$

where $V(\gamma)$ is a random variable with uniform distribution on $[0, A]$. The solution is

$$Y(t) = y_o e^{-V(\gamma)t}, \quad 0 \le t < \infty$$

For each t, Y is a exponential function of $V(\gamma)$, so

$$f(y,t) = \begin{cases} \frac{1}{Ayt} &, \quad y_o e^{-At} \le y \le y_o \\ 0 &, \quad y \notin \left[y_o e^{-At}, y_o\right] \end{cases} \tag{4.269}$$

Of course, if the initial condition is random then the formula (4.266) can be applied. For example, if $Y(0) = Y_o(\gamma)$, where $Y_o(\gamma)$ has normal distribution $N(0,1)$, and it is independent of $V(\gamma)$, then we have

$$\begin{aligned} &Y(t,\gamma) = g(t, t_o = 0, Y_o, V) = Y_o(\gamma)e^{-V(\gamma)t} \\ &V(\gamma) = V(\gamma) \end{aligned}$$

It is clear that

$$Y_o(\gamma) = h(t, Y, V) = Y(t)e^{V(\gamma)t}$$

Since $J = e^{vt}$, we get the result $(f_o(y_o, v_o) = f_{Y_o}(y_o) f_V(v))$

$$\begin{aligned} f(y, v; t) &= f_o\left[h(y, v, t)\right] e^{vt} = \\ &= \frac{1}{A\sqrt{2\pi}} e^{vt - \frac{1}{2}y^2 e^{2vt}} \end{aligned} \tag{4.270}$$

Distribution $f(y;t)$ of the solution is obtained by integration with respect to v.

2. Let us consider a system of n equations (in vector form)

$$\frac{d\mathbf{Y}(t)}{dt} = [\mathbf{A} + \mathbf{V}(\gamma)]\,\mathbf{Y}(t)$$
$$\mathbf{Y}(0) = \mathbf{Y}_o(\gamma) \tag{4.271}$$

where $\mathbf{A}$ is a deterministic matrix, $\mathbf{V}(\gamma)$-random matrix and $\mathbf{Y}_o(\gamma)$-vector of random initial conditions (independent of $\mathbf{V}(\gamma)$). The solution has the form

$$\mathbf{Y}(t,\gamma) = \mathbf{Y}_o(\gamma)e^{[\mathbf{A}+\mathbf{V}(\gamma)]t} \tag{4.272}$$

Averaging yields

$$m_{\mathbf{Y}} = m_{\mathbf{Y}_o} E\left\{e^{[\mathbf{A}+\mathbf{V}(\gamma)]t}\right\} \tag{4.273}$$

Let us assume that the components of $\mathbf{V}(\gamma)$ are mutually independent random variables with normal distribution $N(m_{ij}, D_{ij})$. In this case

$$E\left\{e^{[a_{ij}+V_{ij}(\gamma)]t}\right\}$$
$$= \frac{e^{a_{ij}t}}{(2\pi D_{ij})^{\frac{1}{2}}}\int_{-\infty}^{+\infty} e^{v_{ij}t-\frac{(v_{ij}-m_{ij})^2}{2D_{ij}}}\,dv_{ij}$$
$$= exp\left\{(a_{ij}+m_{ij})t + \frac{1}{2}D_{ij}t^2\right\}$$

Therefore,

$$m_{\mathbf{Y}}(t) = m_{\mathbf{Y}_o} e^{(\mathbf{A}+m_{\mathbf{V}})t+\frac{1}{2}D_{\mathbf{V}}t^2} \tag{4.274}$$

where $D_{\mathbf{V}}$ is the matrix of variances of $\mathbf{V}(\gamma)$. Let us differentiate both sides of (4.274) with respect to time. We get

$$\frac{dm_{\mathbf{Y}}(t)}{dt} = [\mathbf{A} + m_{\mathbf{V}} + D_{\mathbf{V}}t]\,e^{[\mathbf{A}+m_{\mathbf{V}}]t+\frac{1}{2}D_{\mathbf{V}}t^2} m_{\mathbf{Y}_o} \tag{2.75}$$

Making use of (4.274) we have

$$\frac{dm_{\mathbf{Y}}(t)}{dt} = (\mathbf{A} + m_{\mathbf{V}} + D_{\mathbf{V}}t)m_{\mathbf{Y}}(t)$$
$$m_{\mathbf{Y}}(0) = m_{\mathbf{Y}_o} \tag{4.276}$$

which is a first order linear system. If (4.276) is represented in the form

$$\frac{dm_{Y_i}(t)}{dt} + \sum_{j=1}^{n}(-a_{ij} - m_{ij} - D_{ij}t)m_j = 0 \tag{4.276'}$$

we conclude that for for $a_{ij} < 0$, $m_{ij} < 0$ the system (4.276) is stable if (the coefficients of (4.276′) have to be positive)

$$|a_{ij} + m_{ij}| - D_{ij}t > 0. \tag{4.277}$$

For

$$t > \frac{|a_{ij} + m_{ij}|}{D_{ij}} \qquad i, j, 1, \ldots, n$$

the mean value of the solution tends to infinity. This loss of stability (in the mean) with increase of time stems from the fact that in the case of Gaussian coefficients there exists significant probability of the event that random parameters $V_{ij}(\gamma)$ will take nagative values greater in modulus than deterministic parameters a_{ij}.

B.

Another possible approach to analysis of systems with random uncertainty in parameters is based on conditioning (cf. Kotulski, Sobczyk [192]). This means that first the conditional characteristsics of a solution (obtained for fixed parameters) are derived and then an integration with respect to an appropriate distribution (of parameters is performed. Such approach seems to be especially advisable when the systems considered have not only random parameters but also are subjected to random excitation (cf. [192]).

Consider, for illustration, a linear, time-invariant system subjected to a stationary random excitation $X(t)$. An "idealized" model can be written as follows

$$L_t(\mathbf{a}_o)Y(t) = X(t, \gamma) \tag{4.278}$$

where $L_t(\mathbf{a}_o)$ is a linear differential operator with constant coefficients represented symbollically by vector $\mathbf{a}_o$. Often, to be more realistic, a system in question should be described by the equation with random uncertainty in coefficients, that is by equation

$$L_t\left(\mathbf{a}_o, \mathbf{V}(\gamma)\right) Y(t) = X(t, \gamma) \tag{4.279}$$

where $\mathbf{V}(\gamma)$ is a random vector. Randomness involved in $\mathbf{V}(\gamma)$ causes departure of the response of the model (4.279) from that described by equation (4.278).

In order to estimate the quality of a given real system (e.g. reliability, stability, accuracy of the functioning etc.) it is convenient to introduce a *quality measure* Q, a quantity which characterizes the specific property of a system. Such a measure depends on the system parameters and excitations. Let the quality measures associated with models (4.278) and (4.279) be denoted by $Q^{(i)}(\mathbf{a}_o, X)$ and $Q^{(r)}(\mathbf{a}_o, X; V)$, respectively. The quantity

$$q = Q^{(r)}(\mathbf{a}_o, X; \mathbf{V}) - Q^{(i)}(\mathbf{a}_o, X) \tag{4.280}$$

characterizes a change of quality of a system due to random uncertainty in parameters. Various quality measures (e.g. mean-square of the response $Y(t)$, expected rate of crossing a level y_o by process $Y(t)$, average number of maxima of $Y(t)$ per unit time, etc.) have been studied for deterministic systems subjected to random stationary excitation. However, these results can be used to determine $Q^{(r)}$. To do this one should integrate (with respect to the probability distribution of $\mathbf{V}(\gamma)$ the expressions obtained for fixed values of the parameters. In general, we have the formula

$$Q^{(r)}(\mathbf{a}_o, X; \mathbf{V}) = \int Q^{(i)}(\mathbf{a}_o, X; \mathbf{V}(\gamma) = v)\, dF(v) \tag{4.281}$$

where $F(v)$ is the probability distribution of the random vector $\mathbf{V}(\gamma)$. In paper [192] various quality measures have been studied for vibratory systems with random parameters by use of (4.281).

C.

Let us denote by $w(\mathbf{y}, t|\mathbf{v})$ the conditional probability density of the solution $\mathbf{Y}(t)$ at time t (when $\mathbf{V}(\gamma)$ assumes a fixed value). It can be easily shown (cf. Sec. 25.2) that $w(\mathbf{y}, t|\mathbf{v})$ is an integral invariant of motion of the system described by (4.262)–(4.262′) and, therefore, it satisfies the Liouville equation

$$\frac{\partial w}{\partial t} + \sum_{i=1}^{n} \frac{\partial}{\partial y_i}(F_i w) = 0 \tag{4.282}$$

Multiplying the above equation by the probability density $u(\mathbf{v}) = u(v_1, \ldots, v_m)$ of the random variables $V_1(\gamma), \ldots, V_m(\gamma)$ and taking into account that

$$u(\mathbf{v})w(\mathbf{y}, t|\mathbf{v}) = f(\mathbf{y}; \mathbf{v}, t)$$

we obtain the Liouville equation for the one-dimensional (joint) distribution of the solution and coefficients

$$\begin{gathered} \frac{\partial f}{\partial t} + \sum_{i=1}^{n}(F_i f) = 0, \\ f(\mathbf{y}, \mathbf{v}, t_o) = f_o(\mathbf{y}^o, \mathbf{v}) = f_o(y_1^o, \ldots, y_n^o, v_1, \ldots, v_m) \end{gathered} \tag{4.283}$$

The general solution of equation (4.283) is obtained by the Lagrange method (cf. Sec. 25.2).

Example.

Let a motion of a dynamical system be governed by the equation

$$\ddot{Y}(t) + V(\gamma)Y - \varphi(Y) = 0 \tag{4.284}$$

where $\mathbf{V}(\gamma)$ is a random variable. The above equation can be written as

$$\dot{Y}_1(t) = Y_2$$
$$\dot{Y}_2(t) = \varphi(Y_1) - V(\gamma)Y_1$$

In this case

$$F_1 = y_2, \quad F_2 = \varphi(y_1) - vy_1$$

The Liouville equation for the density $f(y_1, y_2, v; t)$ is

$$\frac{\partial f}{\partial t} + y_2 \frac{\partial f}{\partial y_1} + [\varphi(y_1) - vy_1] \frac{\partial f}{\partial y_2} = 0 \tag{4.285}$$

This equation is equivalent to the following system of ordinary equations

$$\frac{dt}{1} = \frac{dy_1}{y_2} = \frac{dy_2}{\varphi(y_1) - vy_1}$$

which can be solved by approximate methods. In special case one can find a stationary solution.

Let $\varphi(Y) = \beta Y^3$, $\beta < 0$. In this case a stationary solution $\left(\frac{\partial f}{\partial t} = 0\right)$ of the Liouville equation (4.285) is

$$f(y_1, y_2, v) = f(y, \dot{y}, v) = \Psi\left(\beta \frac{y^4}{4} - \frac{vy^2}{2} - \frac{\dot{y}^2}{2}\right) \tag{4.286}$$

where Ψ is an arbitrary differentiable function satisfying given conditions.

It is worth noting that the above method (of the Liouville equation) can also be used in the anlysis of systems whose parameters are described by the time-dependent random variables represented in the form of deterministic function of time and random variables (e.g. random polynomials, trigonometric polynomials with random coefficients etc.).

An application of the above equations with random constant parameters to some problems of engineering interest can be found, for example, in the papers [238],[251].

28.3. White noise parametric excitation

If the multiplicative noise process is a Gaussian white noise $\xi(t)$ then the equations (4.261) are interpreted as the Itô or Stratonovich equations and the analysis is based on the theory of diffusion Markov processes. In the case of a linear system we have the following Itô equation

$$d\mathbf{Y}(t) = \boldsymbol{\alpha}(t)\mathbf{Y}(t)dt + \boldsymbol{\beta}(t)\mathbf{Y}(t)d\mathbf{W}(t) \tag{4.287}$$

where $\boldsymbol{\alpha}(t)$ and $\boldsymbol{\beta}(t)$ are given deterministic metrices, and $\mathbf{W}(t)$ is a k-dimensional Wiener process.

The moment equations associated with (4.287) can easily be derived by use of Itô formula. If $f(y_1, y_2, \ldots, y_n) = y_i y_j$ we have

$$d\left[Y_i(t)Y_j(t)\right] = Y_i(t)dY_j(t) + Y_j(t)dY_i(t)$$
$$+ \sum_{r=1}^{k} [\boldsymbol{\beta}_r(t)\mathbf{Y}(t)]_i [\boldsymbol{\beta}_r(t)\mathbf{Y}(t)]_j \, dt$$
$$= \left\{ Y_i(t) [\boldsymbol{\alpha}(t)\mathbf{Y}(t)]_j + Y_j(t) [\boldsymbol{\alpha}(t)\mathbf{Y}(t)]_i \right\} dt$$
$$+ \sum_{r=1}^{k} \left\{ \left(Y_i(t) [\boldsymbol{\beta}_r(t)\mathbf{Y}(t)]_j \right.\right.$$
$$\left. + Y_j(t) [\boldsymbol{\beta}_r(t)\mathbf{Y}(t)]_i \right) dW_r(t) +$$
$$\left. + [\boldsymbol{\beta}_r(t)\mathbf{Y}(t)]_i [\boldsymbol{\beta}_r \mathbf{Y}(t)]_j \, dt \right\}$$

Writing the above relation in the integral form (according to the defintion of a stochastic differential) and evaluating the average gives the following system of differential equations for the second order moments

$$\frac{dm_{ij}(t)}{dt} = \sum_{l=1}^{n} \left\{ \alpha_{jl}(t)m_{jl}(t) + \alpha_{jl}(t)m_{jl}(t) \right.$$
$$\left. + \sum_{s,l=1}^{n} \sum_{r=1}^{k} [\beta_{ir}(t)]_s [\beta_{jr}(t)]_l \, m_{sl}(t) \right\}, \tag{4.288}$$
$$i, j = 1, 2, \ldots, n$$

where

$$m_{ij}(t) = E\left[Y_i(t)Y_j(t)\right]$$

Since $m_{ij}(t) = m_{ji}(t)$ system (4.288) constitutes a set of $\frac{n(n+1)}{2}$ independent linear equations. The equation for $m_1(t) = E\left[Y(t)\right]$ is obtained directly when we take the average of both sides of the integral form of (4.285). It has the form:

$$\frac{d\mathbf{m}(t)}{dt} = \boldsymbol{\alpha}(t)\mathbf{m}_1(t),$$
$$\mathbf{m}(t_o) = \mathbf{Y}_o \tag{4.289}$$

It is seen that system (4.287) is stable in the mean if the deterministic system obtained from (4.287) by neglecting the fluctuating term is stable.

System (4.287) is stable in the mean-square if and only if the deterministic system (4.288) is stable. If the coefficients $\alpha(t) = \alpha^\circ$, $\beta(t) = \beta^\circ$ are constant then the stochastic system (4.287) is stable in the mean-square if and only if the roots of the equation

$$|\lambda \mathbf{I} - \mathbf{\Lambda}| = 0$$

have negative real parts; $\mathbf{I}$ is the unit matrix $(n \times n)$ and $\mathbf{\Lambda}$ is the matrix of system (4.288). As it is known, the necessary and sufficient conditions for this can be given in the form of inequalities satisfied by the elements of the matrix $\mathbf{\Lambda}$ (the Routh-Hurwitz criterion).

To illustrate the above statements let us consider the following *example*:

$$\ddot{Y}(t) + [a_\circ + a_1\xi(t)]\dot{Y}(t) + [b_\circ + b_1\xi_2(t)]Y(t) = 0 \ , t \geq 0 \tag{4.291}$$

where $a_\circ, a_1, b_\circ, b_1$ are constants and $\xi_1(t)$ and $\xi_2(t)$ are two independent Gaussian white noises. Equation (4.291) can be written in the form of a system of Itô equations

$$d\mathbf{Y}(t) = \begin{bmatrix} 0 & 1 \\ -b_\circ & -a_\circ \end{bmatrix} \mathbf{Y}(t)dt + \begin{bmatrix} 0 & 0 \\ 0 & -a_1 \end{bmatrix} \mathbf{Y}(t)dW_1(t)$$
$$+ \begin{bmatrix} 0 & 0 \\ -b_1 & 0 \end{bmatrix} \mathbf{Y}(t)dW_2(t)$$

where

$$\mathbf{Y}(t) = \begin{bmatrix} Y(t) \\ \dot{Y}(t) \end{bmatrix}$$

The mean value satisfies the equation

$$\dot{\mathbf{m}}(t) = \begin{bmatrix} 0 & 1 \\ -b_\circ & -a_\circ \end{bmatrix} \mathbf{m}(t)$$

and it is asymptotically stable if and only if both $a_\circ$ and $b_\circ$ are positive. Equations (4.288) yield the following equations for the second order moments

$$\dot{m}_{11}(t) = 2m_{12}$$
$$\dot{m}_{12}(t) = -b_\circ m_{11} - a_\circ m_{12} + m_{22}$$
$$\dot{m}_{22}(t) = b_1^2 m_{11} - 2b_\circ m_{12} + (a_1^2 - 2a_\circ)m_{22}$$

The matrix $\mathbf{\Lambda}$ of this system is

$$\mathbf{\Lambda} = \begin{bmatrix} 0 & 2 & 0 \\ -b_\circ & -a_\circ & 1 \\ b_1^2 & -2b_\circ & a_1^2 - 2a_\circ \end{bmatrix}$$

and its characteristic equation (4.290) is

$$-det\,[\mathbf{\Lambda} - \lambda\mathbf{I}] = \lambda^3 + \lambda^2(3a_o - a_1^2) + \lambda(4b_o + 2a_o^2 - a_o a_1^2) + 2(2a_o b_o - b_o a_1^2 - b_1^2) = 0$$

Therefore, the real parts of the roots of the above equation are negative if and only if (from the Routh-Hurwitz criterion)

$$a_1^2 < 2a_o, \quad b_1^2 < (2a_o - a_1)b_o \tag{4.292}$$

So, for fixed a_o and b_o, the intensity of the random parametric excitation must not exceed a certain value if the (exponential) stability in the mean-square sense is to be a property of a system.

28.4. Real random parametric excitation

28.4.1. Almost sure asymptotic stability

Here we wish to show briefly how the stability of the system can be investigated in the situation where a random parametric excitation is a non-white Gaussian process (cf. Ariaratnan [147].[148], Kozin [196],[197]).

Let us consider the problem of the stability of a solution of equation (cf. [196])

$$\ddot{Y}(t) + 2\zeta\omega_o\dot{Y}(t) + \left[\omega_o^2 - \alpha(t)\right]Y(t) = 0 \tag{4.293}$$

Writing the above equation in the form of the integral equation

$$Y(t) = p(t)y_o + \int_0^t p(t-\tau)\alpha(\tau)Y(\tau)d\tau \tag{4.294}$$

where $p(t)$ is the impulse response function, and taking absolute values of both sides, we obtain the inequality

$$|Y(t)| \leq |p(t)|\,|Y_o| + \int_0^t |p(t-\tau)|\,|\alpha(\tau)|\,|Y(\tau)|\,d\tau \tag{4.295}$$

Suppose that positive constants a, b can be found such that

$$|p(t)| \leq ae^{-bt} \tag{4.296}$$

Then inequality (4.295) can be written as

$$|Y(t)| \leq ae^{-bt}\,|Y_o| + a\int_0^t e^{-b(t-\tau)}\,|\alpha(\tau)|\,|Y(\tau)|\,d\tau \tag{4.297}$$

or

$$|Y(t)|\,e^{bt} \le a\,|Y_0| + \int_0^t |Y(\tau)|\,e^{b\tau} a\,|\alpha(\tau)|\,d\tau \tag{4.298}$$

Let us make use of the Gronwall-Bellman lemma (cf. Appendix), which asserts that: if functions $u(t) \ge 0$, $v(t) \ge 0$, $t \in [0, \infty)$ are continuous in every finite interval, and if for some constant $C \ge 0$

$$u(t) \le C + \int_0^t u(\tau)v(\tau)d\tau, \qquad t \ge 0$$

then

$$u(t) \le C exp\left[\int_0^t v(\tau)d\tau\right], \qquad t \ge 0 \tag{*}$$

We obtain

$$|Y(t)| \le a\,|Y_0|\,exp\left[-bt + a\int_0^t |\alpha(\tau)|\,d\tau\right]$$

or

$$|Y(t)| \le a\,|Y_0|\,exp\left[-b + a\frac{1}{t}\int_0^t |\alpha(\tau)|\,d\tau\right]t \tag{4.299}$$

If $\alpha(t)$ is an ergodic Gaussian process with mean m_α and variance σ_α^2, then with probability one

$$\begin{aligned}\lim_{t\to\infty}\frac{1}{t}\int_0^t |\alpha(\tau)|\,d\tau &= E\left[|\alpha(t)|\right] = \\ &= \frac{1}{\sigma_\alpha\sqrt{2\pi}}\int_{-\infty}^{+\infty} |\alpha| exp\left[-\frac{(\alpha - m_\alpha)^2}{2\sigma_\alpha^2}\right] d\alpha = \\ &= \sqrt{\frac{2}{\pi}}\sigma_\alpha exp\left[-\frac{m_\alpha^2}{2\sigma_\alpha^2}\right] + \sqrt{\frac{2}{\pi}}\frac{m_\alpha}{\sigma_\alpha}\int_{-m}^{\infty} exp\left[-\frac{\alpha^2}{2\sigma_\alpha^2}\right] d\alpha\end{aligned}$$

If $m_\alpha = E\left[\alpha(t)\right] = 0$, then

$$\lim_{t\to\infty}\frac{1}{t}\int_0^t |\alpha(\tau)|\,d\tau = E\left[|\alpha(t)|\right] = \sigma_\alpha\sqrt{\frac{2}{\pi}}$$

Hence, when $t \to \infty$, (4.299) becomes with probability one

$$\lim_{t\to\infty}|Y(t)| \le \lim_{t\to\infty}\left\{a|Y_0| exp\left[-\left(b - \sigma_\alpha a\sqrt{\frac{2}{\pi}}\right)\right]t\right\}$$

Therefore, for

$$\sigma < \frac{b}{a}\sqrt{\frac{\pi}{2}} \tag{4.300}$$

$|Y(t)|$ tends to zero with probability one as t approaches infinity. This means that condition (4.300) is a condition for almost sure aymptotic stability of the solution. For $0 < \zeta < 1$, taking

$$|p(t)| = \frac{1}{\omega_o\sqrt{1-\zeta^2}}e^{-\zeta\omega_o t}\left|\sin\sqrt{1-\zeta^2}\omega_o t\right| \leq \frac{1}{\omega_o\sqrt{1-\zeta^2}}e^{-\zeta\omega_o t}$$

we have

$$a = \frac{1}{\omega_o\sqrt{1-\zeta^2}}, \qquad b = \zeta\omega_o$$

and the stability conditions (4.300) becomes

$$\sigma_\alpha < \sqrt{\frac{\pi}{2}}\zeta\omega_o^2\sqrt{1-\zeta^2} \tag{4.300'}$$

It is seen that for values of ζ which are near to one the above condition is too restrictive. In this case the following estimate can be adopted (cf. [307])

$$|p(t)| = te^{-\zeta\omega_o t}\left|\frac{\sin\sqrt{1-\zeta^2}\omega_o t}{\sqrt{1-\zeta^2}\omega_o t}\right| \leq \frac{1}{\omega_o}e^{-(\zeta-\frac{1}{e})\omega_o t}$$

and the stabilty condition is

$$\sigma_o < \sqrt{\frac{\pi}{2}}\left(\zeta - \frac{1}{e}\right)\omega_o^2 \tag{4.300''}$$

The investigation of almost sure asymptotic stability of linear systems with ergodic Gaussian parametric noises in a more general setting can be found in the Kozin paper [196]. An interesting discussion of the relations between almost sure stability and the moment properties of the solution is giveen in the paper [197]. Other stability problems, including these of engineering importance are discussed by Wedig (cf. [253],[254]).

28.4.2. Use of stochastic averaging method

The stochastic averaging method presented in Sec. 21.4. can be applied to general class of systems including ones with random parametric excitations. As a matter of fact, this has already been shown by the example in Sec. 27.2.2. Here we consider briefly a linear system with randomly varying parameter excitation . Let

(4.301)
$$\frac{d^2Y}{dt^2} + 2\zeta\omega_o\frac{dY}{dt} + \omega_o^2\left[1 + \varepsilon X(t,\gamma)\right]Y = 0$$

where $X(t,\gamma)$ is a stationary random process with zero mean (satisfying the assumptions of the Khasminskii limit theorem.)

Using transformation (4.183), (4.184) yields the following equations for amplitude $A(t)$ and phase $\Psi(t)$: (*let* $\zeta = \varepsilon^2\bar{\zeta}$)

(4.302)
$$\frac{dA(t)}{dt} = \left\{ -2\varepsilon^2\omega_o\bar{\zeta}A(t)\sin\Phi + \varepsilon\omega_o X(t)A(t)\cos\Phi\right\}\sin\Phi,$$
$$\frac{d\Psi(t)}{dt} = \frac{1}{A(t)}\left\{ -2\varepsilon^2\omega_o\bar{\zeta}A(t)\sin\Phi + \varepsilon\omega_o X(t)A(t)\cos\Phi\right\}\cos\Phi$$

By virtue of the Khasminskii limit theorem, process $[A(t),\Psi(t)]$ converges weakly to a Markov diffusion process whose drift and diffusion coefficients are defined by formulae (3.41), (3.42). In our case these coefficients are (cf. [148],[182])

(4.303)
$$a_A = -\frac{1}{\Theta}\int_0^\Theta E\left[2\omega_o\bar{\zeta}A\sin^2\Phi\right]dt +$$
$$+\frac{1}{\Theta}\int_0^\Theta ds\Bigg\{\int_{-\infty}^0 \Big[E\left(\{\omega_o X(s)\cos\Phi\}\{\omega_o X(s+\tau)A\cos(\Phi+\omega_o\tau)\}\right)\sin\Phi\sin(\Phi+$$
$$+\omega_o\tau) + \frac{1}{A}E\left(\{-\omega_o X(s)A\sin\Phi\}\{\omega_o X(s+\tau)A\cos(\Phi+\omega_o\tau)\}\right)\sin\Phi\sin(\Phi+$$
$$\omega_o\tau)\Big]d\tau + \frac{1}{A}\int_{-\infty}^0 E\left(\{\omega_o X(s)A\cos\Phi\}\{\omega_o X(s+\tau)A\cos(\Phi+\omega_o\tau)\}\right)\cos\Phi\cos(\Phi+$$
$$+\omega_o\tau)d\tau\Bigg\} = -\omega_o\bar{\zeta}A + \frac{3}{16}\omega_o^2 A g_X(2\omega_o)$$

where $g_X(\omega)$ is the spectral density of $X(t)$. By analogous integration we get

(4.304)
$$a_\Psi = \frac{\omega_o^2}{4}\int_{-\infty}^0 K_X(\tau)\sin 2\omega_o\tau d\tau,$$

(4.305)
$$b_{A,A} = \frac{1}{\Theta}\int_0^\Theta ds\int_{-\infty}^{+\infty} E\left[\{\omega_o X(s)A\cos\Phi\}\{\omega_o X(s+t)A\cos(\Phi+\omega_o\tau)\}\right]\times$$
$$\times\sin\Phi\sin(\Phi+\omega_o\tau)d\tau = \frac{1}{8}\omega_o^2A^2g_X(2\omega_o),$$

(4.306)
$$b_{\Psi,\Psi} = \frac{\omega_o^2}{4}\left[g_X(0) + \frac{1}{2}g_X(2\omega_o)\right]$$

Therefore, the appropriate (limiting) process $[A(t), \Psi(t)]$ is governed by the following Itô equations

(4.307)
$$dA = \varepsilon^2\left[-\omega_o\bar{\zeta}A + \frac{3}{16}\omega_o^2 A g_X(2\omega_o)\right]dt + \varepsilon\omega_o\left[\frac{g_X(2\omega_o)}{8}\right]^{1/2} A dW_1(t)$$
$$d\Psi = \varepsilon^2\left[\frac{\omega_o^2}{4}\int_{-\infty}^{0} K_X(\tau)\sin 2\omega_o\tau d\tau\right]dt + \varepsilon\frac{\omega_o}{2}\left[g_X(0) + \frac{1}{2}g_X(2\omega_o)\right]^2 dW_2(t)$$

where $W_1(t)$ and $W_2(t)$ are independent Wiener processes with unit intensity.

The above equations imply that the amplitude and phase are independent diffusion processes. The transition probability density $p_A(a,t;a_o,0)$ is governed by the Fokker-Planck equation

(4.308)
$$\frac{\partial p_A}{\partial t} = -\frac{\partial}{\partial a}\left\{\varepsilon^2\left(-\omega_o\bar{\zeta}a + \frac{3}{16}\omega_o^2 a g_X(2\omega_o\right)p_A\right\} + \frac{\varepsilon^2\omega_o^2}{16}g_X(2\omega_o)\frac{\partial^2}{\partial a^2}(a^2 p_A)$$

with initial condition

(4.309)
$$p_A(a,t;a_o,0) \longrightarrow \delta(a-a_o) \quad \text{as} \quad t \to 0$$

The solution of (4.308), obtained by Stratonovich, has the form

(4.310)
$$p_A(a,t;a_o,0) = = \frac{1}{a\sqrt{2\pi\mu t}}exp\left\{-\frac{1}{2\mu t}\left[ln\frac{a}{a_o} - \alpha t\right]^2\right\}$$

where

$$\mu = \frac{1}{8}\varepsilon^2\omega_o^2 g_X(2\omega_o),$$
$$\alpha = \varepsilon^2\left[\frac{1}{8}\omega_o^2 g_X(2\omega_o) - \omega_o\bar{\zeta}\right]$$

From (4.307) or (4.308) we obtain the following differential equations for the first and second order moments $m_k = E\left[A^k(t)\right]$, $k = 1,2$:

(4.311)
$$\frac{dm_1}{dt} = \varepsilon^2\left[\frac{3}{16}\omega_o^2 g_X(2\omega_o) - \omega_o\bar{\zeta}\right]m_1$$
$$\frac{dm_2}{dt} = 2\varepsilon^2\left[\omega_o^2 g_X(2\omega_o) - \omega_o\bar{\zeta}\right]m_2$$

The conditions for the stability of these moments are, respectively

$$\bar{\zeta} > \frac{3}{16}\omega_\circ g\chi(2\omega_\circ),$$

$$\bar{\zeta} > \frac{1}{2}\omega_\circ g\chi(2\omega_\circ). \tag{4.312}$$

It is clear that from Itô equation (4.307) we can obtain directly the conditions for the sample stability.

28.4.3. Pertubation techniques

In the situations where the random parametric noises are "weak" or "small' one usually introduces into the equations a small parameter $\varepsilon << 1$. Then, the solution is represented in the form of a series expansion with respect to the powers of ε. This is the essence of various pertubation techniques which have been widely used in engineering.

A.

In order to illustrate the procedure along with its deficiencies let us consider a class of equations represented symbolically as

$$L(\gamma)Y = g \tag{4.313}$$

where $L(\gamma)$ is a linear stochastic differential operator (in ordinary or partial derivatives) and the element g is here assumed to be deterministic. Let us assume that the operator L depends on a small parameter ε, i.e. $L = L(\gamma, \varepsilon)$ and that for $\varepsilon = 0$ it reduces to the deterministic operator $L_\circ$. Expanding the operator $L(\gamma, \varepsilon)$ in powers of ε, we can represent equation (4.313) in the form

$$\left[L_\circ + \varepsilon L_1(\gamma) + \varepsilon^2 L_2(\gamma) + \ldots\right] Y = g \tag{4.313'}$$

The operators $L_1(\gamma),\ L_2(\gamma), \ldots$ represent random pertubations of $L_\circ$. We look for the solution in the form of an analogous expansion

$$Y = Y_\circ + \varepsilon Y_1 + \varepsilon^2 Y_2 + \ldots \tag{4.314}$$

By substituting this solution representation into equation (4.313′) and equating terms of the same order of ε, we obtain a system of differential equations:

$$\begin{aligned} L_\circ Y_\circ &= g \\ L_\circ Y_1 &= -L_1(\gamma)Y_\circ \\ L_\circ Y_2 &= -L_1(\gamma)Y_1 - L_2(\gamma)Y_\circ \\ &\vdots \\ L_\circ Y_k &= -\left[L_1(\gamma)Y_{k-1} + L_2(\gamma)Y_{k-2} + \ldots + L_k(\gamma)Y_\circ\right] \end{aligned} \tag{4.315}$$

$k = 1, 2, \ldots$.

It is seen that equation (4.313) or (4.313′) with random coefficients has been replaced by a sequence of equations with constant coefficients and with random inhomogeneous term. If L_o is linear, what we have assumed, its inverse operator is well defiend (it is an integral operator whose kernel is the Green function associated with L_o).. For example, if the original equation is an ordinary differential equation for the process $Y(t)$, then

$$L_o^{-1}\varphi(t) = \int_{t_o}^{t} p(t-\tau)\varphi(\tau)d\tau \tag{4.316}$$

If only particular solutions of (4.315) are considered, they can be written in the forms

$$\begin{aligned} Y_o &= L_o^{-1} g \\ Y_k &= -L_o^{-1}\sum_{i=1}^{k} L_i(\gamma)Y_{k-i}, \quad k = 1, 2, \ldots \end{aligned} \tag{4.317}$$

Therefore, the solution Y as represented by (4.314) can formally be written as

$$\begin{aligned} Y = {} & Y_o - \varepsilon L_o^{-1} L_1(\gamma) Y_o + \\ & + \varepsilon^2 L_o^{-1}\left[L_1 L_o^{-1} L_1 - L_2(\gamma)\right] Y_o + 0(\varepsilon^3) \end{aligned} \tag{4.318}$$

The above explicit representation of the solution (in terms of coefficients and the excitation) can serve as a base for determining its statistical properties. However, in general, the calculations are not easy to perform. In special cases (and under some simplifying assumptions) the first-order moments can be obtained.

The problem that naturally occurs is the convergence of the pertubation expansion of the form (4.314).

As examples show, power pertubation expansions are generally either divegent or convergent too slowly for large values of t (when the ordinary differential equations are considered). In general, this disadvantage is usually accepted since the approximate solution obtained by "cutting off" the formal series at the k-th term proves sufficient in many applications in the asymptotic sense. This means that it approximates an exact solution not when k increases but when k is fixed and $\varepsilon \to 0$. Unfortunately, a second difficulty arises; it turns out that pertubation series contain so called *secular terms*, i.e., terms proportional to a certain positive power of t, and thus tending to infinity as $t \to \infty$. The error arising when expansion (4.314) is substituted into differential equation (4.313) decreases as $\varepsilon \to 0$ and t is fixed, but this decrease is not uniform with respect to t; in other words, finite order approximations are not uniformly valid. To illustrate this deficiency, we shall consider the equation

$$\begin{aligned} & \ddot{Y} + [1 + \varepsilon X(\gamma)]\, Y = 0, \\ & Y(0) = a_1, \quad \dot{Y}(0) = a_2 \end{aligned} \tag{4.319}$$

where ε is a small parameter and $X(\gamma)$ is a random variable. The exact solution of this equation is

$$Y(t,\gamma;\varepsilon) = a_1 \cos(1+X(\gamma))^{\frac{1}{2}} t + (1+X(\gamma))^{-\frac{1}{2}} a_2 \sin(1+X(\gamma))^{\frac{1}{2}} t \tag{4.320}$$

The above solution is a periodic function of t.

Let us try now to look for the solution of (4.319) in the form of pertubation expansion, i.e.

$$Y(t,\gamma;\varepsilon) = \sum_{k=0}^{\infty} \varepsilon^k Y_k(t,\gamma) \tag{4.321}$$

Substituting (4.321) into equation (4.319) and equating terms of the same powers of ε, we obtain a sequence of equations for $Y_k((t,\gamma)$, $k = 0,1,\ldots$. The first two equations are

$$\begin{aligned} \ddot{Y}_o + Y_o &= 0 \\ \ddot{Y}_1 + Y_1 &= -X(\gamma)Y_o(t) \end{aligned} \tag{4.322}$$

The particular solutions of these equations are given by

$$Y_o(t) = a_1 \cos t + a_2 \sin t \tag{4.323}$$

$$Y_1(t,\gamma) = \frac{1}{2} a_2 X(\gamma) t \cos t - \frac{1}{2} a_1 X(\gamma) t \sin t \tag{4.324}$$

It is easily seen that the function $Y_1(t,\gamma)$ oscillates with an increasing amplitude as $t \to \infty$. Both terms in (4.324) are secular. The expression for $Y_n(t,\gamma)$ will involve the products of the form $t^n \cos t$ and $t^n \sin t$. Thus, the asymptotic behaviour of $Y_n(t,\gamma)$ as $t \to \infty$ becomes progressively worse as n increases.

In general, situation is similar when coefficients are stochastic processes. Hence, the use of classical pertubation expansions in the analysis of real random process (described by differential equations) needs special care and good feeling of the physics of the problem. In spite of the deficiencies described above the pertubation technique turned out to be efficient in studying various problems of random vibrations of mechanical systems (cf. [154],[232]) and in stochastic wave analysis (cf. [234]).

28.4.4. Other methods. Remarks

Integral equation formulation

As we have shown in Sec. 26.3.4. an initial-value problem for linear differential equation (with time-varying coefficients) can be represented in the form of an Volterra integral equation.

If coefficients $a_i(t)$ in equation (4.68) are stochastic processes then in the corresponding Volterra integral equation (4.72) the kernel (4.73) and inhomogeneous term (4.74) are random. Like in Sec. 26 the solution is represented in the form (4.76) where $\Lambda = \Lambda(t, \tau, \gamma)$ is given by the Neumann series of iterated kernels. The characterization of solutions requires, in specific problems, that the convergence of the corresponding Neumann series be established and the resolvant kernel calculated. It can be shown (cf. [68]) that in the case when only one coefficient $a_n(t)$ is random a uniform convergence of the Neumann series holds.

Initial value problem for non-linear stochastic systems can also be formulated in the form of stochastic (non-linear) integral equations (cf. [135]). The stochastic integral formulation provides a model which is suitable for consistent numerical analysis.

Equations for moments

A.

Let us consider the following stochastic equation

$$L_o Y(t) + X(t,\gamma) Y(t) = g(t) \tag{4.325}$$

with deterministic initial conditions. The differential operator L_o and function $g(t)$ are deterministic and $X(t,\gamma)$ is a given stochastic process. It is of interest to obtain an equation for the moments of the solution process $Y(t)$.

Averaging both sides of (4.325) gives

$$L_o E[Y] + E\left[X(t,\gamma)Y(t,\gamma)\right] = g \tag{4.326}$$

It is seen that the above equation includes not only $E[Y]$ but also the moment $E\left[X(t,\gamma)Y(t,\gamma)\right]$. To find the equation for $E[XY]$ we multiply both sides of equation (4.325) by $X(t_1,\gamma)$ and take the average. The resulting equation is

$$L_o E\left[X(t_1,\gamma)Y(t,\gamma)\right] + E\left[X(t_1,\gamma)X(t,\gamma)Y(t,\gamma)\right] = E\left[X(t_1,\gamma)\right] g \tag{4.327}$$

which, however, contains the third order moment, so it is necessary to find an equation for it, etc. Finally, we obtain an infinite set of equations for the moments; they are called the *hierarchy equations.* To obtain a finite closed system of equations with a consistent number of unknowns one is forced to introduce simplifying assumptions;

they are called the *truncation* (or *closure*) procedures. One of the procedures postulates that moments of a certain order can be expressed by the lower-order moments; if this is done in such a way as for Gaussian processes the truncation scheme is called the Gaussian closure. Various truncation procedures are considered in the paper by Richardson [219] and Adomian [145]. In the context of stochastic wave propagation the hierarchy equations for moments and the associated closure procedures are discussed in the author's book [234].

B.

There exists an interesting (not white noise) case when one can obtain a closed set of equations for moments of a solution. We have in mind the systems whose coefficients are random telegraph processes (cf. Exercises in Chapter I).

Let us consider a system of linear equations

$$\frac{d\mathbf{Y}(t)}{dt} = \mathbf{A}\mathbf{Y} + \mathbf{B}X(t,\gamma)\mathbf{Y} \qquad (4.328)$$
$$\mathbf{Y}(0) = \mathbf{Y}_0$$

where $\mathbf{Y}(t)$ is an unknown n-dimensional process, $\mathbf{A}$ and $\mathbf{B}$ are $(n \times n)$ matrices and $X(t,\gamma)$ is a random scalar telegraph process, that is

$$X(t) = a(-1)^{N(t)}, \quad X(0) = a, \quad X^2(t,\gamma) = a^2 \qquad (4.329)$$

where a is a constant and $N(t)$ is a homogeneous Poisson process with intensity ν.

Let $R_t[X(\tau)]$ be a functional on the values of $X(\tau)$ for $\tau < t$. It has been shown (cf. [229]) that

$$\frac{d}{dt}E\{X(t)R_t[X]\} = E\left\{X(t)\frac{d}{dt}R_t[X]\right\} - 2\nu E\{X(t)R_t[X]\} \qquad (4.330)$$

The above formula for splitting correlations allows to obtain a closed system of equations for moments. Indeed, taking the average of both sides of equation (4.328) we have

$$E\left[\frac{d\mathbf{Y}}{dt}\right] = \mathbf{A}E[\mathbf{Y}] + \mathbf{B}[X(t)\mathbf{Y}(t)] \qquad (4.331)$$

To find the equation for $E[X(t)\mathbf{Y}(t)]$ we make use of formula (4.330). It should be noticed that $\mathbf{Y}(t)$ as a solution of (4.328) depends only on values of $X(\tau)$ for $\tau \leq t$. By virtue of (4.330) we have

$$\frac{d}{dt}E[X(t)\mathbf{Y}(t)] = E\left[X(t)\dot{\mathbf{Y}}(t)\right] - 2\nu E[X(t)\mathbf{Y}(t)]$$

Substituting $\dot{\mathbf{Y}}(t)$ from the governing equations (4.328) gives a closed equation for $E[Z(t)] = E[X(t)\mathbf{Y}(t)]$.
Thus, we have the following system of equations for the mean of the solution

$$\begin{aligned}\frac{d}{dt}E[\mathbf{Y}] &= \mathbf{A}E[\mathbf{Y}] + \mathbf{B}E[Z] \\ \frac{d}{dt}E[Z] &= (\mathbf{A} - 2\nu\mathbf{I})E[Z] + a^2\mathbf{B}E[\mathbf{Y}]\end{aligned} \tag{4.332}$$

where $\mathbf{I}$ is $(n \times n)$ unit matrix. The initial conditions associated with system (4.332) are

$$\begin{aligned}\mathbf{Y}(0) &= \mathbf{Y}_o, \\ Z(0) &= E[X(0)\mathbf{Y}(0)] = a\mathbf{Y}_o\end{aligned} \tag{4.333}$$

Formula (4.330) makes it also possible to obtain a closed system of equations for the second order moments $E\left[\mathbf{Y}\mathbf{Y}^T\right]$. This system has the form $\left(\zeta = X(t)\mathbf{Y}(t)\mathbf{Y}^T(t)\right)$:

$$\begin{aligned}\frac{d}{dt}E\left[\mathbf{Y}\mathbf{Y}^T\right] &= \mathbf{A}E\left[\mathbf{Y}\mathbf{Y}^T\right] + E\left[\mathbf{Y}\mathbf{Y}^T\right]\mathbf{A}^T + \mathbf{B}E[\zeta] + E[\zeta]\mathbf{B}^T, \\ \frac{d}{dt}E[\zeta] &= \mathbf{A}E[\zeta] + E[\zeta]\mathbf{A}^T - 2\nu E[\zeta] + a^2\mathbf{B}E\left[\mathbf{Y}\mathbf{Y}^T\right] + a^2E\left[\mathbf{Y}\mathbf{Y}^T\right]\mathbf{B}^T\end{aligned} \tag{4.334}$$

where the superscript "T" denotes the transposition of the matrix concerned.

The deterministic systems (4.332) and (4.334) form the base for deriving the conditions of mean and mean-square stability. It is worth noting that equations (4.332), (4.334) hold for general case where matrices $\mathbf{A}$ and $\mathbf{B}$ depend on time (cf. Kotulski, Sobczyk [193]). In paper [193] formula (4.330) has been extended to systems of partial differential equations of the form

$$\frac{\partial u_i(\mathbf{x},t)}{\partial t} = \sum_{j=1}^{k} A_{ij}(\mathbf{x})u_j + X(t,\gamma)\sum_{j=1}^{k} B_{ij}(\mathbf{x})u_j \tag{4.335}$$

where $\mathbf{x} \in R_n$, $t \in [0,\infty)$, $i = 1,2,\dots,k$ and $A_{ij}(\mathbf{x})$, $B_{ij}(\mathbf{x})$ are differential operators with respect to $\mathbf{x}$.

Let us add that in book [94] one can find the analysis of one-dimensional differential equation of the form

$$\frac{dY}{dt} = h(Y(t)) + X(t,\gamma)g(Y(t)) \tag{4.336}$$

where $h(y)$ and $g(y)$ are sufficiently smooth functions and $X(t,\gamma)$ is a random telegraph signal. An integro-differential equation for the probability density of $Y(t)$ has been derived.

29. STOCHASTIC PARTIAL DIFFERENTIAL EQUATIONS

29.1. Use of Hilbert space formulation

Let us consider the following stochastic initial-boundary problem (cf. Kotulski, Sobczyk [108]):

$$\frac{\partial^2 u}{\partial t^2} = \frac{\partial^2 u}{\partial x^2} + \eta(t,x,\gamma)u = f(t,x,\gamma) \tag{4.337}$$

$$u(t,0) = u(t,1) = 0$$

$$u(0,x) = u_o(x,\gamma), \quad \left.\frac{\partial u}{\partial t}\right|_{t=0} = v_o(x,\gamma) \tag{4.338}$$

where $\eta(t,x,\gamma)$ is a given stochastic process, and $u_o(x,\gamma)$, $v_o(x,\gamma)$ and $f(t,x,\gamma)$ are random initial data and random excitation, respectively.

The problem (4.337), (4.338) can be represented in the form of stochastic abstract equation (3.173)

$$\frac{dY}{dt} = AY + B(\eta,t)Y + f(t) \tag{3.173'}$$

where $(v = \frac{\partial u}{\partial t})$:

$$Y = \begin{bmatrix} u \\ v \end{bmatrix}, \quad f = \begin{bmatrix} 0 \\ f \end{bmatrix}, \quad A = \begin{bmatrix} 0 & 1 \\ \frac{\partial^2}{\partial x^2} & 0 \end{bmatrix}, \quad B = \begin{bmatrix} 0 & 0 \\ \eta(t,x,\gamma) & 0 \end{bmatrix} \tag{4.339}$$

The suitable Hilbert space H is taken as

$$H = H_o^1(0,1) \times L_2(0,1)$$

where $H_o^1(0,1)$ is the Sobolev space of square integrable functions possessing square integrable first order derivatives and the inner product on H is [7]:

$$(Y_1,Y_2)_H = \int_0^1 \frac{\partial}{\partial x}u_1(x)\frac{\partial}{\partial x}u_2(x)dx + \int_0^1 v_1(x)v_2(x)dx$$

$$Y_1 = \begin{bmatrix} u_1 \\ v_1 \end{bmatrix}, \quad Y_2 = \begin{bmatrix} u_2 \\ v_2 \end{bmatrix}, \quad Y_1, Y_2 \in H$$

A semigroup of operators generated by A is:

$$K(t)\begin{bmatrix} u(t) \\ v(t) \end{bmatrix} = \begin{bmatrix} \sum_{k=1}^{\infty} 2\left[\{u(t),e_k\}\cos k\pi t + \frac{1}{k\pi}\{v(t),e_k\}\sin k\pi t\right] e_k \\ \sum_{k=1}^{\infty} 2\left[-k\pi\{u(t),e_k\}\sin k\pi t + \{v(t),e_k\}\cos k\pi t\right] e_k \end{bmatrix}$$

where $e_k = e_k(x) = \sin k\pi x$ and

$$\{u(t), e_k\} = \int_0^1 u(t,x)e_k(x)dx$$

The adjoint operators introduced in the method (presented in Sec. 23.3) are:

$$A^* = \begin{bmatrix} 0 & -1 \\ \dfrac{\partial^2}{\partial x^2} & 0 \end{bmatrix},$$

$$B^*Y(t) = \begin{bmatrix} -\int_0^x \int_0^y \eta(t,z,\gamma)v(t,z)dzdy \\ 0 \end{bmatrix}$$

Denoting

$$S(t) = [\varphi(t), \psi(t)]^T, \quad \lambda(t) = [\lambda_1(t), \lambda_2(t)]^T$$

we obtain a specific form of equation (3.177) for $S(t)$:

$$\varphi(t,x) + \int_t^T \sum_{k=1}^{\infty} 2 \int_0^x \int_0^y \eta(t,z)e_k(z)dzdy \Big[k\pi \{\varphi(s), e_k\} \sin k\pi(s-t)$$
$$+ \{\psi(s), e_k\} \cos k\pi(s-t)\Big] ds = \lambda_1(t,x)$$
$$\psi(t,x) = \lambda_2(t,x)$$

If $\Phi_0[a, b, z(t)]$ is a joint characteristic functional of $u_0, v_0, f(t)$ then the characteristic functional of the solution $[u(t,x), v(t,x)]^T$ according to (3.186) takes the form

$$\Phi[\lambda_1, \lambda_2] = \int \Phi_0 \Bigg[\int_0^T \sum_{k=1}^{\infty} 2 \Big[\{\varphi(t), e_k\} \cos k\pi t$$
$$- \frac{1}{k\pi} \{\lambda_2(t), e_k\} \sin k\pi t \Big] e_k dt,$$
$$+ \int_0^T \sum_{k=1}^{\infty} 2 \Big[k\pi \{\varphi(t), e_k\} \sin k\pi t$$
$$+ \{\lambda_2(t), e_k\} \cos k\pi t \Big] e_k dt,$$
$$\int_0^T \sum_{k=1}^{\infty} 2 \Big[k\pi \{\varphi(s), e_k\} \sin k\pi(s-t)$$
$$+ \{\lambda_2(s), e_k\} \cos k\pi(s-t) e_k ds \Bigg] \nu(d\eta) \tag{4.340}$$

where the integration is performed with respect to the probability measure ν defining process $\eta(t)$.

If $u_0 = v_0 = 0$ and $\eta(t,x,\gamma) \equiv 0$, the characteristic functional of the solution $Y = [u,v]^T$ takes the form

$$\Phi[\lambda_1,\lambda_2] = \tilde{\Phi}_0\left[\int_0^T \sum_{k=1}^{\infty} 2\Big[k\pi\{\lambda_1(s),e_k\}\sin k\pi(s-t) + \{\lambda_2(s),e_k\}\cos k\pi(s-t)\Big]e_k ds\right] \tag{4.341}$$

where $\tilde{\Phi}_0$ is the characteristic functional of $f(t)$.

Let us assume now that

$$f(t,x,\gamma) = g(x)X(t,\gamma) = g(x)\sum_i a\delta(t - t_i(\gamma)) \tag{4.342}$$

where $t_i(\gamma)$ are random times of Poisson stream of pulses with average rate Θ and a denotes a constant pulse strength; $g(x)$ is determinstic function. In this case

$$\tilde{\Phi}_0[z(t)] = exp\left(\Theta\int_0^T\left[exp\left\{ia\int_0^T \delta(t-t')\{g,z(t)\}\,dt\right\}-1\right]dt'\right)$$

and

$$\Phi[\lambda_1,\lambda_2] = exp\Bigg\{\Theta\int_0^T\Bigg(exp\Bigg\{ia\int_0^T\sum_{k=1}^{\infty}2\{g,e_k\}H(t-t')\times \Big[k\pi\{\lambda_1(t),e_k\}\sin k\pi(t-t') + \{\lambda_2(t),e_k\}\cos k\pi(t-t')\Big]dt\Bigg\}-1\Bigg)dt'\Bigg\} \tag{4.343}$$

where $H(t)$ is the Heaviside function.

Series expansion of the second *exp* in (4.343) followed by differentiation leads to explicit expressions for moments of the solution process. For further details as well as for analysis of random heat conduction equation the reader is referred to paper [108].

29.2. Stochastic KdV equation

The Korteweg de Vries equation

$$\frac{\partial u}{\partial t} + u\frac{\partial u}{\partial x} + \beta\frac{\partial^3 u}{\partial x^3} = 0 \tag{4.344}$$

where β is a constant constitutes a fruitful model for many various non-linear wave processes in physics. A characteristic feature of this equation is the existence of so called stationary wave (which does not change its form during propagation). This solution has the form $u(\zeta) = u(x - ct)$ and is called the *solitary wave* (or *soliton*). The soliton solution of equation (4.344) has the form

$$u(x,t) = u_0 sech^2\left(\frac{\zeta}{l}\right) \tag{4.345}$$

where u_0 is the amplitude of the initial disturbance and

$$\zeta = x - ct, \qquad c = \frac{1}{3}u_0, \qquad l = \left(\frac{12\beta}{u_0}\right)^{1/2}$$

The soliton propagates without change of its form with constant velocity c; the width of the soliton l decreases with increase of its amplitude. The solitons with greater amplitude propagate quicker.

Introducing new variables

$$x \longrightarrow \beta^{1/3}x, \qquad t \longrightarrow t, \qquad u \longrightarrow -6\beta^{1/3}u$$

leads equation (4.344) to the form

$$\frac{\partial u}{\partial t} - 6u\frac{\partial u}{\partial x} + \frac{\partial^3 u}{\partial x^3} = 0 \tag{4.346}$$

The soliton solution of (4.346) takes the form

$$u(x,t) = -2k^2 sech^2(kx - 4k^3 t) \tag{4.347}$$

where k can be easily expressed in terms of u_0 and l.

It is of interest to study the effect of random external noise on the behaviour of solution (4.347). Let us consider the equation (Orłowski, Sobczyk [214]).

$$\frac{\partial u}{\partial t} - 6\frac{\partial u}{\partial x} + \frac{\partial^3 u}{\partial x^3} = \zeta(t,\gamma) \tag{4.348}$$

where $\zeta(t,\gamma)$ is a given Gaussian stochastic process.

Let us consider the following transformation of variables

$$\begin{aligned} &t' - t, \\ &X = x + \psi(t), \quad \psi(t) = 6\int_0^t W(s)ds, \quad W(s) = \int_0^s \zeta(\tau,\gamma)d\tau, \\ &U(X,t') = u(x,t) - W(t'). \end{aligned} \tag{4.349}$$

It is clear that $\psi(t)$ and $W(t)$ are also Gaussian processes. In new variables equation (4.348) takes the form

$$\frac{\partial U}{\partial t} - 6U(X,t)\frac{\partial U}{\partial X} + \frac{\partial^3 U}{\partial X^3} = 0 \tag{4.350}$$

In original variables

$$u(x,t,\gamma) = W(t,\gamma) - 2k^2 sech^2\left(kx - 4k^3 t + k\psi(t,\gamma)\right) \tag{4.351}$$

Therefore, the soliton randomized by an external random noise $\zeta(t,\gamma)$ is given by (4.351) and forms the non-linear transformation of the original noise $\zeta(t,\gamma)$. The average of (4.351) is

$$\begin{aligned} E\left[u(x,t,\gamma)\right] &= -2k^2 E\left\{sech^2\left[kx - 4k^3 t + k\psi(t,\gamma)\right]\right\} \\ &= -2k^2 \frac{1}{\left[2\pi\sigma_\psi^2(t)\right]^{1/2}} \int_{-\infty}^{+\infty} sech^2(kx - 4k^3 t + k\psi) e^{-\frac{\psi^2}{2\sigma_\psi^2(t)}} d\psi \\ &= -\frac{2k}{\left[2\pi\sigma_\psi^2(t)\right]^{1/2}} \int_{-\infty}^{+\infty} sech^2 z \quad e^{-\frac{\left[z-(kx-4k^3 t)\right]^2}{2k^2\sigma_\psi^2(t)}} dz \end{aligned}$$

From the above expression we can deduce about the asymptotic (for large t) behaviour of $E[u]$. This behaviour is governed by the integral

$$\int_{-\infty}^{+\infty} exp\left\{-\frac{\left[z-(kx-4k^3 t)\right]^2}{2k^2\sigma_\psi^2(t)}\right\} dz.$$

The greatest contribution to this integral (for large t) is provided by the term

$$exp\left\{-\frac{(kx-4k^3 t)^2}{2k^2\sigma_\psi^2(t)}\right\}$$

Taking into account that the integral of $sech^2 z$ is equal to 2 we obtain the following asymptotic formula for the average solitary wave

$$E\left[u(x,t,\gamma)\right] \approx -\frac{4k}{\left[2\pi\sigma_\psi^2(t)\right]^{\frac{1}{2}}} exp\left\{-\frac{(kx-4k^3 t)}{2k^2\sigma_\psi^2(t)}\right\} \tag{4.352}$$

If, for example, $\zeta(t,\gamma)$ is a Gaussian white noise, that is

$$K_\zeta(\tau) = 2D\delta(\tau)$$

then $\sigma_\psi^2(t) = 24Dt^3$ and

$$E\left[u(x,t,\gamma)\right] \approx -\frac{k}{\sqrt{3\pi D}} \frac{1}{t^{3/2}} exp\left\{-\frac{(x-4k^2 t)}{48Dt^3}\right\} \tag{4.353}$$

This indicates that due to the external white noise a solitary wave is deformed during the motion. Asymptotically, its height and width change as $t^{-3/2}$ and $t^{3/2}$, respectively. Further discussion the interested reader will find in paper [214].

Chapter V

STOCHASTIC DIFFERENTIAL EQUATIONS: NUMERICAL METHODS

30. INTRODUCTORY REMARKS

The presentation in the previous chapter indicates that effective analytical solutions of stochastic differential equations are achievable only in some simpler cases. There exists, therefore, an obvious interest to extend a treatment of stochastic differential equations to what might be called *direct numerical integration*. Although for a long time systematic work on numerical methods for stochastic equations has not kept pace with the analytical studies, at present we observe a burst of activity in "stochastic numerics" what is primarily due to progress in stochastic modelling of complex dynamical systems and, of course, due to the common use of computers.

A general look at numerical analysis of differential equations indicates two wide classes of methods. The first one includes approximative methods which, being in the spirit of analytical constructs, are primarily intended for numerical implementation (i.e. iterative methods, variational procedure-like methods of orthogonal projections, the Ritz method etc.). The second class consists of all these schemes which have been constructed with the objective of performing direct numerical evaluation of the approximate value of a solution at some discrete points.

In the context of stochastic differential equations the general approximate schem es have mostly been investigated for regular stochastic equations (in their stochastic operator setting), whereas the numerical methods for the Itô equations have been focussed on direct numerical schemes and their specific features associated with the Wiener process and the Itô integral.

In the case of stochastic Itô differential equations one can work with deterministic F-P-K equations for the probability density. While such partial (deterministic) differential equations can be, in principle, integrated numerically with standard procedures, in practice there occur great difficulties. So, direct numerical methods for stochastic equations seem to be an important and promising tool in characterization of the behaviour of complex systems governed by stochastic differential equations.

In the literature there exists now a number of papers in which basic numerical methods elaborated for deterministic equations have been adopted to stochastic differential equations. Most likely, Maruyama [281] was the first who proposed a scheme for approximate solutions of stochastic equations; it is a natural stochastic analogue of the Euler scheme for deterministic differential equations, and it converges in the mean-square sense. Another important contribution belongs to Milshtein [284] who improved the order of convergence by adding a second-order term in the approximation of the stochastic Itô integral. Among more recent contributions one should mention those of Wagner and Platen [304] and Platen [294] who have derived a formula of the Taylor type for expanding the solution of a stochastic differential equation about the points of a time partition. In series of papers, Talay and Pardoux (cf. bibliography at the end of the book) performed a systematic study of basic problems in stochastic numerical integration. It is not our intention to provide an exhaustive presentation of the existing results in numerical analysis of stochastic differential equations. We only wish to familiarize the reader with the underlying principles of this very fresh subject.

31. DETERMINISTIC EQUATIONS: BASIC NUMERICAL METHODS

31.1. Some approximate methods

A.

The oldest and most popular approximate methods for the solution of differential equations are known as iterative methods. In an iterative method the solution of the equation is constructed by successive approximations. In most instances, y_n, the n-th approximation to the true solution y, is calculated from y_{n-1} alone, when, of course, y_{n-1} was computed from y_{n-2} and so fourth. The procedure used in going from one step to the next is fixed in advance and is often given by a recursion formula. The recursion formula along with initial conditions on the operator (occuring in equation) constitute together an iterative scheme.

Let us consider a class of equations represented in the form of operator equation

$$AY = X \tag{5.1}$$

where, in general, X, Y are elements of appropriate Banach spaces. To make further formulae similar to classical ones, associated with the Neumann series, let us take $A = I - K$ where I is the identify operator. Instead of (5.1) we have

$$Y - KY = X \tag{5.2}$$

The basic recursion formula for successive approximations is given by

$$\begin{aligned} Y_{n+1} &= KY_n + X = \\ &= K^{n+1}Y_o + \sum_{i=0}^{n} K^i X, \quad Y_o\text{-arbitrary} \end{aligned} \tag{5.3}$$

The second term at the right is a partial sum of the infinite series

$$\sum_{i=0}^{\infty} K^i X \tag{5.4}$$

which is called the Neumann series. Here is the basic theorem (a fixed point theorem) for the method of successive approximations (cf. [293]).

Let K be a linear operator on a Banach space such that $\|K\| < 1$. Then equation (5.2) has a unique solution Y^* for every $X \in B$ given by (5.3); that is, the sequence $\{Y_n\}$ given by (5.3) converges to Y^*.
The above theorem has been extended to more general situations, for example, to non-linear operators (cf. [274]).

Let us consider the integral equation (various problems for differential equations can be represented in integral form)

$$\varphi(t) - \lambda K\varphi = f(t) \tag{5.5}$$

where

$$K\varphi = \int_a^b K(t,s)\varphi(s)ds \tag{5.6}$$

In this case the sequence of approximate solutions constructed via the method of successive approximations has the form

$$\begin{aligned} \varphi_{n+1}(t) &= f(t) + \lambda \int_a^b K(t,s)\varphi_n(s)ds \\ &= f(t) + \lambda K\varphi_n, \quad n = 0, 1, \ldots \end{aligned} \tag{5.7}$$

If the first (or, initial) approximation is taken as free term, i.e. $\varphi_0(t) = f(t)$, then

$$\varphi_n(t) = \sum_{m=0}^{n} \lambda^m K^m f \tag{5.8}$$

If interval $[a, b]$ is finite and kernel $K(t, s)$ is bounded

$$|K(t,s)| \leq M = \text{const}.$$

then the sequence of successive approximations (5.7) is uniformly convergent provided

$$|\lambda| < \frac{1}{M(b-a)}.$$

The process of successive approximations in convergent in $L_2(a,b)$ if

$$|\lambda| < \frac{1}{B}, \quad B^2 = \int_a^b \int_a^b |K(t,s)|^2 \, dtds < \infty.$$

In the case of Volterra equations on finite interval the successive approximations are convergent for arbitrary λ.

Formula (5.7) contains the quadrature depending on t, i.e.

$$K\varphi_n = \int_a^b K(t,s)\varphi_n(s)ds$$

In order to calculate $K\varphi_n$ the following formula of often used

$$\int_a^b F(t)dt = \sum_{i=1}^{n} a_i F(t_i)$$

This leads to the following rule for calculating values of $K\varphi_n$ in points t_i:

$$(K\varphi_n)_{t_i} = \sum_{j=1}^{m} a_i K(t_i, t_j)\varphi(t_j).$$

B.

An important class of iterative methods is known as *gradient methods*. They include the method of *steepest descent* and the method of *conjugate gradients* (which is a special case of the conjugate direction method). These methods have been formulated long time ago for systems of linear algebraic equations (for linear operator equations in R_n). Later on, the extension to operator equations in infinite-dimensional spaces has been provided.

Let H be a Hilbert space and operator A maps H into H. Let us assume that A is positive bounded linear operator (cf. [293]); A can be extended to a self-adjoint operator with range being H. Let us consider the equation

$$AY = X, \quad X \in H \tag{5.9}$$

and assume that it has a unique solution Y^* for every $X \in H$. Let us denote:

$$\begin{aligned} r &= r(Y) = X - AY \\ e &= e(Y) = Y^* - Y \end{aligned} \tag{5.10}$$

the residual vector and error vector, respectively. Let

$$J(Y) = (e, r) = (e, Ae) \tag{5.11}$$

be the error functional. It can be shown that gadient of $J(Y)$ is in the direction of the residual vector r. So, r is often said to have the direction of *steepest descent* for J at Y.

A vector $Y \in H$ is considered to be a solution of equation (5.9) if it minimizes $J(Y)$ on the set of all possible solutions. According to gradient methods we minimize $J(Y)$ in the direction of the linearly independent vectors $p_0, p_1, p_2, \dots$ which are to be determined. Thus, an iterative scheme has the form

$$Y_{n+1} = Y_n + \alpha_n p_n, \quad \alpha_n = \frac{(r_n, p_n)}{(p_n, Ap_n)} \tag{5.12}$$

Various gradient methods differ only in the choice of the sequence $\{p_n\}$. In the method of *steepest descent*, $p_n = r_n$. This means that

$$Y_{n+1} = Y_n + \alpha_n r_n, \quad \alpha_n = \frac{\|r_n\|^2}{(r_n, Ar_n)} \tag{5.12'}$$

In the method of conjugate directions we select the p_n so that $(p_n, Ap_m) = 0$ for $n \neq m$.

If operator A has the properties stated above the sequence of approximations $\{Y_n\}$ defined by (5.12) converges to the unique solution Y^* of (5.9).

C.

The gradient methods have the features of iterated methods and variational ones (the recursive formula is assigned to minimize the error functional). The class of methods which are entirely *variational* is based on the following theorem (cf. [283]).

Let A in equation (5.9) be a positive operator (in general, non-linear). If equation (5.9) has solution, then this solution realizes a minimum of the functional

$$\Phi(Y) = (AY, Y) - 2(Y, X) \tag{5.13}$$

and conversely: if there exists element Y for which functional (5.13) attains minimum, then this element is a solution of equation (5.9). The *Ritz method* gives the most common realization of this idea. According to this method an approximate solution of the variational problem is represented in the form

$$Y_n = \sum_{k=1}^{n} a_k \varphi_k \tag{5.14}$$

where φ_k, $k = 1, 2, \dots$ form a complete set of linearly independent elements of a suitable Hilbert space. Coefficients a_k are constants which should minimize $\Phi(Y)$. These constants are determined by system of algebraic equations (the Ritz system). Usually, this system is solved numerically.

In practical implementation of the Ritz method an important issue is associated with the appropriate choice of functions φ_k; they should assure small error of approximation and facility in numerical work. It is worth emphasizing that a sucessful and effective choice of functions φ_k has been provided by the *finite element method.* In this method, functions φ_k are taken as *piece-wise polygonal.* In the last years the finite element method has been applied to various stochastic problems in engineering described by partial stochastic differential equations. However, a rigorous mathematical treatment of the method including its variational onset has not attracted much attention.

31.2. Basic numerical schemes

A.

To illustrate the basic numerical schemes for ordinary differential equations let us consider an initial value problem for a single first order differential equation

$$\frac{dy}{dt} = F(t,y), \quad y(t_o) = y_o, \quad a \le t \le b \tag{5.15}$$

where y_o is known initial condition and function $F(t,y)$ satisfies the conditions for existence and uniqueness of a solution.

The idea of numerical solution of (5.15) is as follows: starting from the given initial condition y_o we wish to evaluate the (approximate) values of the solution for a discrete sequence $\{t_n\}$, $n = 1,2,\ldots$ of values of $t \in [a,b]$. Let y_n be an approximate value of the true solution at $t = t_n$. The basic questions of numerical analysis of initial-value problem (5.1) are:

(i) how the discretization of the interval $[a,b]$ by points $t_1 < t_2 < \cdots < t_n \ldots$ should be performed?

(ii) how the approximations $y_1, y_2, \ldots$ should be calculated?

(iii) how can we find out that y_n is a "good" approximation of the exact unknown $y(t_n)$?

The quantity $h_n = t_n - t_{n-1}$ $(n = 1,2,\ldots)$ is called the *steplength.* If for all points $t_1, t_2, \ldots, t_n - t_{n-1} = h =$const. then we say that calculations are performed with constant steplength, and $t_n = a + nh$ $(n = 1,2,\ldots)$.

The approximations $y_1, y_2, \ldots$ are calculated successively, one after another for ordered points $t_1 < t_2 < \ldots$. Let $y_1, y_2, \ldots, y_{n-1}$ be approximations calculated for $t_1, t_2, \ldots, t_{n-1}$. The calculation of $y_n = y(t_n)$ for the next point t_n is termed as *one step* in calculations.

By a *numerical method* (or, numerical scheme) we understand a specific procedure (usually expressed by analytical formula) for calculating an approximated value in one step. Numerical integration of a differential equation consists in multiple application of a numerical scheme (in subsequent points $t_1, t_2, \ldots$).

An example of one-step numerical method is well known **Euler scheme.** In this method the value y_n is calculated from the Taylor expansion for $y(t_n + h)$ about t_n:

$$y(t_n + h) = y(t_n) + hy^{(1)}(t_n) + \frac{h^2}{2!}y^{(2)}(t_n) + \dots$$

where

$$y^{(q)}(t_n) = \left.\frac{d^q y}{dt^q}\right|_{t=t_n}, \quad q = 1, 2, \dots$$

If we truncate this expansion after two terms and replace $y^{(1)}(t)$ by the right-hand side of equation (5.15), we have

$$y(t_n + h) \approx y(t_n) + hF(t_n, y(t_n))$$

This is an approximate relation between exact values of the solution of (5.15). If we interpret it as an *exact* relation between *approximate* values of the solution of (5.15) we can write down

$$y_{n+1} = y_n + hF(t_n, y_n) \tag{5.16}$$

Formula (5.16) defines the Euler's rule for numerical integration of differential equation (5.1).

The most important class of numerical methods for ordinary differential equations are the *Runge-Kutta methods* and *linear multistep methods.*

The general form of the **m-stage Runge-Kutta methods** is as follows (cf. [277],[278])

$$y_{n+1} = y_n + h\Phi(t_n, y_n, h), \quad n = 0, 1, 2, \dots \tag{5.17}$$

$$\Phi(t_n, y_n, h) = \sum_{i=1}^{m} \alpha_i k_i,$$

$$k_1 = F(t_n, y_n), \quad c_i = \sum_{s=1}^{i-1} a_{is}$$

$$k_i = F(t_n + c_i h, y_n + h\sum_{j=1}^{i-1} a_{ij} k_j) \quad i = 1, 2, \dots, m \tag{5.18}$$

where $m \geq 1$, and α_i, c_i, a_{ij} are known constants. It should be noted that m-stage Runge-Kutta methods involve m function evaluation per step. Each of the functions $k_i(t, y, h)$, $i = 1, 2, \dots, m$ may be interpreted as an approximation to the derivative $\dot{y}(t)$ and the function $\Phi(t, y, h)$ as a weighted mean of these approximations.

Special cases.

1) If $m = 1$, $\alpha_1 = 1$ we obtain the Euler method (5.16).

2) If $m = 2,\ \alpha_1 = 0,\ \alpha_2 = 1,\ c_2 = \frac{1}{2}$ we have

$$y_{n+1} - y_n = hF\left(t_n + \frac{1}{2}h; y_n + \frac{1}{2}hF(t_n, y_n)\right) \tag{5.19}$$

which is original *Runge scheme* known also as the *modified Euler* (or, the improved polygon method) *rule*.

3) If $m = 2,\ \alpha_1 = \frac{1}{2},\ \alpha_2 = \frac{1}{2},\ c_2 = 1$, we obtain the formula

$$\begin{aligned} y_{n+1} - y_n &= \\ &= \frac{1}{2}h\left\{F(t_n, y_n) + F\left(t_n + h, y_n + hF(t_n, y_n)\right)\right\} \end{aligned} \tag{5.20}$$

which is known as the *improved Euler method*.

Of all Runge-Kutta methods the following *fourth-order method* is the most popular, namely

$$\begin{aligned} y_{n+1} &= y_n + \frac{h}{6}(k_1 + 2k_2 + 2k_3 + k_4), \\ k_1 &= F(t_n, y_n), \\ k_2 &= F\left(t_n + \frac{1}{2}h, y_n + \frac{1}{2}k_1\right), \\ k_3 &= F\left(t_n + \frac{1}{2}h, y_n + \frac{1}{2}k_2\right), \\ k_4 &= F(t_n + h, y_n + k_3). \end{aligned} \tag{.21}$$

The Runge-Kutta methods are the main representant of one-step methods (in order to calculate y_n we need only one approximation obtained in the previous step).

The linear multi-step methods include all schemes in which calculations of one step require several previously obtained approximations. The *general linear multi-step (k-step) method* has the form

$$\begin{aligned} y_n &= \alpha_1 y_{n-1} + \ldots + \alpha_k y_{n-k} \\ &+ h(\beta_1 F_{n-1} + \ldots + \beta_k F_{n-k}) + h\beta_o F(t_n, y_n) \end{aligned} \tag{5.22}$$

where $F_{n-k} = F(t_{n-k}, y_{n-k})$, $\alpha_1, \ldots, \alpha_k$ and $\beta_o, \beta_1, \ldots, \beta_k$ are known numbers and $n = k, k+1, \ldots$.

We say that the method (5.22) is *explicit* if $\beta_o = 0$, and *implicit* if $\beta_o \neq 0$. It is seen that the calculations with use of the explicit method can be performed in straightforward way. In the case of implicit method the situation is more involved. Substitution of approximations $y_{n-k}, y_{n-k+1}, \ldots, y_{n-2}, y_{n-1}$ to (5.22) leads to nonlinear algebraic equation of the form

$$y_n = c_n + h\beta_o F(t_n, y_n) \tag{5.23}$$

where c_n denotes the sum of terms including coefficients $\alpha_1, \ldots, \alpha_k$, $\beta_1, \ldots, \beta_k$. This means that y_n is a solution of equation (5.23). In numerical practice, in order to solve (5.23) the Newton method or the method of simple iterations is used. Of course, if we deal with system of differential equations we have to solve a system of non-linear algebraic equations. In numerical calculations the implicit multi-step methods are realized with use of special, separate procedures known in the literature as the *predictor-corrector methods.*

It is worth noticing that the Euler method is also a special case of scheme (5.22) for $k = 1$, $\alpha_1 = 1$, $\beta_1 = 1$, $\beta_0 = 0$. Another example of linear explicit multi-step method is the following *three-step Adams-Bashforth* scheme:

$$(5.24) \qquad y_n = y_{n-1} + h\left(\frac{23}{12}F_{n-1} - \frac{16}{12}F_{n-2} + \frac{5}{12}F_{n-3}\right), \quad n = 3, 4, \ldots$$

One of the simplest implicit one-step method is known as *trapezoidal rule*:

$$(5.25) \qquad y_n = y_{n-1} + \frac{1}{2}h(F_{n-1} + F_n)$$

If $t_{n-1}, y_{n-1} = y(t_{n-1})$, t_n are known, then (5.25) is the equation for determining $y_n = y(t_n)$.

All procedures presented above have been specified for a single first-order differential equation. In the case of system of differential equations, the formulae have the same form, only the scalar quantities should be replaced by their vectorial counterparts. For example, for implicit multi-step method, instead of equation (5.23) we have a system of non-linear algebraic equations

$$(5.23') \qquad \mathbf{y}_n = c_n + h\beta_0 \mathbf{F}(t_n, \mathbf{y}_n).$$

B.

A basic property which we should demand of an acceptable numerical method is that the solution y_n generated by the method *converges*, in some sense, to true solution $y(t)$ as the steplength tends to zeo.

Two other characteristics of any numerical method are: the *rate of convergence* and the *error of approximation*. To explain these notions let us take the Runge-Kutta method determined by formulae (5.17), (5.18).

The approximation y_n is calculated step by step. To estimate the error in one step, let us assume that instead of y_n we substitute in (5.17) the exact value $y(t_n)$. If now

$$k_1(h) = f\left(t_n, y(t_n)\right)$$

$$k_i(h) = F\left(t_n + c_i h, y(t_n) + h\sum_{j=1}^{i-1} a_{ij} k_j(h)\right)$$

then the value

(5.26) $$y(t_n) + h\sum_{i=1}^{m} \alpha_i k_i(h) \equiv y^*_{n+1}$$

approximates the exact value $y(t_n + h)$ of the solution as the approximation from the previous step were not burden by error. Therefore, the difference

(5.27) $$r_n(h) = y(t_n + h) - y^*_{n+1}$$

is a *measure of accuracy* of the method in $(n+1)$-th step. Since, as a starting point for computing $y(t_n + h)$ was taken exact value $y(t_n)$ then it is natural to expect that approximation y^*_{n+1} defined by (5.26) should yield the closest value to its exact counterpart $y(t_n + h)$. How to measure the difference $r_n(h)$?

Let us treat $r_n(h)$ as a function of h and expand it into the Taylor series

(5.28) $$r_n(h) = r_n(0) + hr'_n(0) + h^2\frac{r''_n(0)}{2!} + h^3\frac{r'''_n(0)}{3!} + \dots$$

It is clear that value of $r_n(h)$ diminishes when more and more derivatives $r'_n(0)$, $r''_m(0), \dots$ are equal to zero. Hence, we come to the following definition.

The *approximation error* is the difference $r_n(h)$. We say that numerical method is of order p, if $r_n(0) = 0, \dots, r_n^{(p)} = 0$ and $r_n^{p+1}(0) \neq 0$. If the method is of order p, then

(5.29) $$r_n(h) = h^{p+1}\frac{r_n^{p+1}(0)}{(p+1)!} + \dots$$

Therefore, the practical postulate is as follows: numerical method should be of the highest possible order. The values of derivatives $r_n^{(p)}(0)$ $(p = 1, 2\dots)$ depend on the values of coefficients occuring in the definition of the method (e.g. α_i, c_i, a_{ij} in (5.18)).Optimal selection of these coefficients leads to the highest order of the method.

The approximation error $r_n(h)$ characterizes only an error generated by the method in one step, and what must be emphasized — only under the simplifying and unrealistic assumption that the errors from previous steps have no effect. From this reason error $r_n(h)$ is often called the *local error* of the method; expression (5.29) is called the *principal* local approximation error.

It is clear that in addition to the local error generated by the method we have to consider an error due to multiple application of the procedure (including a transfer of errors). This leads to the *global* (or *accumulated*) *approximation error*:

(5.30) $$e_n = e_n(h) = y(t_n) - y_n(h)$$

where $y(t_n)$ is the true value of the solution at $t = t_n$ whereas y_n is its approximation (depending on h). This error involves all the errors made at each application of the method, and depends in a complicated way on the coefficients of the method and on the initial value problem.

We say that the numerical method is *convergent*, if for each $\varepsilon > 0$, there exists such δ that if $h < \delta$, then for all points $t_n = t_n(h) = a + nh$ we have

$$|e_n(h)| = |y(t_n) - y_n(h)| < \varepsilon \tag{5.31}$$

The proof of convergence of a specific method consists in deriving the appropriate estimate for the approximation error. It has been shown that:
for the Euler method:

$$|r_n(h)| \le ch^2 \tag{5.32}$$

$$|e_n(h)| \le A_0 h \tag{5.33}$$

where c and A_0 are constants;
for the Runge-Kutta methods

$$|r_n(h)| \le c_1 h^{p+1} \tag{5.34}$$

$$|e_n(h)| \le A_1 h^p \tag{5.35}$$

where c_1 and A_1 are constants and p is the approximation order.

The above estimates have asymptotic character, that is they are valid for $h \to 0$ (in practice, for sufficiently small h). To reflect this in explicit way we commonly use the notation

$$f(x) = 0\,(\varphi(x)) \quad \text{as} \quad x \to x_0 \tag{5.36}$$

which means that there exists a positive constant C such that for x sufficiently close to x_0:

$$|f(x)| \le C\,|\varphi(x)|\,. \tag{5.37}$$

The estimates (5.34)–(5.35) are usually expressed in terms of this notation

$$r_n(h) = 0(h^{p+1}),$$

$$e_n(h) = 0(h^p).$$

32. APPROXIMATE SCHEMES FOR REGULAR STOCHASTIC EQUATIONS

32.1. Method of successive approximation

It is clear from the presentation in previous chapter that stochastic differential systems can be represented equivalently in the form of stochastic integral equations. The system of stochastic equations

$$\frac{dY}{dt} = F\left[t, Y(t), \gamma\right], \quad Y(0) = Y_0(\gamma) \tag{5.38}$$

can be written as a stochastic non-linear integral equation of Volterra type

$$Y(t,\gamma) = Y_0(\gamma) + \int_0^t F\left[\tau, Y(\tau), \gamma)\right] d\tau \tag{5.39}$$

In particular, if system (5.38) has the form

$$\frac{dY}{dt} = A(\gamma)Y(t) + f\left[t, Y(t)\right], \quad t \in [0, \infty) \tag{5.40}$$

where $Y = Y(t,\gamma)$ is an n-dimensional random process, $A(\gamma)$ is an $n \times n$ random matrix, and $f(t,y)$ is an n-dimensional vector-valued function, then it can be represented as

$$Y(t,\gamma) - \int_0^t exp\left\{A(\gamma)(t-\tau)\right\} f\left(\tau, Y(\tau,\gamma)\right) d\tau = e^{A(\gamma)t} Y_0(\gamma) \tag{5.41}$$

Denoting

$$X(t,\gamma) = e^{A(\gamma)t} Y_0(\gamma)$$

and

$$K(t,\tau,\gamma) = e^{A(\gamma)(t-\tau)}, \quad 0 \le \tau \le t < \infty$$

equation (5.41) can be written in the form

$$Y(t,\gamma) = X(t,\gamma) + \int_0^t K(t,\tau,\gamma) f\left(\tau, Y(\tau,\gamma)\right) d\tau \tag{5.42}$$

Now, on the basis of paper [290] by Padgett and Tsokos, we shall discuss briefly the problem of obtaining an approximation to a realization of $Y(t,\gamma)$ by method of successive approximations at each $t \in R_+$. To formulate the problem rigorously we need to introduce the appropriate spaces of functions.

Let $C_c = C_c(R_+, L_2(\Gamma, \mathcal{F}, P)$ be the space of all continuous functions from R_+ into $L_2(\Gamma, \mathcal{F}, P)$ such that

$$\left\{\int_\Gamma |Y(t,\gamma)|^2 \, dP(\gamma)\right\}^{\frac{1}{2}} \leq Mg(t)$$

where M is a positive constant and $g(t)$, $t \in R_+$ is a positive continuous function. The topology in C_c is defined by means of the family of semi-norms:

$$\|Y(t,\gamma)\|_n = \sup_{0 \leq t \leq n} \left\{\int_\Gamma |Y(t,\gamma)|^2 \, dP(\gamma)\right\}^{\frac{1}{2}},$$
$$n = 1, 2, \ldots$$

Let $C = C(R_+, L_2(\Gamma, \mathcal{F}, P)$ denote the space of all second-order stochastic processes defined on R_+ which are bounded and continuous in mean-square. Let B and D be two Banach spaces such that $B, D \subset C_c$ and let T be a linear operator mapping C_c into C_c. We say that spaces (B, D) are *admissible* (cf. [282]) with respect to the operator T if and only if $T(B) \subset D$.

A *solution* $Y(t,\gamma)$ of equation (5.42) is understood as a function which belongs to $C_c(R_+, L_2(\Gamma, \mathcal{F}, P)$ and satisfies the equation almost surely.
The following theorem holds [290].
Assumptions:

1) B and D are Banach spaces such that (B, D) is admissible with respect to the operator

$$T(\gamma)Y(t,\gamma) = \int_0^t K(t,\tau,\gamma)Y(\tau,\gamma)d\tau \tag{5.43}$$

where $K(t,\tau,\gamma)$ is continuous with probability one;

2) the mapping $Y(t,\gamma) \to f(t, Y(t,\gamma))$ is a continuous operator defined on

$$S = \left\{Y(t,\gamma) : Y(t,\gamma) \in D, \quad \|Y(t,\gamma)\|_D \leq \rho\right\}$$

for some $\rho \geq 0$ with values in B, satisfying the condition:

$$\|f(t, Y_1(t,\gamma)) - f(t, Y_2(t,\gamma))\|_B \leq k\,\|Y_1(t,\gamma) - Y_2(t,\gamma)\|_D,$$

for $Y_1, Y_2 \in S$ and k a positive constant;

3) $X(t,\gamma) \in D$

4) let $U(\gamma)$ be the mapping from S into D defined as

$$U(\gamma)\,[Y(t,\gamma)] = X(t,\gamma) + \int_0^t K(t,\tau,\gamma) f(\tau, Y(\tau,\gamma))\, d\tau \tag{5.44}$$

5) the distribution of the random variable $X(t,\gamma)$ is known for each $t \in R_+$ or value of $X(t,\gamma)$ can be observed at each $t \in R_+$;
6) let the sequence (of successive approximations) be defined as follows.

$$Y_0(t,\gamma) = X(t,\gamma),$$
$$Y_{n+1}(t,\gamma) = U(\gamma)\left[Y_n(t,\gamma)\right], \quad n \geq 0. \tag{5.45}$$

Then:
a) there exists a unique solution of the stochastic differential equation (5.42) provided

$$k < \frac{1}{N}$$
$$\|X(t,\gamma)\|_D + N\,\|f(t,0)\|_B \leq \rho(1-kN) \tag{5.46}$$

where N is the norm of the operator $T(\gamma)$;
b) for each $t \in R_+$, the sequence of successive approximations $\{Y_n(t,\gamma)\}$ defined by (5.45) converges to the unique solution of equation (5.42) with probability one and in mean-square sense.

Under the assumptions of the above theorem one can also investigate the rate of convergence of the sequence (5.45) and the error of approximation (in the norm of $L_2(\Gamma, \mathcal{F}, P)$. The results are (cf. [290]).

$$\|Y_{n+1}(t,\gamma) - Y_n(t,\gamma)\|_{L_2(\Gamma,\mathcal{F},P)} < (kN)^n \rho(1-kN), \tag{5.47}$$
$$\|Y(t,\gamma) - Y_n(t,\gamma)\|_{L_2(\Gamma,\mathcal{F},P)} < (kN)^n \rho. \tag{5.48}$$

Let us consider the discrete representation (approximation) of equation (5.42) for purposes of numerical integration. Thus, we wish to obtain a solution at each of the discrete points: $0 = t_0 < t_1 < \cdots < \infty$, where $t_i - t_{i-1} = h,\ i = 1,2,\ldots$ and $t_n = t_0 + nh = nh$. For fixed $t = t_n$ the interval from zero to t is devided into n subintervals. As $h \to 0$, then for fixed t such that $t = t_n = hn$, we must have $n \to \infty$.

The integral of equation (5.42) is approximated as:

$$\int_0^t K(t_n,\tau,\gamma) f(\tau, Y(\tau,\gamma))\,d\tau =$$
$$= \sum_{i=0}^{n} w_{n,i} K_{n,i}(\gamma) f_i\left(Y_i(\gamma)\right) \tag{5.49}$$

where

$$K_{n,i}(\gamma) = K(t_n, t_i, \gamma),$$
$$f_i\left(Y_i(\gamma)\right) = f\left(t_i, Y(t_i,\gamma)\right) \tag{5.50}$$

and $w_{n,i}$ are appropriate weights (such as in the composite trapezoidal rule). Therefore, the discrete version of (5.42) can be represented as

$$Y(t_n,\gamma) = X(t_n,\gamma) + \sum_{i=0}^{n} w_{n,i} K_{n,i}(\gamma) f_i\left(Y_i(\gamma)\right) \equiv UY_n \tag{5.51}$$

and recurrent formula for successive approximations $\tilde{Y}_n(\gamma)$ of $Y_n(\gamma)$ for each $t = t_k$ takes the form

$$\begin{aligned} \tilde{Y}_0(t_k,\gamma) &= Y_0(t_k) = X(t_k,\gamma), \\ \tilde{Y}_{n+1}(t_k,\gamma) &= U\left[Y_n(t_k,\gamma)\right], \quad n \geq 0 \end{aligned} \tag{5.52}$$

It has been proven that the sequence of approximations (5.52) converges in the Banach space D to the true solution $Y(t,\gamma)$ at $t = t_k = kh$ as $n \to \infty$.

Other iterative methods can also be extended to stochastic differential equations. The most common way is to interpret such equations as stochastic operator equations (cf. Sec. 21.3) and apply the probabilistic fixed point theorems. Along such a line the steepest descent method has been "randomized". As far as convergence and error estimates (for each $\gamma \in \Gamma$) are concerned the analysis is similar as in the case of deterministic operator equations. New aspects are concerned with measurability and probabilistitic properties of the approximate solutions (cf. Nashed and Engl [286]).

32.2. Approximation and simulation

Numerical analysis of stochastic differential equations consists in various approximations of stochastic processes in question. On the other hand, in situations when we deal with real physical systems the digital simulation of random processes appears most often to be the only feasible approach. Although the subjects mentioned are rather vast, and it is not our intention to discuss them thorouhgly, we wish to throw some light on these problems.

A.

It is known that one of the most common ways of approximation of functions in deterministic analysis is the polynomial approximation grounded in the Weierstrass theorem (cf. [259]). The Weierstrass-type theorems can also be formulated for stochastic processes (and random functions of several variables) — cf. [288].

A random polynomial of degree n is defined as:

$$P_n(t,\gamma) = \sum_{k=0}^{n} a_k(\gamma) t^k \tag{5.53}$$

where $a_k(\gamma)$ are random variables.

A possible extension of the Weierstrass approximation theorem to random processes is the following (cf. [288]):
if $X(t, \gamma)$ is a random process, continuous in probability in the interval $I \subset R_1$, then there exists a family of random polynomials $\{P_n(t, \gamma)\}$ converging uniformly in probability to $X(t, \gamma)$, that is for $\varepsilon > 0$, $\eta > 0$ there exists an integer $N = N(\varepsilon, \eta)$ such that, for all $n \geq N$ and any $t \in I$

$$P\{\gamma : |X(t, \gamma) - P_n(t, \gamma)| \geq \varepsilon\} \leq \eta$$

Let $X(t, \gamma)$ be a random process defined on $I = [0, 1]$, then the *random Bernstein polynomial* is defined as

$$B_n(t, \gamma) = \sum_{k=0}^{n} c_n^k (1-t)^{n-k} X\left(\frac{k}{n}, \gamma\right) \tag{5.54}$$

where $c_n^k = \binom{n}{k}$. If $X(t, \gamma)$ is continuous in probability on I and is bounded, i.e.

$$\sup_{t \in I} |X(t, \gamma)| \leq M < \infty$$

then $\{B_n(t, \gamma)\}_{1 \leq n < \infty}$ converges uniformly in probability to $X(t, \gamma)$.

The most common procedures in simulation of real random processes are based on trygonometric representation of a stationary Gaussian process.

Let $X(t, \gamma)$ be a real-valued, zero-mean stationary Gaussian process. The well-known spectral representation of $X(t, \gamma)$ has the form (cf. Sec. 5):

$$X(t, \gamma) = \int_{-\infty}^{+\infty} e^{i\omega t} d\Phi(\omega) \tag{5.55}$$

where $\Phi(\omega)$, called the spectral process, has orthogonal increments, i.e. $d\phi(\omega_1)$ and $d\Phi(\omega_2)$ are uncorrelated $\omega_1 \neq \omega_2$. Since process $X(t, \gamma)$ is real, representation (5.55) becomes

$$X(t, \gamma) = \int_0^{\infty} \cos \omega t dU(\omega) + \int_0^{\infty} \sin \omega t dV(\omega) \tag{5.56}$$

where $U(\omega)$ and $V(\omega)$ are two independent Gaussian processes with independent increments.

The above formulae are valid for $-\infty < t < +\infty$. When we restrict $X(t, \gamma)$ to the finite interval $[-T, T]$ we do not have direct counterpart of this what in deterministic theory is known as the Fourier series, i.e.

$$x(t) = \sum_{n=0}^{\infty} \left(a_n \cos \frac{n\pi t}{T} + b_n \sin \frac{n\pi t}{T}\right)$$

How to approximate general stationary processes (with continuous spectrum) by "discrete" trygonometric representation?
Theoretically, we have the following assertion (originating from Slutcki):
for arbitrary stationary process $X(t,\gamma)$, for any $\varepsilon > 0$ and arbitrary T (sufficiently large) there exist pairwise uncorrelated random variables $a_1(\gamma),\ldots,a_n(\gamma)$ and $b_1(\gamma),\ldots,b_n(\gamma)$ and real numbers $\omega_1,\ldots,\omega_n$ such that for arbitrary $t\in[-T,T]$

$$E\left[X(t,\gamma)-\sum_{k=1}^{n}(a_k(\gamma)\cos\omega_k t+b_k(\gamma)\sin\omega_k t)\right]^2\le\varepsilon.$$

However, for simulation purposes we would like to have a recipe for calculating the ω_k, a_k and b_k. In this context it is of interest the following result by Pakula [291].

Let us assume that Gaussian stationary process $X(t,\gamma)$ has a spectral density $g_X(\omega)$ satisfying

$$\int_{-\infty}^{+\infty}\frac{\log g_X(\omega)}{1+\omega^2}d\omega > -\infty,$$

then for $T>0$ there exists another Gaussian process $Y(t,\gamma)$ having the discrete spectrum $\{\omega_k\}$ such that $Y(t,\gamma)$ and $X(t,\gamma)$ are identical in probability low on the interval $[-T,T]$; in fact

$$Y(t,\gamma)=\sum_k\alpha_k(\gamma)e^{it\omega_k} \tag{5.57}$$

where $\alpha_k(\gamma)$ are independent Gaussian variables, the series being convergent in the mean-square sense and with probability one.
Although in paper [291] a recipe for calculating the ω_k and α_k is provided, the construction is complicated even for the simple case where $g(\omega)$ is a rational function.

Practical digital simulations are most often accomplished by use of finite sum of cosine functions with random phase angles (cf. Shinzuka [297], [298] and Yang [305]. The reasoning leading to such an approximation is as follows.

By virtue of the property of process $\Phi(\omega)$ in representation (5.55), and process $U(\omega)$, $V(\omega)$ in (5.56) we have

$$E\left|d\Phi(\omega)\right|^2=g_X(\omega)d\omega,$$
$$E\left[dU(\omega)\right]^2=E\left[dV(\omega)\right]^2=\tilde g_X(\omega)d\omega$$

where $\tilde g_X(\omega)$ is the so called one-sided spectral density: $\tilde g_X(\omega)=2g_X(\omega)$. If one defines

$$dU(\omega_k)=[2\tilde g_X(\omega_k)\Delta\omega]^{\frac12}\cos\Psi_k=\sqrt2A_k\cos\Psi_k,$$
$$dV(\omega_k)=[2\tilde g_X(\omega_k)\Delta\omega]^{\frac12}\sin\Psi_k=\sqrt2A_k\sin\Psi_k, \tag{5.58}$$

where Ψ_k are mutually independent random variables distributed uniformly on $[0, 2\pi]$ and

$$A_k = [\tilde{g}_X(\omega_k)\Delta\omega]^{\frac{1}{2}}, \quad \omega_k = \left(k - \frac{1}{2}\right)\Delta\omega \tag{5.59}$$

then a discrete approximation of (5.56) is assumed in the form

$$\begin{aligned} X(t,\gamma) &= \sum_{k=1}^{N}\left[\cos\omega_k t dU(\omega_k) + sin\omega_k t dV(\omega_k)\right] \\ &= \sqrt{2}\sum_{k=1}^{N}\left[A_k \cos\omega_k t \cos\Psi_k + A_k \sin\omega_k t \sin\Psi_k\right] \\ &= \sqrt{2}\sum_{k=1}^{N} A_k \cos(\omega_k t - \Psi_k(\gamma)) \end{aligned} \tag{5.60}$$

where ω_k are realized values of frequency distributed according to the density $g_X(\omega)$. The degree of the approximation depends on whether the interval of frequencies $[0, \omega^*]$ taken into account in discretization is large enough and whether $\Delta\omega$ in (5.59) is small enough so that (cf. Shinozuka [298]):

$$A_k^2 = \int_{(k-1)\Delta\omega}^{k\Delta\omega} \tilde{g}_X(\omega_k)\Delta\omega = \tilde{g}_X(\omega_k)\Delta\omega$$

is valid. In simulation practice, formula (5.60) is used with Ψ_k being replaced by their realized values ψ_k.

It is clear that the larger number N used in simulation is, the better accuracy of the simulated process will result. However, we should keep in mind, that too large number N, for instance, 10^5, could be impractical from economical point of view. Therefore, it is of vital importance to investigate the accuracy of simulation within the practical range of N, such as $10^2 - 10^3$. It is also desirable to investigate the sensitivity of the accuracy of simulation with respect to the number, N, of cosine terms in the approximation formula. Recently, with the advancement in the numerical technique of fast Fourier transform, it has become possible to investigate the accuracy of simulation with respect to N (cf. [305]).

B.

In the approach to simulation described above the process under consideration is represented by a finite number of cosine functions and the stochastic aspects are reduced to the selection of amplitudes and phases. The process itself is discrete in the frequency domain and continuous in time. All frequency components contribute at any given time, and the amount of computation in each time step grows rapidly with the required frequency band. Although the use of the fast Fourier transform algorithm improves the situation, but then the full time history must be calculated simultaneously.

In the recent years sequential simulation algorithms have been introduced in the form of *Moving-Average* (MA), *Auto-Regressive* (AR) and the combined *Auto-Regressive Moving-Average* process (ARMA). These processes are discrete in time and continuous in the frequency domain. The general form of the ARMA model is (cf. Kozin [273], Krenk [276]):

$$X_n + \sum_{k=1}^{N} a_k X_{n-k} = \sum_{k=0}^{M} b_k \xi_{n-k} \tag{5.61}$$

where $\{a_k\}, \{b_k\}$ are constant coefficients, and $\xi_n, \xi_{n-1}, \ldots, \xi_{n-M}$ is a sequence of independent identically distributed random variables (most often taken to be Gaussian variables). The sequence $\{X_n, X_{n-1}, \ldots\}$ denotes the discrete values (observations in discrete instants of time) of the process to be simulated; it is called a time series.

The model (5.61) is of *order* (N, M) and all variables are scalar quantities. This corresponds to a so-called single input-single output linear model. ARMA models can be extended to multi-input and multi-output linear systems, in which case the observations X_k and the random terms ξ_k become vectors, and the coefficients $\{a_k\}, \{b_k\}$ become matrices.

As it is seen from the formula (5.61), ARMA model corresponds to passing a white noise through a discrete filter, and available methods of analysis for digital filters can be used to describe the properties of the simulated sequence (cf. e.g. Oppenheim and Schafer [289]).

The MA process corresponds to $N = 0$ in (5.61); in this case the value of process at time $t = n$ is characterized by a linear combination of Gaussian random variables. The AR process corresponds to $M = 0$; this means that X_n is generated from a linear combination of N previous values of the process plus a single independent random variable.

The basic problem of modelling of the time series $\{X_n\}$ by the ARMA process consists in estimating the parameters $\{a_k\}, \{b_k\}$. There is also the question of determining the best model order to fit the observed data (best choice of N and M) — cf. [273].

The ARMA processes have many advantages and they received considerable attention in a variety of disciplines (e.g. simulation of load processes such as wind, sea waves and earthquakes). It is worth noting that the AR representation is equivalent to a model of a time series which has maximum entropy. Furthermore, the ARMA process turned out to be particularly suited for analysis of random vibration of linear systems; it can be shown that the ARMA process with $N = M + 1 = 2n$ is an exact representation of the discretely sampled response of an n-degree of freedom linear system to white noise excitation (cf. Gersch, Liu [263]). For further and more detailed information about ARMA process the reader is referred to [273],[276],[300].

Other questions concerned with simulation of stochastic processes are discussed by Franklin [262]; simulation study of stochastic Van der Pol equation is provided by Spanos in paper [301].

33. NUMERICAL INTEGRATION OF ITÔ STOCHASTIC EQUATIONS

33.1. Preliminaries

In this section we shall be concerned with the Itô stochastic equation

$$(5.62)\qquad \begin{aligned} dY_t &= m(t, Y_t)dt + \sigma(t, Y_t)dW_t \\ Y_{t_0} &= Y_0 \end{aligned}$$

or in integral form

$$(5.63)\qquad Y_t = Y_{t_0} + \int_{t_0}^{t} m(s, Y_s)ds + \int_{t_0}^{t} \sigma(s, Y_s)dW(s)$$

When we think on various approximations of the solution process, the first method which, most likely,comes to the mind is the method of successive approximations. It has been used in Sec. 22.1 for proving the existence of the solution and, in principle, can be adopted for numerical calculations. The recursive formula for successive approximations is:

$$(5.64)\qquad Y_t^{(n)} = Y_0 + \int_{t_0}^{t} m(s, Y_s^{(n-1)})ds + \int_{t_0}^{t} \sigma(s, Y_s^{(n-1)})dW(s)$$

where $n = 1, 2, \ldots$. As a zero approximation we take Y_0.
As we know, under the conditions of the existence and uniqueness theorem the sequence $\{Y_t^{(n)}\}$ converges uniformly with probability one on the considered interval $[t_0, T]$ to a solution Y_t of equation (5.62).

In numerical implementation of this method the values of the first integral in (5.64) can be calculated by use of any known method of numerical integration. For calculation of values of the second integral in (5.64) one has to use the definition of the integral with respect to the Wiener process. If we adopt the Stratonovich definition we obtain:

$$(5.65)\qquad \int_{t_0}^{t} \sigma(s, Y_s^{(n-1)})dW(s) \approx \sum_{i=0}^{n-1} \sigma\left(s, \frac{Y_{t_i}^{(n-1)} + Y_{t_{i+1}}^{(n-1)}}{2}\right) \Delta W_i$$

where $\Delta t = t_{i+1} - t_i = h$ is the discretization step, and $\Delta W_i = W_{t_{i+1}} - W_{t_i}$. In specific calculations the increments ΔW_i constitute a sequence a mutually independent Gaussian random variables.
The method of successive approximations is, however, not popular in numerical practice since it consumes too much computing time (to calculate the approximation of the process at time t one has to have in the disposal the values of the previous approximation for all $s \in [t_o, t]$). Due to this fact, like as for deterministic equations, various one-step difference methods are used.

In contrast to the deterministic case, where different numerical methods converge (if they are convergent) to the same solution, in the case of stochastic equation (5.62) different schemes can converge to *different* solutions (for the same noise sample and initial condition). In addition, we can consider various types of approximations; the most common are:

(i) mean-square approximation (in L_2-norm),
(ii) pathwise (or, trajectory) approximation,
(iii) approximation of moments $E[f(Y_t)]$.

Another essential feature of numerical analysis of stochastic Itô equations manifests itself in additional methodical difficulties when we wish to deal with the multidimensional equations; for instance, in this case the solution process can not be, in general, expressed as a continuous functional of the Wiener process alone.
In general, the main problem of numerical integration of the Itô stochastic equations can be verbalized as follows:

we wish to approximate (in the appropriate sense) a solution of equation (5.62) by a time-discretized recurrent formula

$$Y_{t_{i+1}} = f(Y_{t_i}, W_s; t_i \le s \le t_{i+1}); \tag{5.66}$$

the question associated is:
which of the convergent discretization schemes (5.66) can be considered as "good" approximation to (5.62)?
A goodness of a scheme depends primarily on the type of chosen approximation (cf. (i), (ii), (iii)). In the sequel we shall discuss briefly the basic procedures.

We consider here the scalar Itô equation (5.62) for $t \in [t_o, T]$, $Y_{t_o} = Y_o$, independent of $W_t - W_{t_o}, t > t_o$ and $E(Y_o^2) < \infty$, where the drift $m(t, y)$ and diffusion $\sigma(t, y)$ satisfy all conditions guaranteeing the existence and uniqueness of a solution.

Let the discretization of the interval $[t_o, T]$ be defined as

$$\tau_n = \{t_i : 0, 1, \ldots, n; \quad t_o < t_1 < \cdots < t_n = T\}$$

Let us introduce the following notation:
the time increment: $\Delta t_i = t_{i+1} - t_i = h$,
the increment of the standard Wiener process: $\Delta W_i = W_{t_{i+1}} - W_{t_i}$,
the approximation process $Y_t^{(n)} = \overline{Y}_t$, and $\overline{Y}_{t_i} = \overline{Y}_i$.

Let us consider the integral counterpart of equation (5.62) on the interval $[t_i, t_{i+1}]$:

$$Y_{t_{i+1}} = Y_{t_i} + \int_{t_i}^{t_{i+1}} m(s, Y_s)ds + \int_{t_i}^{t_{i+1}} \sigma(s, Y_s)dW_s \tag{5.67}$$

"Good" approximate schmemes should be constructed in such a way as to approximate as well as possible the integral terms in (5.67). If we are interested in the ***mean-square approximation*** this means that for each $t \in [t_0, T]$

$$E(Y_t - \overline{Y}_t)^2 \to 0 \quad \text{as} \quad h \to 0$$

and, moreover, that for fixed small h (i.e. for large n), $E(Y_n - \overline{Y}_T)^2$ be as small as possible. In other words, one is looking for a scheme such that for all $t \in [t_0, T]$:

$$E(Y_t - \overline{Y}_t)^2 = 0(h^r) = 0\left(\frac{1}{n^r}\right)$$

with r as large as possible.

33.2. Stochastic Euler scheme

The simplest numerical scheme for the Itô equation (5.62) is the stochastic analogue of the Euler method. The stochastic Euler scheme is obtained from (5.67) by replacing $m(s, Y_s)$ by $m(t_i, Y_i)$ for $s \in [t_i, t_{i+1}]$ and $\sigma(s, Y_s)$ by $\sigma(t_i, Y_{t_i})$ for $s \in [t_i, t_{i+1}]$. As a result we obtain

$$\overline{Y}_{i+1} = \overline{Y}_i + m(t_i, \overline{Y}_i)h + \sigma(t_i, \overline{Y}_i)\Delta W_i \tag{5.68}$$

It was shown by Maruyama [281] that the Euler scheme (5.68) converges uniformly in mean-square sense to Y_t governed by (5.62) as $h \to 0$. It has also been shown (cf. [292],[296]) that the order of convergence of the Euler scheme (5.68) is $0(h)$, that is, for each $t \in [t_0, T]$

$$E(Y_t - \overline{Y}_t)^2 = 0(h) \tag{5.69}$$

The order of approximation (5.69) of the Euler scheme is too low; numerical calculations with use of this scheme show significant departure of the approximate values from the exact ones (cf. [272]).
It is worth noting that in the particular case: $m(t, y) = 0$, $\sigma(t, y) = 1$ one has exactly

$$E\left|Y_T - \overline{Y}_T\right|^2 = e^T - \left(1 + \frac{T}{n}\right)^n \approx \frac{e^T T^2}{2n}. \tag{5.70}$$

33.3. Milshtein scheme

If we want to reduce the error of approximation we may replace $\sigma(s, Y_s)$ by

$$\sigma(t_i, Y_{t_i}) + \sigma'(t_i, Y_{t_i})\sigma(t_i, Y_{t_i})[W_s - W_{t_i}]$$

where $\sigma'(t, y) = \frac{\partial \sigma}{\partial y}(t, y)$. Noting that

$$\int_{t_i}^{t_{i+1}} (W_s - W_{t_i})dW_s = \frac{1}{2}\left[(\Delta W_{i+1})^2 - h\right]$$

we obtain the *Milshtein scheme*: $(i = 0, 1, \dots)$

$$\begin{aligned}
\overline{Y}_{i+1} &= \overline{Y}_i + \left[m(t_i, \overline{Y}_i) - \frac{1}{2}\sigma(t_i, \overline{Y}_i)\sigma'(t_i, \overline{Y}_i)\right] h + \\
&\quad + \sigma(t_i, \overline{Y}_i)\Delta W_i + \frac{1}{2}\sigma(t_i, \overline{Y}_i)\sigma'(t_i, \overline{Y}_i)[\Delta W_i]^2, \qquad (5.71)\\
\overline{Y}_0 &= Y_{t_0}
\end{aligned}$$

The above scheme can be derived from the Taylor expansion of $m(t, Y)$ and $\sigma(t, Y)$ when we keep all terms of order Δt (and remember that $\Delta W_i \approx 0(\Delta t)^{1/2}$). We have

$$\begin{aligned}
Y_{t_{i+1}} - Y_{t_i} &= m(t_i, Y_{t_i})\Delta t + \sigma(t_i, Y_{t_i})\Delta W_i + \\
&\quad + \int_{t_i}^{t_{i+1}} [Y_s - Y_{t_i}]\sigma'(t_i, Y_{t_i})dW_s + \dots \\
&= m(t_i, Y_{t_i})\Delta t + \sigma(t_i, Y_{t_i})\Delta W_i \\
&\quad + \sigma'(t_i, Y_{t_i}) \int_{t_i}^{t_{i+1}} \left[\int_{t_i}^{s} \sigma(\tau, Y_\tau)dW_\tau\right] dW_s + \dots \\
&= m(t_i, Y_{t_i})\Delta t + \sigma(t_i, Y_{t_i})\Delta W_i + \\
&\quad + \sigma(t_i, Y_{t_i})\sigma'(t_i, Y_{t_i}) \int_{t_i}^{t_{i+1}} (W_s - W_{t_i})dW_s + \dots
\end{aligned}$$

The Milshtein scheme can also be obtained as the Euler scheme for the Stratonovich version of equation (5.62) — cf. [292].

An important result associated with the Milshtein scheme (5.71) can be stated as follows (Milshtein [284], Talay [303]):

If $E(Y_0^4) < \infty$ and:

(a) m and σ are of class C^2,

(b) m, m', σ and σ' are unformly Lipschitz, then, if $\overline{Y}_{i+1}$ is defined by (5.71), for each $t \in [t_0, T]$:

$$E\left(|Y_t - \overline{Y}_t|^2\right) = 0(h^2).$$

It should be noticed that the Euler and Milshtein schemes described above produce for each regular partition of the interval an approximation of Y_t that depends on the values of the forcing Brownian motion process only at the dividing points of the partition. In other words, we say that the methods produce a P_n-measurable approximation, where P_n denotes the "partition" σ-field generated by $W_{\frac{iT}{n}}$, $i = 0, 1, \ldots$, $(t_o = 0)$.

It has been shown (cf. [261]) that the spead of convergence of the Milshtein scheme is, in a sense, optimal within a large class of P_n-measurable approximations; this problem is also studied in the paper by Newton [287].

It is worth emphasizing that one can construct other types of algorithms which yield higher order convergence. For example, Wagner and Platen [304] and Platen [294] developed a family of discretization methods with any prescribed order of accuracy for the mean-square error. However, the approximations of this type depend in very complicated way on the Wiener process and are not P_n-measurable. One of simpler schemes which include (in addition to ΔW_i) the integral of $W_s - W_{t_i}$ is (cf. Platen [294])

$$
\begin{aligned}
\overline{Y}_{i+1} = \overline{Y}_i + m_i h + \sigma_i dW_i + \frac{1}{2}\sigma_i \sigma_i' \left[(\Delta W_i)^2 - h)\right] + \\
+ \sigma_i m_i' \Delta Z_i + \left(m_i \sigma_i' + \frac{1}{2}\sigma_i^2 \sigma_i'' \right) (\Delta W_i h - \Delta Z_i) + \\
+ \frac{1}{2}\left[\sigma_i'' \sigma_i + (\sigma_i')^2\right] \sigma_i \Delta W_i \left[\frac{1}{3}(\Delta W_i)^2 - h \right] + \\
+ \left(m_i m_i' + \frac{1}{2}\sigma_i^2 m_i'' \right) \frac{h^2}{2},
\end{aligned}
\tag{5.72}
$$

where

$$
m_i = m(t_i, \overline{Y}_i), \quad \sigma_i = \sigma(t_i, \overline{Y}_i),
$$

$$
\Delta Z_i = \int_{t_i}^{t_{i+1}} (W_s - W_{t_i}) ds
$$

The rate of convergence of scheme (5.72) is $0(h^3)$.

33.4. Stochastic Runge-Kutta schemes

It has always been tempting to extand the Runge-Kutta methods widely elaborated in deterministic numerical analysis to stochastic equations. Such an extension has been explicitly displayed in a number of papers (cf. Helfand [267], Greenside and Helfand [265], Koeden and Pearson [272], Rumelin [296], Klauder and Peterson [271]).

A stochastic counterpart of the two-stage Runge-Kutta method (called Heun-method — [260]) has the form

$$\overline{Y}_{i+1} = \overline{Y}_i + \frac{1}{2}h\left[m(t_i, \overline{Y}_i) + m(t_{i+1}, \hat{Y}_i)\right] + \\ + \frac{1}{2}\left[\sigma(t_i, \overline{Y}_i) + \sigma(t_{i+1}, \hat{Y}_i)\right] dW_i \tag{5.73}$$

where

$$\hat{Y} = \overline{Y}_i + m(t_i, \overline{Y}_i)h + \sigma(t_i, \overline{Y}_i)dW_i \tag{5.74}$$

It has been proven by McShane [52] that this scheme converges in mean-square sense to the solution of the Itô equation

$$dY_t = \left[m(t, Y_t) + \frac{1}{2}\sigma(t, Y_t)\sigma'(t, Y_t)\right] dt + \sigma(t, Y_t)dW_t \tag{5.75}$$

that is, to the solution of equation (5.62) in the Stratonovich interpretation. A general m-stage stochastic Runge-Kutta method can be defined as follows (cf. formulae (5.17), (5.18)):

$$\begin{aligned} &\overline{Y}_0 = Y_0 \\ &\overline{Y}_{i+1} = \overline{Y}_i + h\Phi(t_i, \overline{Y}_i, h) + \Psi(t_i, \overline{Y}_i, h)\Delta W_i \end{aligned} \tag{5.76}$$

where

$$\begin{aligned} \Phi(t_i, \overline{Y}_i, h) &= \sum_{j=1}^{m} \alpha_j K_j, \quad m \geq 1 \\ \Psi(t_i, \overline{Y}_i, h) &= \sum_{j=1}^{m} \beta_j G_j \end{aligned} \tag{5.77}$$

where

$$\begin{aligned} &K_1 = m(t_i, \overline{Y}_i), \quad G_1 = \sigma(t_i, \overline{Y}_i), \\ &K_j = m(t_i + c_j h, Y_i^{(j)}), \\ &G_j = \sigma(t_i + c_j h, Y_i^{(j)}), \quad j = 2, \ldots, m \\ &Y_i^{(j)} = \overline{Y}_i + h\sum_{k=1}^{j-1} a_{jk}K_k + \sum_{k=1}^{j-1} b_{jk}G_k\Delta W_i \end{aligned} \tag{5.78}$$

and

$$\sum_{j=1}^{m} \alpha_j = \sum_{j=1}^{m} \beta_j = 1.$$

Coefficients α_j, β_j, c_j, a_{jk}, b_{jk} are constant which are selected in similar way as in determinstic case.

The following **theorem** has been proven by Rumelin [296].

Let

$$m, \sigma, \frac{\partial m}{\partial t}, \frac{\partial m}{\partial y}, \frac{\partial \sigma}{\partial t}, \frac{\partial^2 \sigma}{\partial t^2}, \frac{\partial^2 \sigma}{\partial t \partial x}, \frac{\partial^2 \sigma}{\partial x^2}$$

be bounded. Then the approximation $\overline{Y}_t$, $t \in [t_o, T]$ defined by the m-stage Runge-Kutta scheme (5.76)–(5.78) converges uniformly in the mean-square sense to the solution Y_t of the following equation

$$dY_t = \left[m(t, Y_t) + \lambda\sigma(t, Y_t)\frac{\partial \sigma}{\partial y}(t, Y_t)\right] dt + \sigma(t, Y_t) dW_t \tag{5.79}$$

with correction factor $\lambda = 0$ for $m = 1$ and with

$$\lambda = \sum_{j=2}^{m} \beta_j \sum_{k=1}^{j-1} b_{jk} \tag{5.80}$$

for $m \geq 2$.

Let us notice, that $\lambda = 0$ corresponds to the Euler scheme, which gives convergence to the solution of the Itô equation. In the case of scheme (5.73) when $m(t, y) \equiv 0$, $\sigma = \sigma(y)$ we have $b_{21} = 1$ and $\lambda = \beta_2 = \frac{1}{2}$ which reproduces the Stratonovich interpretation of equation (5.62). The Runge-Kutta scheme in its general form (5.76)—(5.78) allows us to obtain a large spectrum of stochastic equations whose solutions are the limits (in mean-square) of Runge-Kutta approximations with various combinations of parameters.

As far as convergence of the stochastic Runge-Kutta schemes is concerned one can easily show (by reducing to the Milshtein scheme) that the Heun scheme (5.73)–(5.74) has the order of mean-square convergence $0(h^2)$; its one step approximation error is of order $0(h^3)$. A general result is as follows cf. [296]).

Given one-dimensional stochastic equation (5.62) with m and σ and their partial derivatives up to the fourth order bounded an continuous; let the values of the driving Wiener process $W(t)$ be known at discrete time $\{t_o, t_1 \dots, t_n\}$; let the function $S = S(t, y)$ be defined as

$$S = \sigma\frac{\partial m}{\partial y} - \frac{\partial \sigma}{\partial t} - m\frac{\partial \sigma}{\partial y} - \frac{1}{2}\sigma^2\frac{\partial^2 \sigma}{\partial y^2} \tag{5.81}$$

Then:

(i) if coefficients of equation (2.62) are such that $S(t, y) \not\equiv 0$ then the best integration scheme for (2.62) has a mean-square approximation order $0(h^2)$;

(ii) if $S(t, y) \equiv 0$, the best possible order of (global) mean-square approximation is at least $0(h^4)$; this order can be realized by the stochastic counterpart of the classical fourth order Runge-Kutta scheme and $\lambda = \frac{1}{2}$.

Further investigation of the stochastic Runge-Kutta method can be found in the paper by Klauder and Peterson [271]. The authors consider a two-stage Runge-Kutta algorithm for vector Itô stochastic differential equations with multiplicative noise with some examples illustrating the method. These examples indicate that an error estimate for the first two cumulants is very sensitive to the number of sample paths used in calculations.

33.5. Multi-dimensional systems

Let us consider a system (in R_n) of Itô stochastic differential equations

$$d\mathbf{Y}_t = \mathbf{m}(t, \mathbf{Y}_t)dt + \sum_{j=1}^{m} \boldsymbol{\sigma}_j(t, \mathbf{Y}_t)dW_j(t)$$

$$\mathbf{Y}_{t_o} = \mathbf{Y}_o \tag{5.82}$$

where $\mathbf{Y}_t = [Y_1(t), \ldots, Y_n(t)]$, $\mathbf{W}(t) = [W_1(t), \ldots, W_m(t)]$ and $W_j(t)$ are mutually independent Wiener processes and $\mathbf{m}$ and $\boldsymbol{\sigma}_j$ are Lipschitz continuous functions from $[t_o, T]$ into R_n. The above Itô equation may be rewritten in Stratonovich form as:

$$d\mathbf{Y}_t = \tilde{\mathbf{m}}(t, \mathbf{Y}_t)dt + \sum_{j=1}^{m} \boldsymbol{\sigma}_j(t, \mathbf{Y}_t) \circ dW_j(t) \tag{5.83}$$

where

$$\tilde{\mathbf{m}} = \mathbf{m} - \frac{1}{2}\sum_{j=1}^{m} \boldsymbol{\sigma}_j \boldsymbol{\sigma}'_j \tag{5.84}$$

and $\boldsymbol{\sigma}'_j$ denotes the matrix whose elements are equal to: $\frac{\partial \sigma_{j,i}}{\partial y_k}$.

In general, the results presented for one-dimensional case are not extendable directly to the multi-dimensional case. Only in the situation when the multi-dimensional equation includes one scalar Wiener process such an extension is straightforward. In this case, the term $\sigma'\sigma$ occuring in numerical schemes above should be replaced by $(\nabla_y \boldsymbol{\sigma})\boldsymbol{\sigma}$.

When $n > 1$ and $\mathbf{W}(t)$ is the vector process, it is not possible to express, as above, the solution of Itô equation as a continuous functional of the processes $W_j(t)$, $j = 1, \ldots, m$. In certain cases, Y_t can be expressed as a continuous functional of the processes

$$\left\{ W_j(t), \int_{t_o}^{t} W_i(s)dW_j(s), \ldots, \quad i, j = 1, \ldots, m \right\}$$

but not of the processes $W_j(t)$ alone. The possibility of expressing the solution as a continuous functional of a given set of processes (and the complexity of such functional) depends on the structure of the Lie algebra associated with vectors$\boldsymbol{\sigma}_j$, $j = 1, 2, \ldots, p$; cf [275]. Due to this difficulty we introduce the following **restriction**:

(5.85) $$\boldsymbol{\sigma}_i'(t, \mathbf{y})\boldsymbol{\sigma}_j(t, \mathbf{y}) \equiv \boldsymbol{\sigma}_j'(t, \mathbf{y})\boldsymbol{\sigma}_i(t, \mathbf{y})$$

Under usual regularity conditions on the coefficients two basic schemes have the form:
the *Euler scheme*:

(5.86) $$\overline{\mathbf{Y}}_{i+1} = \overline{\mathbf{Y}}_i + \mathbf{m}(t_i, \overline{\mathbf{Y}}_i)h + \sum_{j=1}^{m} \boldsymbol{\sigma}_j(t_i, \overline{\mathbf{Y}}_i)\,[W_j(t_{i+1}) - W_j(t_i)]\,;$$

the *Milshtein scheme*:

(5.87) $$\begin{aligned} \overline{\mathbf{Y}}_{i+1} &= \overline{\mathbf{Y}}_i + \tilde{\mathbf{m}}(t_i, \overline{\mathbf{Y}}_i)h + \sum_{j=1}^{m} \boldsymbol{\sigma}_j(t_i, \overline{\mathbf{Y}}_i)\,[W_j(t_{i+1}) - W_j(t_i)] + \\ &+ \frac{1}{2} \sum_{j,k=1}^{m} \boldsymbol{\sigma}_j'(t, \overline{\mathbf{Y}}_i)\boldsymbol{\sigma}_k(t, \overline{\mathbf{Y}}_i)\,[W_j(t_{i+1}) - W_j(t_i)]\,[W_k(t_{i+1}) - W_k(t_i)] \end{aligned}$$

with $\overline{\mathbf{Y}}_o = \mathbf{Y}_t$ in both schemes.

The convergence rates differ depending on whether the commutativity condition (5.85) is satisfied or not. It has been shown (cf. [284],[296]) that:

if the coefficients of equation (5.82) do not satisfy condition (5.85) then the global mean-square approximation error of schemes (5.86) and (5.87) is $0(h)$; if condition (5.85) holds then a possible achievable error is $0(h^2)$; this order is realized by the Milshtein scheme (5.87), or by "corrected" n-dimensional Heun scheme.
Let us notice, for example, that the condition (5.85) does not hold for the system

$$\begin{aligned} dY_1 &= dW_1(t) \\ dY_2 &= Y_1(t)dW_2(t) \end{aligned}$$

Indeed,

$$\boldsymbol{\sigma}_1'\boldsymbol{\sigma}_2 = \begin{pmatrix} 0 \\ 0 \end{pmatrix}, \quad \boldsymbol{\sigma}_2'\boldsymbol{\sigma}_1 = \begin{pmatrix} 0 \\ 1 \end{pmatrix}.$$

Remark.

All procedures discussed in this section give approximations of the solution in the mean-square sense. In various situations, we are given (e.g. by a measurement device) the actual realization of the driving process (either in continuous or in discrete time) and we wish to compute an approximation of the corresponding trajectory

$\{Y_\gamma(t),\ t \in [t_0, T],\ \gamma \in \Gamma\}$. In such cases the numerical scheme should converge pathwise, i.e. for each ω with probability one. This kind of approximation is discussed by Pardoux and Talay in paper [292].

Another problem which will not be discussed here arises when we wish to approximate numerically certain functionals defined on the solution process. For instance, we may wish to compute the quantity $E[f(Y_t)]$. Usually in practice, in order to approximate such averages one chooses n which is large enough and calculates the quantity

$$\frac{1}{n}\sum_{i=1}^{n} f(\overline{Y}^{(h)})$$

Then, an important problem is to minimize the error

$$\left| \int f(y)dP(y) - \frac{1}{n}\sum_{i=1}^{n} f\left(\overline{Y}^{(h)}\right) \right|$$

A scheme which is good for approximation of the solution itself has not to have similar efficiency in approximation of $E[f(Y_t)]$. As we have said in the introduction to this section, a goodness of an approximation scheme depends on the problem and the criterion (cf.[285]).

33.6. Approximation and simulation

A.

So far we have discussed various approximate schemes for stochastic Itô equations. In numerical practice an additional approximation is necessary; the Wiener process has to be replaced by its suitable simulation. This, however, can be done in different ways.

For example, we may approximate the trajectories $W(t)$ by a sequence $\left\{W_\gamma^{(n)}(t)\right\}$ of continuous functions of bounded variation, so that for almost all $\gamma \in \Gamma$ and all $a \in [t_0, T]$

$$\sup_{t_0 \le t \le a} \left| W_\gamma^{(n)}(t) - W_\gamma(t) \right| \longrightarrow 0, \quad n \to \infty$$

This is the case of the *polygonal approximation* at the points $t_0 < t_1 < \cdots < t_n = a$:

$$W^{(n)}(t) = W_{t_k} + (W_{t_{k+1}} - W_{t_k})\frac{t - t_k}{t_{k+1} - t_k}, \qquad t_k \le t \le t_{k+1} \tag{5.88}$$

and $\Delta_n = \max(t_{k+1} - t_k) \to 0$.
In the equation

$$Y_t^{(n)} = Y_0 + \int_{t_0}^{t} m(s, Y_s^{(n)})ds + \int_{t_0}^{t} \sigma(s, Y_s^{(n)})dW^{(n)}(s) \tag{5.89}$$

the last integral is now an ordinary Riemann-Stieltjes integral for particular trajectories. As we already said in Sec. 22.2., under certain conditions on functions m and σ, the sequence of $Y_t^{(n)}$ converges with probability one, uniformly on $[t_\circ, a]$ to the solution of the Stratonovich equation (equivalent to the Itô equation (5.75)).

Another possible approximation of the Wiener process can be accomplished by the sequence of so called *transport processes* (cf. Griego et al. [266] and Gorostiza [264]).

Let $Z_n(t)$, $n = 1, 2, \ldots$ be a sequence of continuous, piecewise linear functions with alternating slopes n and $-n$, $Z_n(0) = 0$; the times between consecutive slope change are independent, exponentially distributed with parameter n^2. The processes $Z_n(t)$, called uniform transport processes were shown (cf. [266]) to converge to $W(t)$ uniformly on finite intervals with probability one. Let us consider a family of differential equations

$$\begin{aligned} &dY_n(t) = m(t, Y_n(t))\, dt + \sigma(t, Y_n(t))\, dZ_n(t) \\ &Y_n(t_\circ) = Y_\circ \end{aligned} \tag{5.90}$$

It has been shown in [264] that under certain assumptions on $m(t,y)$ and $\sigma(t,y)$ such as continuity and boundedness with first derivatives

$$P\left\{ \sup_{t_\circ \le t \le T} |Y_n(t) - Y(t)| > \alpha n^{p-\frac{1}{2}} (\log n)^{5/2} \right\} = o(n^{-q})$$

as $n \to \infty$, for $0 < p < \frac{1}{2}$ and all $q > 0$, where α is positive constant and $Y(t)$ is the solution of stochastic equation (5.75).

So, the process $Y_n(t)$ is an approximate solution of the stochastic Stratonovich equation. Trajectories of $Y_n(t)$ are obtainable by solving equation (5.90); this can be done by simulating the transport process $Z_n(t)$ in a computer, and solving equation (5.90) for each generated $Z_n(t)$-trajectory using standard schemes for deterministic differential equations (with their own degree of accuracy).

Although the use of polygonal approximation (5.88) and the transport process approximation of the Wiener process for creating the approximàte equations of type (5.90) is methodically appealing, a digital simulation of the solution of the Stratonovich equation through these equations has to be performed with special care.

To indicate a possible risk, let us try to solve numerically equation (5.89). To do this, one has to evaluate the integrals by numerical techniques; this, however, involves another discretization (the first one there is in the polygonal approximation (5.88)). Though the value of the Riemann integrals is independent of the choice of evaluating points of the approximating sums, the mixing up of these two limiting procedures might lead to the result which not coincides with the solution of equation (5.75).

To see things better let us consider $Y_t^{(n)}$ as it is represented by (5.89) at discrete points $t = t_i$, then (5.88) and (5.89) give

$$Y^{(n)}_{t_{i+1}} = Y^{(n)}_{t_i} + \int_{t_i}^{t_{i+1}} m(s, Y^{(n)}_s)ds + \\ + (W_{t_{i+1}} - W_{t_i})\frac{1}{h}\int_{t_i}^{t_{i+1}} \sigma(s, Y^{(n)}_s)ds \tag{5.91}$$

Now, let us use the Euler formula for computing the ordinary integrals occuring in (5.91), i.e.

$$\int_{t_i}^{t_{i+1}} \sigma(s, Y^{(n)}_s ds \approx \sigma(t_i, Y^{(n)}_{t_i})h$$

Substituting this into (5.91) leads to the stochastic Euler scheme (5.68) which converges to the solution of Itô stochastic equation (5.62) and not to the solution (5.75) as the Wong-Zakai theorem would predict.
However, the use of the trapezoidal rule

$$\int_{t_i}^{t_{i+1}} \sigma(s, Y^{(n)}_s ds \approx \frac{1}{2}\left[\sigma(t_i, Y^{(n)}_{t_i}) + \sigma(t_{i+1}, \hat{Y}^{(n)}_{t_i})\right] h$$

with the predictor $\hat{Y}$ given by formula (5.74) leads to the stochastic Heun scheme (5.73) which, as stated in Sec. 33.4., converges to equation (5.75), that is, to the correct limit.

More general problems associated with approximation of the Wiener process in stochastic integrals and stochastic differential equations are discussed in the papers by Ikeda et al. [268] and Mackevicius [280].

B.

One of the basic questions which may bother us is the practical implementation of the numerical schemes presented above. In particular, what should be the appropriate values of h leading to acceptable final error? It is also of interest to have information about the computational efficiency of particular algorithms and their comparative accuracy (e.g. does the accuracy gained by using the more sophisticated schemes really balance their complexity?).

Of course, in order to gain a reliable insight into the problems like these just mentioned one has to perform large variety of numerical computations. The example reported in paper [292] is as follows.

Let Y_0 be a $N(0,1)$ random variable, and $\{W_t,\ t \geq 0\}$ a standard Wiener process independent of Y_0. Let us consider the following equation

$$\begin{aligned} &dY_t = (2 - \log Y_t)Y_t dt + 2Y_t dW_t \\ &Y_{t_0} = Y_0 \end{aligned} \tag{5.92}$$

Let us take the transformation

$$Y_t = exp\left(\sqrt{2}Z_t\right)$$

The use of Itô formula gives

$$Z_t = e^{-t}Z_0 + \sqrt{2}e^{-t}\int_0^t e^s dW_s. \tag{5.93}$$

For $T = 10$ and $n = 100$, one trajectory of W_t, $0 \le t \le T$ is simulated (by simulating independent increments on time intervals of length $10^{-1} \times \frac{T}{n}$). This simulation is first introduced into (5.93) (after integrating by parts of the stochastic integral, and discretization of the integral with mesh of size $10^{-1} \times \frac{T}{n}$). Taking the exponential, we obtain an Y-trajectory, which we claim to be "exact". On the other hand, we substitute the same (simulated) Z_0 and W-trajectory into different numerical schemes associated with equation (5.92). Let $\{\overline{Y}^n_k,\ 1 \ge k \ge n\}$ be a sequence of approximations in any verified scheme. The measure of the error assumed is:

$$e = \frac{1}{n}\sum_{k=1}^{n} |Y_{t^n_k} - \overline{Y}^n_k|$$

The numerical calculations (cf. [292]) lead to the following values of e;
for the stochastic Euler method: $e = 0.465$,
for the Milshtein scheme: $e = 0.045$.
for the scheme (5.72): $e = 0.010$.

The difference in accuracy between the above three schemes is significant. The example indicates clearly the importance of the convergence rate of the particular scheme (the Euler scheme is of order $0(h)$, the Milshtein scheme — $0(h^2)$, and the scheme characterized by formula (5.72) converges with order $0(h^3)$).
Other comparative study of different discretization schemes is reported by Liske and Platen [279].

Chapter VI

APPLICATIONS: STOCHASTIC DYNAMICS OF ENGINEERING SYSTEMS

34. INTRODUCTION

34.1. General remarks

The importance of differential equations in mathematics itself and in studying phenomena of the real world can not be over-estimated. The differential models of real phenomena constitute the only form which completely satisfies the modern physicist's demand for causality. A. Einstein (1930) pointed out that "the clear conception of the differential law is one of Newton's greatest achievements". And R. Thom (1973) went even much further saying that: "the possibility of using the differential model is, to my mind, the final justification for the use of quantitative models in the sciences".

The above statements have been expressed in connection with classical deterministic differential equations which constitute the basic language of physics and engineering sciences. The importance of such models becomes much more evident if we think about stochastic differential equations, which in addition to the inherent physical regularity of the processes account for various random interactions.

Today, stochastic differential equations arise in virtually every field of science which deal with time-varying phenomena. The mathematics of stochastic differential equations provides the methodical principles for the *stochastic dynamics*, being a branch of science investigating real dynamical systems with use of the apparatus of stochastic process theory.

Stochastic dynamics is now a greatly advanced field comprising the methods of investigation of various systems subjected to parametric and external random excitations. It includes, for instance, the analysis of control systems in the presence of random noises, stochastic vehicle dynamics, stochastic structural dynamics studying the response to earthquake, wind load, sea waves etc. Although the first studies and subsequent rapid developments in stochastic dynamics of engineering systems have been concerned with control and aerospace problems, at present the methods of stochastic dynamics are widely accepted in many other fields including such areas of scientific activity as population dynamics, groundwater (pollution) transport, biophysics etc.

In order to show the possibilities of the methods presented in the previous parts of this book we devote this chapter to some selected problems of stochastic dynamics of real engineering systems.

34.2. Underlying models for stochastic dynamics

In order to perform theoretical (or, mathematical) analysis of a real system we introduce its *mathematical model*, which constitutes a description (in the formal language of mathematics) of the mutual relationships between the most essential quantities characterizing the system at time. The motion of a dynamical system can be stimulated by various factors of external or internal origin.

Although each real system is subjected to large variety of external excitations, not all of them are equally essential, and in the mathematical modelling we select those which (in the situation under consideration) influence the system in the most significant way; these external factors are usually called the *inputs*, or *excitations*. The response of a system to prescribed excitations is characterized mathematically by a certain transformation, which is usually called the *system operator*. For very wide class of real dynamical systems the relationship between excitations and response is characterized by the differential equations; these equations are called the *equations of motion*, or the *governing equations*. The governing equations can be linear or non-linear, with constant or variable coefficients, ordinary or partial derivatives. They can be deterministic or stochastic.

It is clear that the "world" of mathematical models of dynamical systems is enormous. Nevertheless, in this huge variety of dynamical models one can distinguish the models which turned out to be fundamental in investiagtion of real physical and engineering processes. They reflect the most essential features in behaviour of many real dynamical systems of various physical nature. In what follows we shall provide a compendium of such underlying models with a special emphasis on these which are commonly used in stochastic dynamics of vibratory engineering systems.

1. Harmonic oscillator

It is well known that the fundamental relation in modelling of a motion of material particle of mass m subjected to the action of force $\mathbf{F}$ is the second Newton law

$$m\ddot{\mathbf{r}} = \mathbf{F} \tag{6.1}$$

where $\ddot{\mathbf{r}} = \mathbf{a}$ is the acceleration of the particle and force $\mathbf{F}$ is treated as a given function of time; this function comes, in each situation, from the experimental data or from the physics of the phenomenon.

Let $\mathbf{F} = k\mathbf{r}$ where k =const.. Such a force is called the *elastic force*; with good approximation it can be realized by springs. A material point subjected to the elastic force is called the *harmonic oscillator*; its motion in time describes the harmonic vibrations. For x-component of a motion, equation (6.1) gives the following differential equation

$$\ddot{x}(t) + \omega^2 x(t) = 0, \quad \omega^2 = \frac{k}{m} \tag{6.2}$$

which describes the free linear harmonic vibrations.
When the initial displacement $x(t_\circ) = x_\circ$, and the initial velocity $\dot{x}(t_\circ) = v_\circ$ are given, the solution of (6.2) is

$$x(t) = x_\circ \cos\omega(t - t_\circ) + \frac{v_\circ}{\omega} \sin\omega(t - t_\circ) \tag{6.3}$$

Let $F_x = -kx - c\dot{x}$, in this case the motion is governed by the equation: $m\ddot{x}(t) + c\dot{x} + kx(t) = 0$. The more common form is:

$$\ddot{x}(t) + 2h\dot{x}(t) + \omega^2 x(t) = 0, \quad h = \frac{c}{2m} \tag{6.4}$$

The above equation turns out to be very useful in modelling and analysis of various physical processes. It appears everywhere where occurs the "restoring" force kx proportional to $x(t)$ and the "resistant" force $c\dot{x}$ proportional to velocity.

The solution of (4.6) has the form:
for $\omega^2 - h^2 = \Omega^2 > 0$

$$x(t) = e^{-h(t-t_\circ)}\left[x_\circ \cos\Omega(t - t_\circ) + \frac{v_\circ + hx_\circ}{\Omega} \sin\Omega(t - t_\circ)\right] \tag{6.5}$$

for $\omega^2 - h^2 < 0$

$$\begin{aligned} x(t) &= \frac{\lambda_2 x_0 - v_0}{\lambda_2 - \lambda_1} e^{\lambda_1(t-t_\circ)} + \frac{\lambda_1 x_0 - v_0}{\lambda_1 - \lambda_2} e^{\lambda_2(t-t_\circ)}, \\ \lambda_{1,2} &= -h \pm \sqrt{h^2 - \omega^2}. \end{aligned} \tag{6.6}$$

The term $2h\dot{x}$ in equation (6.4) accounts for the dissipation of energy: the equation (6.4) implies that

$$\frac{d}{dt}\left(\frac{\dot{x}^2}{2} + \frac{\omega^2}{2}x^2\right) = -2h\dot{x}^2 \tag{6.7}$$

where the term in brackets characterizes the total energy of the oscillator. If $h > 0$, then this energy decreases at time until $\dot{x}$ reaches zero. This means that the linear oscillator (6.4) when $h > 0$, describes the *damped oscillations*. If $h < 0$ the linear oscillator describes a process with increasing amplitude.

Let $F_x = -kx - c\dot{x} + f(t)$, where $f(t)$ is an external force acting on the point considered. In this case, the motion commonly termed as *excited vibration* of damped linear harmonic oscillator — is governed by the equation

$$m\ddot{x}(t) + c\dot{x}(t) + kx(t) = f(t) \tag{6.8}$$

The above equation is widely used in characterization of vibrations of one-degree of freedom systems subjected to various external excitations. It is used, for example, in modelling of a ship motion (treated as a motion of a rigid body) on the surface of the sea, in study of various giroscopic systems, in vehicle dynamics and in many non-mechanical phenomena.

2. Non-linear oscillators

The equations given above are based on the assumption that interactions in systems considered are linear. In many situations such a simplification can not be adopted and the modelling has to rely on non-linear equations of the form

$$F(x, \dot{x}, \ddot{x}, t) = f(t) \tag{6.9}$$

where F is a non-linear function satisfying suitable conditions.

One of the simplest cases is the model (conservative oscillator)

$$\ddot{x}(t) + 2h\dot{x} + Q(x) = 0 \tag{6.10}$$

In mechanical process such a model accounts for the non-linear relationship between the elastic force and displacement. In the case, where $h = 0$:

$$\begin{aligned} &\frac{\dot{x}^2}{2} + V(x) = \text{const.}, \\ &V(x) = \int_{t_0}^{t} Q(\xi)d\xi \end{aligned} \tag{6.11}$$

It is interesting that in the case where $\frac{dQ}{dx} > 0$ and $Q(0) = 0$ the behaviour of the trajectories of non-linear oscillator (6.10) on the plane $(x, \dot{x})$ is the same as that of oscillator (6.2).

A popular representation of (6.10) is the *Duffing oscillator*

$$\ddot{x}(t) + 2h\dot{x}(t) + ax(t) + bx^3(t) = 0 \tag{6.12}$$

In this case, the behaviour of the trajectories on the phase space (plane $(x, \dot{x})$) depends on the sign of the coefficients a and b and is, in general, complicated. Model (6.12) has been widely used for characterization of vibratory elements in the machine dynamics.

A particular case of the model (6.10):

$$\ddot{\varphi}(t) + \frac{mgl}{J} \sin \varphi = 0 \tag{6.13}$$

is known as the physical pendulum; m, g, l and J are constant parameters. In this case the phase space is a cylinder.

An important class of models included in the general equation (6.9) is characterized by

$$\ddot{x}(t) + G(x, \ddot{x}) = 0 \tag{6.14}$$

The best known representant is the *Van der Pol oscillator*

$$\ddot{x}(t) - 2h\dot{x}(t)\left[1 - \alpha x^2(t)\right] + \omega^2 x(t) = 0 \tag{6.15}$$

3. Non-autonomous models

The oscillatory models presented above are autonomous, i.e., they do not depend explicitly on time. The non-autonomous vibratory systems can be characterized by equations of the form

$$\ddot{x}(t) + 2H(x, \dot{x}, t)\dot{x}(t) + Q(x, t) = 0 \tag{6.16}$$

It is seen that the models belonging to this class account for non-linear interactions and for time-varying parametric excitation. The best known representants of such models is the *Hill equation*

$$\ddot{x}(t) = [\mu + \varphi(t)]\, x(t) = 0 \tag{6.17}$$

and its special form — the *Mathieu equation*

$$\ddot{x}(t) + [\mu + \nu \sin t]\, x(t) = 0 \tag{6.18}$$

In equation (6.17), $\varphi(t)$ can be arbitrary periodic function. The behaviour of systems governed by the equations (6.17), (6.18) depends crucially on the mutual relationship between parameters μ and parameters characterizing the periodic excitation (e.g. ν); the basic problems in the analysis are concerned with *parametric resonances* and *instability* of the system. In Chapter IV we discussed a stochastic extension of these models, where instead of $\varphi(t)$ we had a stochastic process.

4. Non-linear hysteretic models

It turns out that structural elements when subjected to severe dynamic loading (especially, random) may become inelastic, i.e. the restoring force may become non-linear and depends on the time history of the response. The restoring force of an ineleastic system may be characterized by its hereditary behaviour and possible degradation with time. A differential equation model for such systems was first proposed by Bouc [314] and then generalized by Wen (cf. [384]) and others. This model is constructed as follows.

The basic governing equation is:

$$\ddot{x}(t) + G(x, \dot{x}, t) = f(t) \tag{6.19}$$

where

$$G(x, \dot{x}, t) = g(x, \dot{x}) + h(x) \tag{6.20}$$

in which g is a non-hysteretic component (an algebraic function of the instantaneous x and $\dot{x}$), and h is a hysteretic component, a function of the time history of x. Most often, the restoring force of a nearly elasto-plastic system may be modelled by

$$G(x, \dot{x}, t) = G(x, t) = \alpha k x + (1 - \alpha) k z \tag{6.20'}$$

in which k is the pre-yielding stiffness, α is the ratio of post-yielding stiffness to pre-yielding stiffness and $(1 - \alpha)kz$ characterizes the hysteretic part of the restoring force in which z is described by the following non-linear differential equation

$$\dot{z}(t) = A\dot{x}(t) - \nu \left[\beta \left| \dot{x}(t) \right| \left| z(t) \right|^{n-1} z(t) - \delta \dot{x}(t) \left| z(t) \right|^{n} \right] \tag{6.21}$$

where paramters A, β, ν, δ and n characterize the amplitude and shape of the hysteretic loop as well as the smoothness of transition from elastic to inelastic ranges. Proper choices of parameters give various softening as well as hardening of the material (of structural element in question).

The original Bouc model has the form $(n = 1,\ \nu = 1)$

$$\begin{aligned} &\ddot{x}(t) + 2\zeta\omega_o \dot{x}(t) + \alpha\omega_o^2 x(t) + (1 - \alpha)\omega_o^2 z = f(t), \\ &\dot{z}(t) = A\dot{x}(t) - \beta \left| \dot{x}(t) \right| \left| z(t) \right| - \delta \dot{x}(t) \left| z(t) \right|. \end{aligned} \tag{6.22}$$

5. Systems of interacting oscillators

The dimensions of models discussed so far has not been higher than three. Of course, there exist situations where highly multi-dimensional systems have to be studied. A possible model in such situation is a system of *weakly interacting oscillators* of the form

$$\ddot{x}_k(t) + \omega_k^2 x_k(t) = \varepsilon F(x_1, \ldots, x_n; \dot{x}_1, \ldots, \dot{x}_n) \qquad k = 1, 2, \ldots, n \tag{6.23}$$

and ε is a small parameter. If $\varepsilon = 0$, the model reduces to the system of n independent harmonic oscillators with solutions

$$x_k(t) = A_k \cos(\omega_k t + \varphi_k) \tag{6.24}$$

If $\varepsilon \neq 0$ but is sufficiently small, the solution is usually represented in the form (6.24) but A_k and φ_k are not constants but slowly varying functions of time ($\dot{A}_k(t)$ and $\dot{\varphi}_k(t)$ are of order ε).

Since the external force $f(t)$ in the majority of real situations is very irregular (cf. a profile — imposed excitation of road vehicles, pressure of the sea waves, earthquake excitation, wind etc.) a more adequate model for large variety of real systems should assume that $f(t)$ is a stochastic process. Such modelling leads to stochastic differential equations.

6. Models of continuous systems

Analysis of dynamic behaviour of structural systems such as beams, plates, shells constitutes, in general, more difficult problems in comparison with the cases of systems with discrete distribution of mass (discussed above). This is mainly due to the fact that the dynamics of systems is described by partial differential equations. It turns out, however, that a wide class of vibration problems for linear structures can be studied with the use of simple equation of form (6.8).

Let a linear vibratory system be governed by the equation

$$L_o u(\mathbf{r}, t) + c\dot{u}(\mathbf{r}, t) + m\ddot{u}(\mathbf{r}, t) = p(\mathbf{r}) f(t) \tag{6.25}$$

where L_o is a linear operator in the spatial variables, m is the mass density, c — damping coefficient and $p(\mathbf{r})$, $f(t)$ are the spatial and temporal components of the external loading, respectively; in general $\mathbf{r} = (x, y, z)$.

On the basis of the *normal mode approach* we represent $u(\mathbf{r}, t)$ in the form

$$u(\mathbf{r}, t) = \sum_j q_j(t) \varphi_j(\mathbf{r}) \tag{6.26}$$

where $q_i(t)$ and $\varphi_j(\mathbf{r})$ denote the generalized coordinates and the normal modes, respectively. The generalized coordinates $q_j(t)$ are governed by the set of equations

$$\ddot{q}_j(t) + 2\zeta_j\omega_j\dot{q}_j(t) + \omega_j^2 q_j(t) = \beta_j f(t) \tag{6.27}$$

where

$$\beta_j = \int_D \varphi_j(\mathbf{r})p(\mathbf{r})dr \Big/ \int_D m\varphi_j^2(\mathbf{r})dr \tag{6.28}$$

and ω_j denotes the natural frequency of j-the vibrational mode, ζ_j — the damping ratio, $2\zeta_j\omega_j = \frac{c}{m}$.

It is seen that when the normal modes $\varphi_j(\mathbf{r})$ are given, the problem reduces to the analysis of harmonic oscillators governed by uncoupled equations (6.27). These equations can also constitute a basis for performing a stochastic analysis when $f(t)$ in (6.25) is assumed to be a stochastic process.

The normal mode approach can be formulated in a more general form. The governing equation is represented in the vector form and interpreted with the use of the language of Hilbert space theory. For example, instead of equation (6.25) we can consider the system

$$A\frac{\partial^2 \mathbf{u}}{\partial t^2} + B\frac{\partial \mathbf{u}}{\partial t} + C\mathbf{u} = \mathbf{f}(\mathbf{x}, t) \tag{6.29}$$

where $\mathbf{u}(\mathbf{x}, t)$ and $\mathbf{f}(\mathbf{x}, t)$ are the functions defined for $\mathbf{x} \in G \subset R_n$, $t \in [0, \infty)$. For fixed t, $\mathbf{u}$ and $\mathbf{f}$ are the elements of a suitable Hilbert space H_1 and H_2; and A, B, C are linear operators from H_1 into H_2.

It is worthy to add that equations of the form (6.27) occur in the earthquake engineering. Namely, if $\mathbf{u}(\mathbf{r}, t)$ represents a displacement of a point $\mathbf{r}$ of a building (measured in a coordinate system moving jointly with the base) and $\varphi_j(\mathbf{r})$ are normal modes, then the generalized coordinates $q_j(t)$ are governed by the equations

$$\ddot{q}_j(t) + 2\zeta_j\omega_j\dot{q}_j(t) + \omega_j^2 q_j(t) = -a_j(t) \tag{6.30}$$

where $a_j(t)$ denotes the generalized acceleration of the soil (cf. [312] and Sec. 37 of this Chapter).

35. RANDOM VIBRATIONS OF ROAD VEHICLES

35.1. On road-induced excitation

The road surface overrun by a vehicle is always more or less irregular. Measurements of the roughness of various types of roads indicate that the profile of the road can rationally be described by a stochastic process. At present, a stochastic characterization of the road roughness is widely accepted and used in analysis of a vehicle vibrations (cf. Schiehlen [360]).

The road roughness manifests itself with respect to the four wheeles of a two-axle vehicle and is characterized, in general, by the vector random function of the space coordinate s of the longitudinal motion of the vehicle:

$$\zeta(s) = [\zeta_{fr}(s), \zeta_{fl}(s), \zeta_{rr}(s), \zeta_{rl}(s)] \tag{6.31}$$

where two-dimensional "process" $[\zeta_{rr}(s), \zeta_{rl}(s)]$ of the rear axle is only delayed by the axle distance $\Delta s = l$ from $[\zeta_{fr}(s), \zeta_{fl}(s)]$ of the front axle. Numerous measurements have shown that the road roughness can be characterized by a Gaussian and stationary stochastic process with zero mean and with the spectral density estimated from data. Most often the measurment spectra are approximated by rational functions of the frequency. In the space domain the measurements are represented as outputs of some shape filters.

The road-imposed excitation acting on the vehicles wheels takes place at time. Therefore, for a *constant traversal velocity* the time-varying excitation $X(t)$ is defined as $(s = vt)$

$$X(t) = \zeta\left(s(t)\right) = \zeta(vt), \tag{6.32}$$

and

$$K_X(\tau) = K_\zeta(v\tau), \quad g_X(\omega = \frac{1}{v} g_\zeta(\omega). \tag{6.33}$$

If the traversal *velocity is variable*, i.e. $v = v(t)$ $\neq$const., then the relations (6.32), (6.33) are more complicated. Above all, a variable traversal velocity generates non-stationarity of the excitation. Indeed, the excitation acting on wheels of a vehicle at time t is

$$X(t) = \zeta\left(s(t)\right), \quad s = s(t) \tag{6.34}$$

The correlation function of process (6.34) is then

$$\begin{aligned} K_X(t_1, t_2) &= E\left[X(t_1)X(t_2)\right] = E\left[\zeta\left(s(t_1)\right)\zeta\left(s(t_2)\right)\right] \\ &= K_\zeta\left(s(t_2) - s(t_1)\right) \end{aligned} \tag{6.35}$$

For example, if a vehicle moves with variable velocity such that $s(t) = \frac{1}{2}at^2$ where a is constant (constant acceleration), then $X(t) = \zeta\left(\frac{1}{2}at^2\right)$. If, for example

$$K_\zeta(x_2 - x_1) = \sigma_\zeta^2 e^{-\alpha^2(x_2-x_1)^2}\cos\beta(x_2 - x_1)$$

then

$$\begin{aligned} &K_X(t_1, t_2) \\ &= \sigma_\zeta^2 exp\left\{-\alpha^2\frac{a^2}{4}(t_2^2 - t_1^2)^2\right\}\cos\beta\frac{a}{2}(t_2^2 - t_1^2) \\ &= \sigma_\zeta^2 exp\left\{-\alpha_1^2(t_2^2 - t_1^2)^2\right\}\cos\beta_1(t_2^2 - t_1^2), \end{aligned}$$

where

$$\alpha_1^2 = \frac{\alpha^2 a^2}{4}, \quad \beta_1 = \frac{\beta a}{2}.$$

So, the correlation function of the excitation clearly depends on both instants of time t_1 and t_2.

If the traversal velocity varies at time with random fluctuations (e.g. due to unpredictible conditions), that is, it is characterized by a random process $s(t)$ then the excitation acting on wheels of a vehicle is a composition of two random processes $\zeta(s)$ and $s(t)$. In this case, probabilistic properties of process $X(t)$ have to be inferred from the statistics of $\zeta(s)$ and $s(t)$. Characterization of the composition of two stochastic processes is an interesting problem itself.

35.2. Response to random road roughness

A.

The random vibrations of a vehicle (due to road-imposed excitation) affect the ride comfort of passengers as well as the safety (or, reliability) of the whole system. For the analysis of these vibrations (and subsequent estimation of comfort and reliability) vehicles are modelled, in general, as multi-body systems, what means that all parts of the vehicle are considered as rigid bodies with inertia interconnected by rigid bearings. The bodies are subject to additional forces and torques by supports, springs etc. (cf. [359]). Such a multibody system is characterized by n generalized coordinates corresponding to n degrees of freedom (cf. Fig.1).

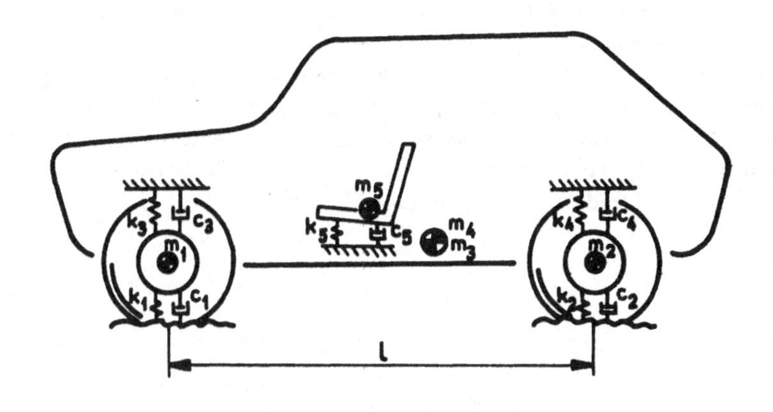

Fig.1. Five-degree of freedom vehicle model [316]

In general, the non-linear differential equations of motion can be written as a vector differential equation of the form

(6.36) $$\mathbf{M}\ddot{\mathbf{y}}(t) + \mathbf{k}(\mathbf{y}, \dot{\mathbf{y}}) = \mathbf{q}(\mathbf{y}, \dot{\mathbf{y}}, \mathbf{z}, t)$$

where $\mathbf{M}$ is the $(n \times n)$ matrix, $\mathbf{k}$ and $\mathbf{q}$ are n-dimensional vectors of coriolis and generalized forces, respectively ($\mathbf{q}$ includes, in particular, the road-induced excitations), $\mathbf{y}(t)$ is the n-dimensional position vector of the generalized coordinates, $\mathbf{z}(t)$ is the p-dimensional vector summarizing state variables of additional forces. Process $\mathbf{z}(t)$ can be represented symbolically by the following equation

(6.37) $$\dot{\mathbf{z}}(t) = \mathbf{G}(\mathbf{y}, \dot{\mathbf{y}}, \mathbf{z}, t)$$

Therefore, equations (6.36), (6.37) can be represented in the form of one system of N equations

(6.38) $$\dot{\mathbf{Y}}(t) = \mathbf{F}\left[\mathbf{Y}, \mathbf{X}(t, \gamma), t\right]$$

where

(6.39) $$\mathbf{Y}(t) = [\mathbf{y}(t), \dot{\mathbf{y}}(t), \mathbf{z}(t)]$$

and $\mathbf{X}(t,\gamma)$ characterizes the excitation due to random roughness of the road. It is seen that (6.38) has the general form of stochastic equation investigated in the previous chapters.

Equation (6.38) can be regarded as a general framework for various particular models used in the vehicle dynamics. The choice of a specific model for a vehicle depends on the technical problem under consideration.

A complete analysis of road vehicles generally requires non-linear and multi-dimensional models, and under some approximations can be performed (analytically or numerically). However, to display some qualitative features of the motion, or certain quantitative specific effects (e.g. the effect of variable traversal velocity), a simple one degree of freedom model can be used (cf. Fig. 2).

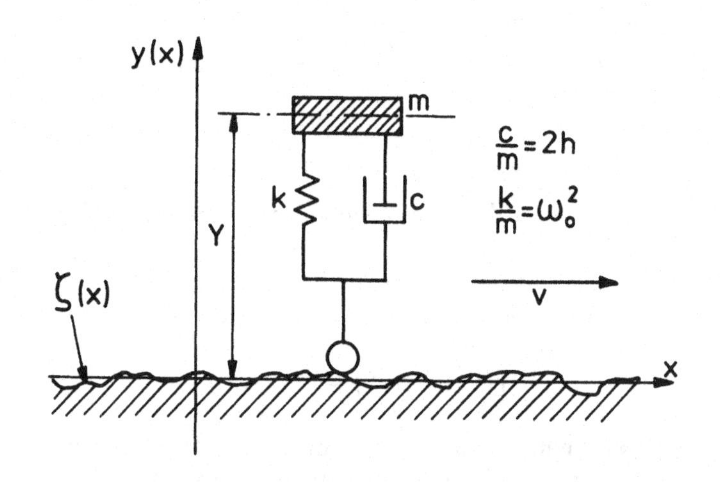

Fig.2 Single-degree of freedom vehicle model [365]

It has the form

$$
\begin{aligned}
&\ddot{Y}(t) + 2h\dot{Y}(t) + \omega_0^2 \mathbf{Y}(t) = \tilde{X}(t,\gamma) \\
&\tilde{X}(t,\gamma) = 2h\dot{X}(t,\gamma) + \omega_0^2 X(t,\gamma)
\end{aligned}
\tag{6.40}
$$

where process $X(t,\gamma)$ is related to the random road profile $\zeta(x,\gamma)$ by formulae (6.32), (6.34). If the spectral density $g_\zeta(\omega)$ of the random profile is such that

$$\int_{-\infty}^{+\infty} \omega^2 g_\zeta(\omega) d\omega < \infty$$

then the derivative $\dot{X}(t,\gamma)$ exists. In the case of constant traversal velocity, according to the spectral method, a stationary response is characterized by

$$
\begin{aligned}
g_Y(\omega) &= \frac{g_{\dot{X}}(\omega)}{|-\omega^2 + 2ih\omega + \omega_0^2|^2} \\
&= \frac{|2hi\omega + \omega_0^2|^2 g_X(\omega)}{(\omega^2-\omega_0^2)^2 + 4h^2\omega^2} = \frac{4h^2\omega^2 + \omega_0^4}{(\omega^2-\omega_0^2)^2+4h^2\omega^2}\frac{1}{v}g_\zeta(\omega)
\end{aligned} \tag{6.41}
$$

The above formula expresses the spectral density of stationary response of the single-degree of freedom vehicle model by the spectral density of the road roughness $g_\zeta(\omega)$, the traversal velocity v, and the system parameters h and ω_0. As is well known $g_{\dot{Y}}(\omega) = \omega^2 g_Y(\omega)$ and $g_{\ddot{Y}}(\omega) = \omega^4 g_Y(\omega)$. Having $g_{\ddot{Y}}(\omega)$ we can easily calculate the mean-square $E[\ddot{Y}]^2$ of the vertical acceleration of the mass. This usually serves for selection of such values of ω_0^2 and h which minimize $E[\ddot{Y}]$, and, therefore, optimize the ride comfort.

The estimation of reliability requires, in general, more sophisticated analysis; such characteristics of $Y(t)$ as the distribution of maximal values and the first-passage time distribution are necessary. For the analysis of road vehicle dynamics modelled by multi-dimensional equations the reader is referred to the papers [316], [361] and to recent issues of the Vehicle System Dynamics (the international journal of vehicle mechanics).

B.

In the analysis of random vibrations of road vehicles it is usually assumed that: (i) the road roughness is statistically homogeneous (stationary) random function, (ii) the traversal velocity of the system is constant.

There is no evidence to doubt stationarity of most profiles (cf. [355]). However, the assumption concerning the uniformity of traversal velocity is evidently not strictly true and it is usually introduced for simplification of the analysis. Often, the actual velocity of travelling systems is variable. Moreover, when the variations are due to imperfection in spead control arrangements and some external factors, the traversal velocity can depart randomly from a constant value. It is therefore of importance to obtain information about the response of a vehicle travelling with variable velocity. It is also of interest to know what kind of probabilistic and statistical problems are produced by variable velocity (cf. Sobczyk, Macvean [365]).

As we have already said, a variable traversal velocity generates non-stationarity in the excitation. Therefore, the random vibrations of a road vehicle travelling with variable (deterministic) velocity are, in one-dimensional case, governed by stochastic differential equation (6.40) with $X(t,\gamma)$ being a non-stationary process with correlation function (6.35). Solution of this equation can be obtained by the methods presented in Sec. 26.3.

Let $t = 0$ be the initial instant of time at which the system begins to travel with variable velocity. Two different cases can be distinguished.

(i) The system starts its traversal motion at $t = 0$ with velocity $v(0) = 0$ and with acceleration $a(t)$, i.e.

$$\ddot{s}(t) = a(t) \quad s(0) = \dot{s}(0) = 0 \tag{6.42}$$

This means that

$$s(t) = \begin{cases} 0, & t \le 0 \\ s_1(t), & t > 0 \end{cases}, \quad v(t) = \begin{cases} 0, & t \le 0 \\ \dot{s}_1(t), & t > 0 \end{cases}, \tag{6.43}$$

where

$$s_1(t) = \int_0^t (t - \tau) a(\tau) d\tau \tag{6.44}$$

In this case:

$$\begin{aligned} X(t) &= \zeta\left((s_1(t)\right), \\ K_X(t_1, t_2) &= K_\zeta\left(s_1(t_2) - s_1(t_1)\right), \quad t_1, t_2 > 0 \end{aligned} \tag{6.45}$$

(ii) The vehicle travels with constant velocity v from the "infinite" past and then at time $t = 0$ it begins to accelerate with acceleration $a(t)$, i.e.

$$s(t) = \begin{cases} vt, & t \le 0 \\ vt + s_1(t), & t > 0, \end{cases}, \quad v(t) = \begin{cases} v, & t \le 0 \\ v + \dot{s}_1(t), & t > 0 \end{cases} \tag{6.46}$$

In this case

$$\begin{aligned} X(t) &= \zeta(vt) \quad , \quad t \le 0 \\ X(t) &= \zeta\left(vt + s_1(t)\right), \quad t > 0 \end{aligned}$$

The correlation function takes the form

$$K_X(t_1, t_2) = \begin{cases} K_\zeta\left(v(t_2 - t_1) + s_1(t_2) - s_1(t_1)\right), & t_1, t_2 > 0 \\ K_\zeta\left(v(t_2 - t_1) - s_1(t_1)\right), & t_1 > 0,\ t_2 < 0 \\ K_\zeta\left(v(t_2 - t_1) + s_1(t_2)\right), & t_1 < 0,\ t_2 > 0 \\ K_\zeta\left(v(t_2 - t_1)\right), & t_1, t_2 < 0. \end{cases} \tag{6.47}$$

If the acceleration $a(t)$ tends to zeo, then formulae (6.32), (6.33) are obtained.

The response in case (i) is characterized as follows

$$
\begin{aligned}
Y(t) &= Y_1(t) + X(0)\psi(t), \\
Y_1(t) &= \int_0^t p(\tau)x(t-\tau)d\tau, \\
\psi(t) &= e^{-ht}\left(\cos\lambda_\circ t - \frac{h}{\lambda_\circ}\sin\lambda_\circ t\right), \quad \lambda_\circ = (\omega_\circ - h^2)^{\frac{1}{2}}, \\
p(\tau) &= \frac{1}{\lambda_\circ}e^{-h\tau}\left[2h\lambda_\circ\cos\lambda_\circ\tau - (2h^2-\omega_\circ^2)\sin\lambda_\circ\tau\right], \\
E\left[Y^2(t)\right] &= E\left[Y_1(t)\right] + 2E\left[X(0)Y_1(t)\right]\psi(t) + \psi^2(t)E\left[X^2(0)\right], \\
E\left[Y_1^2(t)\right] &= \int_0^t\int_0^t p(\tau_1)p(\tau_2)K_\zeta\left(s_1(t-\tau_2) - s_1(t-\tau_1)\right)d\tau_1 d\tau_2, \\
E\left[X(0)Y_1(t)\right] &= \int_0^t p(\tau)K_\zeta\left(s_1(t-\tau)\right)d\tau.
\end{aligned}
\tag{6.48}
$$

The appropriate formulae can be written down for the case (ii). A detailed discussion of these formulae along with numerical illustrations is provided in paper [365]. Other approach to the problem of non-stationary response of road vehicles travelling with variable velocity is represented by Hammond and Harrison [325], [326] and by Narayanan [346]. The problem concerned with randomly varying traversal velocity is considered by Sobczyk, Macvean and Robson in paper [366].

36. RESPONSE OF STRUCTURES TO RANDOM TURBULENT FIELD

36.1. On turbulent-induced excitation

The aerospace and civil engineering structures are subjected to wind loads of various physical origins. One of the most important is the load induced by the atmospheric turbulence.

A wind load acting on structures consists of regular and fluctuating components. For sufficiently rigid structures of civil engineering the regular component is believed to be dominating and the pressure acting on a structure is assumed to be proportional to the square of average wind velocity. If, however, the frequencies of free vibrations of a structure in question are comparable with frequencies of turbulent fluctuations, then the fluctuating component has to be accounted for. This is, as a matter of fact, the case of all flexible structures (tall chimneys, masts, as well as the aircraft panels)

— cf. Fig.3.

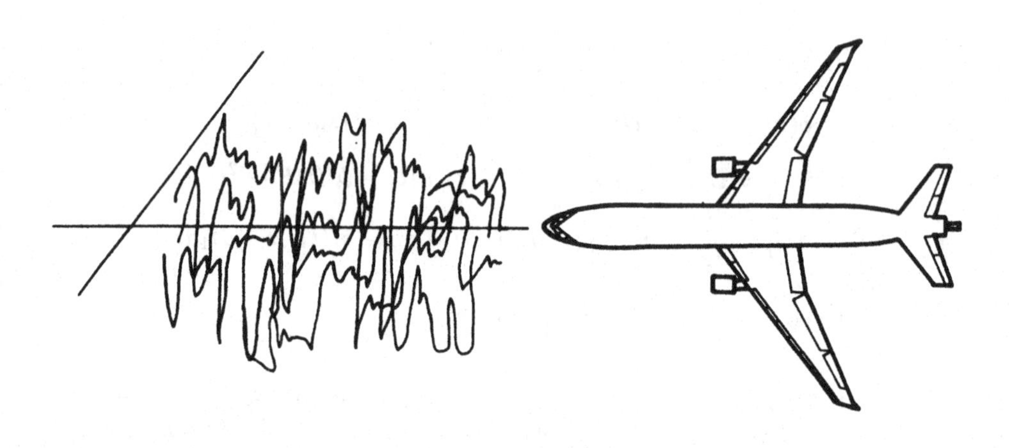

Fig. 3. Airplane in turbulent stream

In regions which are sufficiently far away from the surface of the Earth, random fluctuations of wind pressure are commonly described with the use of the model of homogeneous turbulence (cf. Batchelor [309]). The probabilistic characteristics of turbulent fluctuations depend on the meteorological conditions, the geographical position, the height over the Earth surface etc. Statistical regularities of these fluctuations are not easily predictable. Without going into details, we wish to say here that the engineering analysis of different types of structures assumes some "standard" representations of the spectrum of a turbulent wind (considered to be stationary process). For example, for needs of structural dynamics the following spectral density (of the longitudinal component of wind velocity) is used (cf. [313])

$$g_X(\omega) = \frac{2\sigma_X^2 \lambda_Z}{3v_o} \frac{\tilde{\omega}^2}{(1+\tilde{\omega}^2)^{4/3}}, \quad \tilde{\omega} = \frac{\lambda_Z}{v_o}\omega \tag{6.49}$$

where σ_X^2 is the variance, v_o — the average wind velocity, and λ_Z — scale of vertical heterogeneity. In analysis of aircraft structures, various forms of the spectral density are assumed; one of them is (cf. [313])

$$g_X(\omega) = \frac{\sigma_Z^2 \lambda_X}{\pi v_o} \frac{1+3\tilde{\omega}^2}{(1+\tilde{\omega}^2)^2}, \quad \tilde{\omega} = \frac{\lambda_X}{v_o}\omega \tag{6.50}$$

where σ_Z^2 is the variance of vertical component, λ_X denotes the scale of turbulence in longitudinal direction. Another important type of loads is the load induced by the boundary-layer turbulence. The boundary-layer turbulence results when the space vehicle travels at high speeds. The statistical properties of boundary-layer pressure fields have been sampled by a number of investigators, both on actual vehicles and on wind-tunnel models, and some analytical forms of the correlation function have been proposed (cf. Elishakoff [320], Lin [339]).

36.2. Random vibrations of elastic plate

A.

In order to illustrate the analysis of random vibrations of structures subjected to random turbulent excitation we shall consider a rectanglular flat panel which is a very common element in various engineering applications.

The governing equation for transverse vibrations of elastic plate can be written in the form

$$\nabla^4 u + \frac{\partial^2 u}{\partial t^2} + 2c\frac{\partial u}{\partial t} = p(x, y, t) \tag{6.51}$$

where $u(x, y, t)$ is the transverse displacement; x and y are the cartesian coordinates; t, c, and p denote the "reduced" quantities related to their actual counterparts $\bar{t}, \bar{c}, \bar{p}$ by the formulae

$$t = \left(\frac{D}{\rho}\right)^{\frac{1}{2}} \bar{t} \quad , \quad c = \frac{1}{2}(\rho D)^{-\frac{1}{2}} \bar{c},$$
$$p = \frac{\bar{p}}{D} \quad , \quad D = \frac{h^3 E}{12(1-\nu^2)} \tag{6.52}$$

where E is the Young modulus, ν — the Poisson ratio, h — the plate thickness and ρ is the mass density of the material; D is the flexural rigidity of the plate and $p(x, y, t)$ is an external random load.

To gain a better clarity of further analysis we assume that the external excitation has the form

$$p(x, y, t) = P(x, y)R(t) \tag{6.53}$$

where $P(x, y)$ and $R(t)$ are statistically independent random functions. Additionally, we shall assume (without restriction of generality) that mean values of $P(x, y)$ and $R(t)$ are equal to zero. Of course, equation (6.51) is suplemented by the appropriate initial and boundary conditions.

To characterize the reponse $u(x,y,t)$ we shall use the normal-mode approach characterized briefly in Sec. 34.2. Therefore, we look for a solution in the following form

$$u(x,y,t) = \sum_{m,n} \varphi_{mn}(x,y) q_{mn}(t) \tag{6.54}$$

where $\varphi_{mn}(x,y)$ are eigenfunctions satisfying the equation

$$\nabla^4 \varphi_{mn} - \omega_{mn}^2 \varphi_{mn} = 0 \tag{6.55}$$

and all boundary conditions on the plate; ω_{mn} are the eigenvalues corresponding to φ_{mn}. The functions $\varphi_{mn}(x,y)$ form a complete orthogonal set, which permits us to express $P(x,y)$ in the form of expansion

$$P(x,y) = \sum_{m,n} H_{mn} \varphi_{mn}(x,y). \tag{6.56}$$

Since $P(x,y)$ is a random function, the coefficients H_{mn} are random variables and series (6.56) is convergent in the mean-square sense, if the following series is convergent

$$\sum_{k,l,m,n} E\left[H_{kl} H_{mn}\right] V_{kl}(x,y) V_{mn}(x,y) \tag{6.57}$$

Using the orthogonality of the eigenfunctions we obtain that the coefficients H_{mn} are given by the formula

$$\begin{aligned} H_{mn} &= \frac{1}{\alpha_{mn}^2} \int_S p(x,y) \varphi_{mn}(x,y) dS \\ \alpha_{mn}^2 &= \int_S \varphi_{mn}^2(x,y) dS \end{aligned} \tag{6.58}$$

where S denotes the area of the plate. Taking into account (6.53), (6.54) and (6.56) we obtain the following equations for $q_{mn}(t)$:

$$\ddot{q}_{mn}(t) + 2c\dot{q}_{mn}(t) + \omega_{mn}^2 q_{mn}(t) = H_{mn} R(t) \tag{6.59}$$

which are the linear differential equations with random excitation $H_{mn}R(t)$. According to general principles of Sec. 26.1.

$$q_{mn}(t) = H_{mn} \int_{t_0}^{t} G_{mn}(t-\tau) R(\tau) d\tau \tag{6.60}$$

where $G_{mn}(t-\tau)$ is the impulse function (or, the Green function) and $(0 < c < \omega_{mn})$:

$$G_{mn}(t-\tau) = \frac{1}{\beta_{mn}} e^{-c(t-\tau)} \sin \beta_{mn}(t-\tau)$$

$$\beta_{mn} = (\omega_{mn}^2 - c^2)^{\frac{1}{2}} \tag{6.61}$$

Substituting (6.60) into (6.54) yields

$$u(x,y,t) = \sum_{m,n} H_{mn}\varphi_{mn}(x,y) \int_{t_0}^{t} G_{mn}(t-\tau)R(\tau)d\tau \tag{6.62}$$

The above series is convergent in the mean-square sense if the following series

$$\sum_{k,l,m,n} E\left[H_{kl}H_{mn}\right]\varphi_{kl}(x,y)\varphi_{mn}(x,y) \times \int_{t_0}^{t}\int_{t_0}^{t} K_R(\tau_1,\tau_2)G_{kl}(t-\tau_1)G_{mn}(t-\tau_2)d\tau_1 d\tau_2 \tag{6.63}$$

is convergent.

Therefore, formula (6.62) represents a random displacement field $u(x,y,t)$ as a linear transformation of random variables H_{mn} (defined by (6.58)) and random process $R(t)$. It constitutes a ground for calculating the probabilistic properties of $u(x,y,t)$.

B.

The mean value of $u(x,y,t)$ is equal zero due to our assumptions (we have assumed that the initial conditions are homogeneous). The correlation function is as follows

$$\begin{aligned} K_u(x_1,y_1,t_1;x_2,y_2,t_2) &= E\left[u(x_1,y_1,t_1)u(x_2,y_2,t_2)\right] \\ &= \sum_{k,l,m,n} E[H_{kl}H_{mn}]\varphi_{kl}(x_1,y_1)\varphi_{mn}(x_2,y_2) \\ &\times \int_{t_0}^{t_1}\int_{t_0}^{t_2} K_R(\tau_1,\tau_2)G_{kl}(t-\tau_1)G_{mn}(t-\tau_2)d\tau_1 d\tau_2 \end{aligned} \tag{6.64}$$

where

$$E[H_{kl}H_{mn}] = \frac{1}{\alpha_{kl}^2\alpha_{mn}^2}\int_S\int_S K_P(x_1,y_1;x_2,y_2) \times \varphi_{kl}(x_1,y_1)\varphi_{mn}(x_2,y_2)dx_1dx_2dy_1dy_2 \tag{6.65}$$

In particular, the above formulae give the expression for the spatial correlation function $K_u(x_1, y_1, t; x_2, y_2, t)$ and for the temporal correlation function: $K_u(x, y, t_1, t_2)$.

All what is written above holds for general case, when stochastic process $R(t)$ is non-stationary. But, in numerous real situations process $R(t)$ can be adequately regarded as stationary one; for example, this is the case of the excitation by the atmospheric turbulence.

Let us assume now that $R(t)$ is a *weakly stationary process* with the correlation function $K_R(t_2 - t_1)$ and that we are interested in a stationary response (i.e., when $t \to \infty$). Then

$$u(x,y,t) = \sum_{m,n} H_{mn}\varphi_{mn}(x,y) \int_0^\infty R(\tau)G_{mn}(t-\tau)d\tau \tag{6.66}$$

The temporal correlation function is

$$K_u(x,y,t_1,t_2) = \sum_{k,l,m,n} E\left[H_{kl}H_{mn}\right]\varphi_{kl}(x,y)\varphi_{mn}(x,y) \times \int_0^\infty \int_0^\infty K_R(\tau_2 - \tau_1)G_{kl}(t_2 - \tau_2)G_{mn}(t_1 - \tau_1)d\tau_1 d\tau_2 \tag{6.67}$$

The transformation variables

$$t_1 - \tau_1 = \Theta_1, \quad t_2 - \tau_2 = \Theta_2, \quad t_2 - t_1 = \Theta$$

and the reasoning analogous to that in Sec. 26.2 leads to the result (cf. [131])

$$K_u(x,y,\Theta) = \sum_{k,l,m,n} E\left\{\int_{-\infty}^{+\infty} \Phi_{kl}(x,y,-i\omega)\Phi_{mn}(x,y,i\omega)g_R(\omega)e^{i\omega\Theta}d\omega\right\} \tag{6.68}$$

where

$$\Phi_{rs}(x,y,i\omega) = \frac{H_{rs}\varphi_{rs}(x,y)}{\omega_{rs}^2 - \omega^2 + 2i\omega c} \tag{6.69}$$

It is clear that (6.68) can be written as

$$K_u(x,y,\Theta) = \int_{-\infty}^{+\infty} E\left\{|\Psi_u(x,y,i\omega)|^2\right\} g_R(\omega)e^{i\omega\Theta}d\omega \tag{6.70}$$

where

$$\Psi_u(x,y,i\omega) = \sum_{m,n} \Phi_{mn}(x,y,i\omega) \tag{6.71}$$

Therefore, the spectral density of a stationary response $u(x,y,t)$ in a fixed point (x,y) is expressed as

$$g_u(x,y,\omega) = E\left\{|\Psi_u(x,y,i\omega)|^2\right\} g_R(\omega) \tag{6.72}$$

Using the results for the correlation function and for the spectral density of $u(x,y,t)$ we can easily evaluate the correlation function and the spectral density of the stresses in the plate.

An extension of the problems shortly presented above (beams, sheels etc.) in linear and non-linear formulation can be found in very wide literature on the subject (cf. Bolotin [313], Dimentberg [317], Elishakoff [319], [320], [321], Fiedorov [322], Lin [339], Shinozuka, Yang [362], Witt, Sobczyk [385].

37. RESPONSE OF STRUCTURES TO EARTHQUAKE EXCITATION

37.1. Description of earthquake excitation

During a long time an earthquake action (e.g., the horizontal ground acceleration) was characterized by deterministic functions with various degrees of complextity. Of course, such modelling was too far from the actual records of earthquakes.

At present it is commonly accepted that the earthquake should be described by a stochastic process. The random nature of the earthquake mechanism and, consequently, the resulting strong ground motion was first pointed out by Housner (cf. [327]). Early statistical data concerned with particular characteristics of strong earthquakes were provided in the book by Gutenberg and Richter [324]. Today there exists very extensive literature devoted to both the seismic source mechanisms and the surface ground motion (cf. a review paper by Zerva [389]).

A complicated ground motion caused by tectonic phenomena is a result of complex interaction of seismic waves propagating from the source through inhomogeneous layered media. Multiple scattering of waves at randomly distributed inhomogeneities makes the surface displacement and acceleration fields highly unpredictable (cf. Sobczyk [234]).

One of the methods for obtaining a tractable model for the reliability predictions consists in treating the system transmitting the motion from the source to the ground surface as a suitable filter characterized by a frequency transfer function. This transfer function characterizes approximately the averaged effects of wave propagation through the earth strata. The required frequency transfer function is approximated on the basis of analytical theory of wave propagation and system identification techniques.

Various stochastic models for a strong ground motion have been proposed. Modelling started from uncorrelated impulses (Housner G.W.) and white-noise representations (Bycroft G.N.) and has been developed to account for non-uniform spectra (Kanai K., Tajimi H.) as well as for the temporal non-stationarity of a random seismic action (Bolotin V.V., Amin M., Ang A.H.S.). These investigations lead to a commonly accepted model for (horizontal) ground acceleration having the form of a non-stationary modulated stochastic process

$$\ddot{X}(t,\gamma) = A(t)X_1(t,\gamma) \tag{6.73}$$

where $A(t)$ is a deterministic envelope function imposed on stationary process $X_1(t,\gamma)$.

The most common expression for the spectral density of stationary process $X_1(t,\gamma)$ is known as *Kanai-Tajimi spectrum*:

$$g_{X_1}(\omega) = \frac{\left[1 + 4\zeta_g^2\left(\frac{\omega}{\omega_g}\right)^2\right] g_o}{\left[1 - \left(\frac{\omega}{\omega_g}\right)^2\right]^2 + 4\zeta_g^2\left(\frac{\omega}{\omega_g}\right)^2} \tag{6.74}$$

where g_o is a constant spectral density, ω_g is a characteristic ground frequency (the frequency of a "ground filter"), and ζ_g is a characteristic ground damping ratio. Kanai [331] has suggested the following numerical values: $\omega_g = 15.6$ rads/sec., $\zeta_g = 0.6$ as being representative for firm soil conditions. The validity of these values has been confirmed by Ruiz and Penzien [357]. Of course, other values should be selected when site conditions are significantly different.

More general model has the form (cf. [378])

$$\ddot{X}(t,\gamma) = \sum_k A_k(t)X_k(t,\gamma)H(t) \tag{6.75}$$

where $X_k(t,\gamma)$ are stationary processes, $A_k(t)$ deterministic functions (envelopes) and $H(t)$ is Heaviside function.

Recently, Suzuki and Minai (cf. [374]) elaborated the model:

$$\ddot{X}(t,\gamma) = a_1(t)V(t,\gamma) + a_2(t)\xi_2(t,\gamma) \tag{6.76}$$

where $V(t,\gamma)$ is the output of a time-dependent linear filter driven by Gaussian white noise $\xi_1(t,\gamma)$, namely

$$\begin{aligned} V(t,\gamma) &= L_2\varphi(t) \\ L_1\varphi(t) &= a_3(t)\xi_1(t,\gamma) \end{aligned} \tag{6.77}$$

where the differential operators L_1 and L_2 are given by

$$L_1 = \sum_{i=0}^{m} \alpha_i \frac{d^i}{dt^i}, \quad L_2 = \sum_{j=0}^{l} \beta_j \frac{d^j}{dt^j}; \tag{6.78}$$

in the above expressions ξ_1 and ξ_2 are zero mean independent Gaussian white noises, $a_1(t)$, $a_2(t)$, $a_3(t)$ are deterministic amplitude-modulation functions, and α_i, β_j are, in general, time-dependent coefficients.

37.2. Stochastic seismic response

In general, a response of structures to earthquake excitation is described by the following stochastic equation

$$Lu = -\ddot{X}(t,\gamma) \tag{6.79}$$

where $\ddot{X}(t,\gamma)$ a component of the ground acceleration, is described by one of the models presented above ((6.73), (6.75) or (6.76)) whereas operator L characterizes symbolically a type of structure and its possible dynamic behaviour.
For example, in the linear case, (6.79) can be represented as a sequence of equations (6.30) for the generalized coordinates, that is we have

$$\ddot{q}_j(t) + 2\zeta_j\omega_j\dot{q}(t) + \omega_j^2 q_j(t) = -\ddot{X}(t,\gamma) \tag{6.80}$$

If we are interested in the response of a tall, slender, tower-shaped structure (cf. Zembaty [388]) subjected to horizontal seismic acceleration of the foundation, a possible model is a cantilever beam with the governing equation

$$\begin{aligned} &\frac{\partial^2}{\partial z^2} EJ(z) \left[\frac{\partial^2 u(z,t)}{\partial z^2} + k\frac{\partial^3 u(z,t)}{\partial z^2 \partial t}\right] \\ &+ \mu m \frac{\partial u(z,t)}{\partial t} + m\frac{\partial^2 u(z,t)}{\partial t^2} = -m\ddot{X}(t,\gamma) \end{aligned} \tag{6.81}$$

where $u(z,t)$ is the relative horizontal displacement of the beam structure (for each value of vertical coordinate z and t); $EJ(z)$ is the flexural rigidity, m — the mass per unit length, and k, μ, are constants characterizing a stiffness and damping. Using the normal-mode approach the problem of solving equation (6.81) reduces to analysis of equation (6.80). The methods of analysis of this type of equations were discussed in Sec. 26.3.

More generally, a seismic excitation should be considered as multi-component one; this means that it should be characterized by three-dimensional random process (an earthquake disturbance is usually recorded in three directions, that is, in the east-west, north-south and up-down directions).

Let us consider an elastic structure in the cartesian coordinate system (x_1, x_2, x_3) — cf. Figure 4.

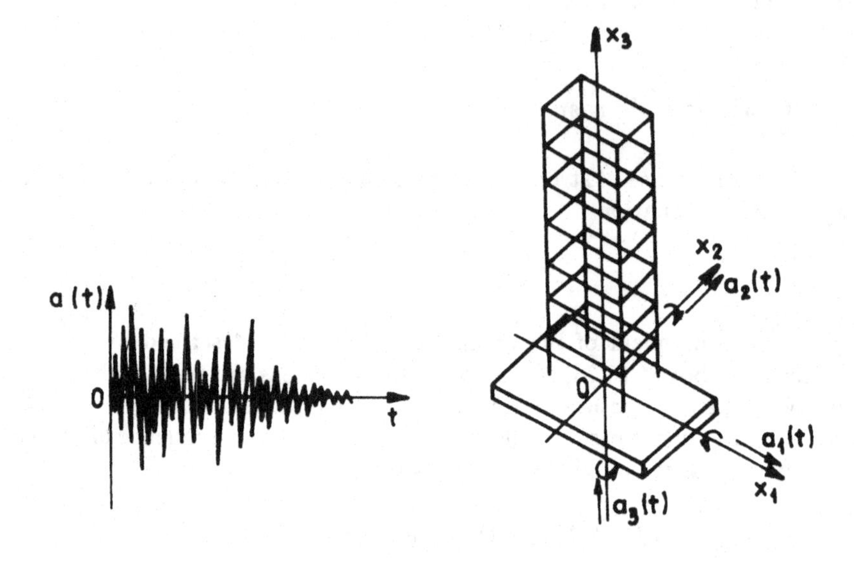

Fig.4 Structure excited by earthquake [312]

Let us denote the components of inertia forces by $F_j(x_1, x_2, x_3)$, and components of the displacement vector by $u_j(x_1, x_2, x_3, t)$. Let the displacements generated by the external static forces be determined by the system of three linear differential equations

$$L_j(u_1, u_2, u_3) = F_j, \quad j = 1, 2, 3 \tag{6.82}$$

Let us denote by $\alpha_j(t)$ a linear-acceleration of the origin of the cartesian coordinate system, and by $\beta_j(t)$ — a rotational acceleration. Then the actual accelerations $a_j(x_1, x_2, x_3, t)$ at a given point are [312]

$$\begin{aligned} \hat{a}_1 &= \alpha_1 + x_3\beta_2 - x_2\beta_3 \\ \hat{a}_2 &= \alpha_2 + x_1\beta_3 - x_3\beta_1 \\ \hat{a}_3 &= \alpha_3 + x_2\beta_1 - x_1\beta_2 \end{aligned} \tag{6.83}$$

The dynamic governing equations have the form (6.82) where

$$F_j = -\rho\left(\hat{a}_j + \frac{\partial^2 u_j}{\partial t^2}\right), \quad j = 1, 2, 3,$$

and ρ is the mass density. Hence, the dynamic behaviour of the structure in question is governed by the system of stochastic equations

$$L_j(u_1, u_2, u_3) + \rho\frac{\partial^2 u_j}{\partial t^2} = -\rho\ddot{X}_j(x_1, x_2, x_3, t; \gamma) \tag{6.84}$$

with $\ddot{X}_j$ defined by (6.83). This system can also be treated by the normal-mode approach (cf. [312]). One may try to treat this system with use of stochastic equations for processes in Hilbert space (cf. Sec. 23 and Sec. 29); this, however, seems to be an open problem.

Many structures exhibit highly non-linear and hysteretic response when exposed to severe earthquake loading, and it is important to include this in the response analysis. In this case the response is governed by the non-linear hysteretic stochastic equations (6.22) in which $f(t)$ in a non-stationary stochastic process characterizing a horizontal component of ground acceleration. A study of such systems has been just started (cf. Noori et al. [348], Suzuki and Minai [374]).

38. RESPONSE OF STRUCTURES TO SEA WAVES

38.1. Description of sea wave excitation

A.

Very complicated nature of wind-generated sea waves and lack of possibility of determining exact initial conditions (for this kind of motion) make that the efforts to describe the associated phenomena deterministically are not satisfactory. Since over last three decades the sea wave dynamics has been performed within the stochastic theory.

Although the sea motion (or, sea state) can be partially characterized by some parameters (e.g., the wave height h_s, the mean wave period T_s, the wave direction Θ), an underlying quantity in stochastic theory of the sea motion is the elevation $\eta(x, y, t)$ which is regarded to be random function of position and time. Probabilistic properties of $\eta(x, y, t)$ are derived partially from the measurement and in part — from hydrodynamic wave theory (cf. [338]). In almost all studies in ocean engineering it is assumed that the sea wave process is *stationary*, *ergodic* and *Gaussian*. Under such hypothesis (confirmed in many situations) the process $\eta(x, y, t)$, for each fixed (x, y), is characterized by the spectral density $g_\eta(\omega)$.

Using the dimensional arguments *Phillips* (1958) showed the evidence that in the high frequency range, where wave breaking is important, the asymptotic form of the sea spectrum ($\omega \to \infty$) is given by

$$g_\eta(\omega) = Aa_g^2\omega^{-5} \tag{6.85}$$

where A is a constant of the order of $1,35 \times 10^{-2}$, and a_g is the acceleration of gravity.

The following form of spectrum (*Striekalov, Massel* [373]) has been estimated from measurements on the Baltic sea:

$$g_\eta(\omega) = \frac{h_s}{\omega_o}\left\{0,38exp\left[-35\left(\frac{\omega}{\omega_o} - 0,8\right)^2\right] + 0,22\left(\frac{\omega}{\omega_o}\right)^{-5} exp\left[-1,34\left(\frac{\omega}{\omega_o}\right)^{-8}\right]\right\} \tag{6.86}$$

where h_s is the average wave height, and ω_o is the reference frequency expressed in terms of the gravity acceleration g and the mean wind velocity.

A number of different forms of $g_\eta(\omega)$ have been proposed as an explicit function of the wind velocity V. For example, the *Darbyshiere spectrum* is as follows (cf. [353])

$$\begin{aligned} g_\eta(\omega) &= Aexp\left\{-\left(\frac{(\omega-\omega_o)^2}{0,054(\omega-\omega_o+0,265)}\right)^{\frac{1}{2}}\right\} \\ &= 0 \qquad \text{when} \qquad \omega - \omega_o < -0,265 \end{aligned} \tag{6.87}$$

where

$$\begin{aligned} A &= 0,186 \times 10^{-5}V^4, \\ \omega_o &= 6,284(1,94V^{\frac{1}{2}} + 2,5 \times 10^{-7}V^4)^{-1} \end{aligned} \tag{6.88}$$

It seems that the most popular in practice is the following *Pierson-Moskowitz spectrum* [351]:

$$g_\eta(\omega) = \frac{Aa_g^2}{\omega^5}exp\left[-B\left(\frac{a_g}{\omega V}\right)^4\right], \quad \omega > 0 \tag{6.89}$$

where A and B are dimensionless constants equal to:
$A = 8.1 \times 10^{-3}$, $B = 0.74$; V is the mean wind velocity at a height of $19.5m$ above the still sea surface. This spectrum turns out to be very sensitive to the values of V; hence, it is practical to introduce the parameters directly related to the observed values.

Having the spectral density of the Gaussian sea elevation $g_\eta(\omega)$, we can characterize the time-scale (the "mean wave period") T_s and the wave height h_s. The time scale T_s defined as a mean time between the upcrossings of the mean still water level is (cf. [344])

$$T_s = 2\pi \left(\frac{\lambda_0}{\lambda_2}\right)^{\frac{1}{2}} \tag{6.90}$$

where λ_0 and λ_2 are the (one-sided) spectral moments,i.e.

$$\lambda_k = \int_0^\infty \omega^k g_\eta(\omega) d\omega, \quad k = 0, 2 \tag{6.91}$$

The wave height h_s is defiend as the expected value of the highest one-third of the wave. If the process $\eta(t)$ can be assumed to be narrow-band, then h_s can be related to the envelope of process $R(t)$, namely

$$h_s = 3\sigma_\eta \int_{r_0}^\infty 2r f_R(r) dr = 4\sigma_\eta \tag{6.92}$$

where $\sigma_\eta = \sqrt{\lambda_0}$ is the standard deviation of $\eta(t)$.
For the Pierson-Moskowitz spectrum

$$T_s = \frac{2\pi}{(B\pi)^{\frac{1}{4}}} \frac{V_{19,5}}{a_g}, \quad h_s = 2\sqrt{\frac{A}{B}} \frac{V_{19,5}^2}{a_g} \tag{6.93}$$

The Pierson-Moskowitz is a special case of a gamma spectrum, i.e. a spectrum of the form (cf. [344]):

$$g_\eta(\omega) = \hat{A}\omega^{-\alpha} exp(-\acute{B}\omega^{-\beta}), \quad \omega > 0 \tag{6.94}$$

It is seen that the Pierson-Moskowitz spectrum corresponds to $\alpha = 5$ and $\beta = 4$.

Typically, wave spectra are unimodal with a limited bandwidth. Under the assumption of normality a simple approximation for the joint probability density of the wave amplitude R and the zero-crossing period T can be obtained (cf. Longuet-Higgins [342], [343]).

Often, in analysis of engineering problems, there is a need for numerical techniques involving the sea wave process. A significant element of these techniques is the proper digital simulation of sea elevation records which are compatible with a specified target power spectrum. Most of the currently used algorithms for numerical generation of sea wave records are based on the superposition of harmonic waves with random amplitudes and phases, and on ARMA models (cf. Sec. 32.2). Of course, some other algorithms can also be adopted (cf. Spanos, Hansen [370]).

Closing this short presentation on the decription of the sea waves we wish to emphasize that the assumption of normality of the sea elevation is not fully satisfactory. One of the most typical examples is the profile of wind-generated waves in finite water depth. Records of waves in such situation show a significant excess of high crests and shallow troughs in contrast to those of waves in deep water. For example, Ochi and Wang (cf. Ochi [349]) analyzed more than 500 wave records measured at various water depths during stage of a storm. The results show a significant departure of histograms of wave profile from the Gaussian distribution. Also the application of the non-linear Stokes wave theory leads to non-Gaussian distribution of the sea surface elevation (cf. Tayfun [376], Huang and Long [328], Ochi [349], Cieślikiewicz [315]).

B.

If we wish to study the response of the structures to sea waves we must be able to specify clearly a random loading (generated by the waves) acting on the structure.

Using a linearized wave theory (cf. [338]) we can obtain a relationship between the fluid particle velocity $u(x, y, z, t)$ and the surface elevation $\eta(x, y, t)$. In particular case when $u = u(x, z, t)$ and $\eta = \eta(x, t)$ this relationship is (cf. [344], [363])

$$u(x, z, t) = \omega \frac{\cos h \quad k(z + d)}{\sin h \quad (kd)} \eta(x, t) \equiv \omega H(z, \omega) \eta(x, t) \tag{6.95}$$

where d is the constant depth of the sea and $H(z, \omega)$ plays a role of a transfer function. So, the spectral density of $u(x, z, t)$ has the form

$$g_u(\omega) = g_u(x, z, \omega) = [\omega H(z, \omega)]^2 g_\eta(x, \omega) \tag{6.96}$$

From the above formula we obtain the variance of the horizontal acceleration $\sigma_{\dot{u}}(z, x)$.

If the characteristic dimension of a structure is small compared with the wave length, then the load consists of two basic components: a drag force proportional to the square of the normal component of the incident particle velocity and an inertia or mass force associated with the normal component of the particle acceleration. These forces are combined in the *Morison formula* for the force per unit length of a fixed cylinder:

$$\mathbf{p} = k_d \mathbf{u} |\mathbf{u}| + k_m \dot{\mathbf{u}} \tag{6.97}$$

where $\mathbf{u}$ is the incident particle velocity normal to the cylinder, and k_d and k_m are given in terms of the drag and mass coefficients C_d and C_m. If cylinder has a diameter D then

$$k_d = \frac{1}{2} C_d \rho D, \quad k_m = \frac{1}{4} C_m \rho \pi D^2$$

where ρ is the mass density of the water.

It should be noticed, that even when the sea surface elevation is Gaussian, the non-linearity of the Morison formula yields a force $p(t)$ which, in general, is non-Gaussian process. The departure from the Gaussian distribution (at a given cross section) depends of the coefficients k_d and k_m. The total Morison force on a fixed vertical cylinder is obtained by an integration of (6.97).

Formula (6.97) does not take into account the possible flexibility of the structure. If the velocity of the structure is not negligible compared with the water particle velocities, the Morison equation should be modified. The *modified Morison equation* is

$$p = \frac{1}{2} C_D \rho D(u - \dot{x})|u - \dot{x}| + (C_m - 1)\rho A(\dot{u} - \ddot{x}) + \rho A\dot{u} \tag{6.98}$$

where D is the diameter of the structural member, A — its cross-sectional area, u is the incident normal particle velocity and $\dot{x}$-the corresponding component of the velocity of the structure.

In addition to the statistical properties of the wave force at a particular cross section at fixed time, it is of interest to have the correlation of wave forces at different positions and different times. Some approximations can be obtained on the basis of the Morison equation.

It is worth emphasizing that the particle velocity and acceleration in the Morison formula are usually assumed as values of the incident wave without any interaction effect. But, they can be replaced by the values obtained from the solution of the wave diffraction problem.

38.2. Ship motion in random sea waves

In mathematical modelling of a ship motion it is usually assumed (cf. [353]) that a ship is a rigid body with motions associated with the six degree of freedom known as:

surge, sway, heave, roll, pitch and yaw (cf. Fig. 5).

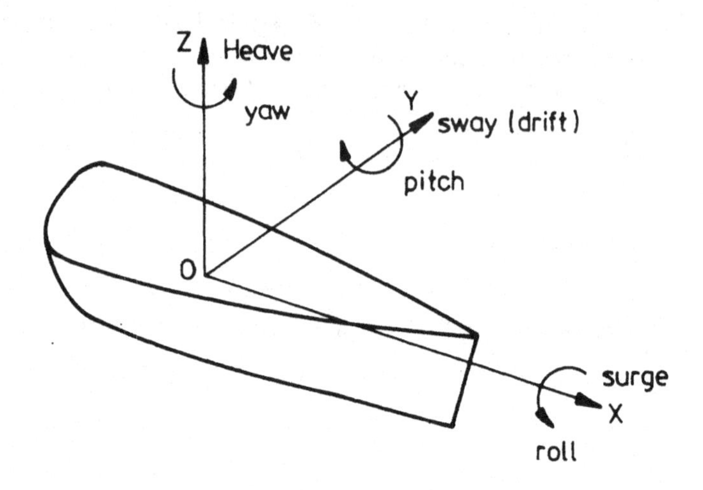

Fig. 5. Ship motion in random sea [353]

The motions excited in the ship are mainly due to the wave disturbance which, in effect, is the input to the system considered.

For a rigid ship travelling with constant speed v at an arbitrary angle to sea waves, the resultant motions in the six degrees of freedom are governed by the set of second order differential equations. In linear formulation they have a standard form (cf. Price and Bishop [353]:

$$(\mathbf{m}+\mathbf{a})\ddot{\mathbf{q}}(t)+\mathbf{b}\dot{\mathbf{q}}(t)+\mathbf{c}\mathbf{q}(t)=\mathbf{Q}(t) \tag{6.99}$$

where

$$\begin{aligned}\mathbf{q}(t)&=[x(t),y(t),z(t),\varphi(t),\Theta(t),\psi(t)]^T\,,\\ \mathbf{Q}(t)&=[X(t),Y(t),Z(t),K(t),M(t),N(t)]^T\,;\end{aligned} \tag{6.100}$$

$X(t)$, $Y(t)$, $Z(t)$ are the amplitudes of surge, sway and heave forces and $K(t)$, $M(t)$, $N(t)$ are the amplitudes of roll, pitch and yaw moments; the components of $\mathbf{q}(t)$ are the corresponding responses. The matrix $\mathbf{m}$ includes as its elements, the mass of the ship (m) and the principal moments of inertia. The elements of the (6×6) matrix $\mathbf{a}$ are a combination of hydrodynamic forces; $\mathbf{b}$ is a damping matrix and $\mathbf{c}$ is a matrix which accounts for the hydrostatic contributions.
The system of equations (6.99) has been written in the coordinate system associated

with a ship (so, the velocity v does not occur in explicit way). Of course, system (6.99) simplifies if we consider only selected motions (e.g. haeve and pitch).

It is clear that system (6.99) is a linear system of differential equations with random excitation and can be analyzed by the methods presented in Chapter IV. The possible difficulties of computational nature are due to the complex interaction of hydrodynamic forces (cf. [353]).

Because of such "factors" as free-surface conditions, viscous effects, geometric properties of the hull, etc. the non-linear description is often necessary. To make the problem tractable only one degree of freedom is often considered. For normal ship motion, rolling is probably the most obviously non-linear [353]. It is also considered to be the motion which can most realistically be treated in isolation (from other degrees of freedom).

Let φ be the roll angle. The equation for φ derived in [353] has the form

$$\ddot{\varphi}(t) + 2k_\varphi\omega_\varphi\dot{\varphi}(t) + \omega_\varphi \sum_{n=1,2} \alpha_{2n-1}\varphi^{2n-1}(t) = X(t) \tag{6.101}$$

where k_φ, ω_φ and α_{2n-1} are appropriate constants.
The analysis of the above equation can be performed by one of the methods presented in Sec. 27, depending on the properties of a stochastic process $X(t)$.

The problem of the ship roll stability has attracted the attention of many authors. The governing equation treated by Roberts [354] is the following

$$\ddot{\varphi}(t) + 2\zeta\omega_o\dot{\varphi}(t) + \omega^2\left[1 + \varepsilon\Theta(t,\gamma)\right]\varphi(t) = \varepsilon X(t,\gamma) \tag{6.102}$$

where $\Theta(t,\gamma)$ represents the pitch angle which is assumed to be a random stationary process, and $X(t,\gamma)$ describes the sea wave random excitation.

Making use of the averaging method (cf. Sec. 28.3) we represent the solution as

$$\begin{aligned} \varphi(t) &= A(t)\cos\Phi(t) \\ \Phi(t) &= \omega_o t + \Psi(t) \end{aligned}$$

The standard equations for the amplitude and phase are

$$\begin{aligned} \dot{A}(t) &= \Big\{ -2\zeta\omega_o A(t)\sin\Phi + \varepsilon\omega_o\Theta(t)\cos\Phi - \frac{\varepsilon}{\omega_o}X(t)\Big\}\sin\Phi \\ \dot{\Psi}(t) &= \frac{1}{A(t)}\Big\{ -2\zeta\omega_o A(t)\sin\Phi + \varepsilon\omega_o A(t)\Theta(t)\cos\Phi - \frac{\varepsilon}{\omega_o}X(t)\Big\}\cos\Phi \end{aligned}$$

Under the assumption of the averaging method, process $[A(t), \Psi(t)]$ converges weakly (as $\varepsilon \to 0$) to a diffusion Markov process $[a(t), \psi(t)]$ governed by the Itô equations

$$\begin{aligned} da &= \left[-\alpha a + \frac{\beta}{2a}\right]dt + \left(\nu a^2 + \beta\right)^{\frac{1}{2}} dW(t), \\ d\psi &= \eta dt + \left(\delta + \frac{\beta}{a^2}\right)^{\frac{1}{2}} dW(t), \end{aligned}$$

where (cf. [354])

$$\alpha = \zeta\omega_\circ - \frac{3}{16}\varepsilon^2\omega_\circ^2 g_\Theta(2\omega_\circ), \quad \beta = \frac{\varepsilon^2}{2\omega_\circ^2} g_X(\omega_\circ),$$
$$\nu = \frac{1}{8}\varepsilon^2\omega_\circ^2 g_\Theta(2\omega_\circ), \quad \delta = \frac{1}{4}\varepsilon^2\omega^2\left[g_\Theta(0) + \frac{1}{2}g_\Theta(2\omega_\circ)\right],$$
$$\eta = \frac{1}{4}\varepsilon^2\omega_\circ^2 \int_{-\infty}^{\circ} K_\Theta(\tau)\sin 2\omega_\circ\tau d\tau.$$

The probability density of the amplitude $a(t)$ is governed by the following F-P-K equation

$$\frac{\partial p(a,t)}{\partial t} = -\frac{\partial}{\partial a}\left[\left(-\alpha a + \frac{\beta}{2a}\right)p\right] + \frac{1}{2}\frac{\partial^2}{\partial a^2}\left[(\nu a^2 + \beta)p\right]$$

whose stationary solution is

$$p(a) = Ca\nu^{1+\mu}/(\beta + \nu a^2)^{1+\mu}, \quad \mu = \frac{1}{2} + \frac{\alpha}{\nu}; \tag{6.103}$$

C is evaluated from the normalization condition:

$$\int_\circ^\infty p(a)da = 1$$

The above integral exists for $\mu > 0$; then $C = 2\mu\left(\frac{\beta}{\nu}\right)^\mu$.
Therefore, $p(a)$ exists for $\mu > 0$; this gives the following condition for the stability of ship rolling:

$$\zeta > \frac{1}{8}\varepsilon^2\omega_\circ g_\Theta(2\omega_\circ) \tag{6.104}$$

38.3. Response of offshore platforms

A.

It is known that the oil exploration under a bottom of the sea provided a strong motivation for the analysis and the construction of special hydroengineering structures which are called the offshore platforms. These structures have to work in severe conditions of sea environment with the complex interactions between the ground, sea water and wind pressure.

A "classical" type of the offshore platform is the *jacket-type platform*. This is a deformable structure with the foundation attached to the sea bed. The largest part of the structure, consisting of welded metallic pipes, and called a jacket, is summerged.

These pipes have circular sections with diameters much smaller than the pipe lengths (Fig. 6)

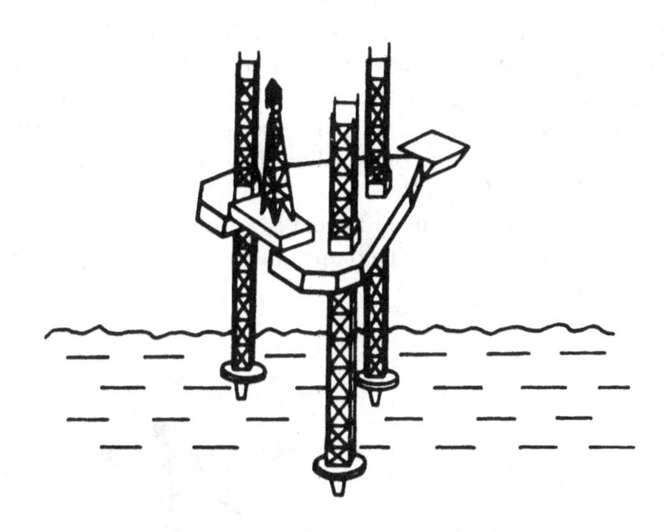

Fig. 6. Fixed jacket-type platform [356]

In the dynamic study of large jackets, the non-linearities are generated by the hydrodynamic viscous drag forces and by the fluid-structure interactions. Sometimes, it is also necessary to consider the non-linearities caused by the behaviour of the foundation in the sea bed (cf. [358]).

In mathematical modelling of the platform structures it is usually assumed that the structural elements are elastic and the appropriate discretization is introduced. In this way the forces are applied to the nodes of the structure, and the governing equations describe the reponse as a function of time at fixed spatial points. Therefore, the mathematical model has the form

$$\mathbf{M}\ddot{\mathbf{Y}}(t) + \mathbf{G}(\dot{\mathbf{Y}}, \mathbf{Y}) = \mathbf{f}(t, \gamma) \tag{6.105}$$

where $\mathbf{M}$ is the mass matrix, $\mathbf{G}$ — the appropriate non-linear vector-function and $\mathbf{f}(t, \gamma)$ is a random stochastic process characterizing the forces acting on the structure (and related to the sea waves by the Morison equation). Often, the existing analyses assume linearized relationships in the structure itself, and the non-linearity is introduced by the wave force $f(t, \gamma)$ — cf. [350]. In each case we have to do with rather complicated differential equations which can be analyzed in significant part by computational procedures.

B.

As an alternative to fixed base platforms, which become, very heavy, difficult to install and uneconomical, a new type of structures, namely the *compliant* offshore structures becomes increasingly important. Prominent examples of this type of structures are the *Tension-Leg platforms* also called *Guyed Towers*. The basic features of these structures which distinguish them from the more traditional fixed jacket platforms, is that they primarily ride with the waves rather than resist them; this explains the name compliant. Numerous authors have developed non-linear models of the tension-leg platforms (cf. [330]).

The tension-leg platform (or, guyed tower) is a slender tower of uniform cross-section resting on a flexible foundation held upright by a number of guylines attached to the tower near its top (Fig. 7).

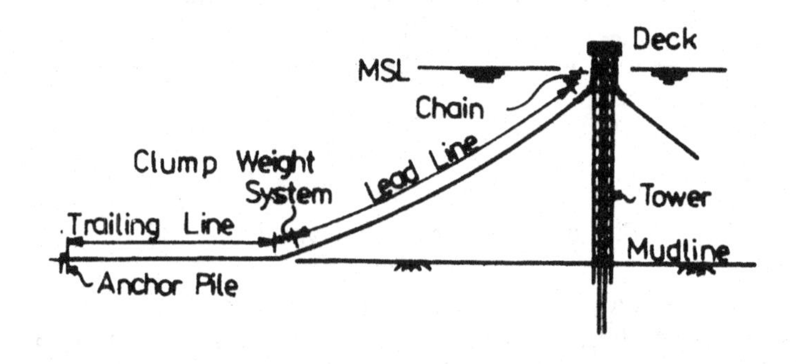

Fig. 7. Guyed tower platform [332]

Each individual guyline is a multi-component mooring line conisting of a lead line, clumpweught and trailing line. Under usual operating conditions, the tower does not move appreciably, but during storms, the clump weight lifts off, making the system "soft" and thus absorbing the wave energy. Non-linear stiffness is provided by the guyline as a result of the geometric non-linerarity, load dependence on position etc.

To indicate a type of equations describing the dynamics of the guyed tower we shall provide the model presented by Kanegaonkar and Haldar [332].

The tower of the platform is modelled as a rigid column with a hinge at the base. The guylines are assumed to be circumferentially symmetric. The hydrodynamics load on the guylines is neglected and the tower is assumed to move in one plane only. The horizontal restoring force is related to the linear and cubic functions of the tower rotation, while the vertical reaction is related to a constant and a second order function of the tower rotation.

The governing equation of motion (for the angle $\Theta = \Theta(t)$ made by the tower with the vertical direction) for the idealized guyed tower system is obtained by taking the moments of all the forces about the base hinge. For relative small displacements, the governing equation becomes (cf. [332])

$$
\begin{aligned}
&J\ddot{\Theta} + cd^2\dot{\Theta} + (z_c c_1 - DW_p - dDw_T - z_c c_1' - F_b z_b)\,\Theta \\
&+ z_c(c_2 - c_2')\Theta^3 = hf(t)
\end{aligned}
\tag{6.106}
$$

where J — is the structural inertia about the base, c — damping constant, d — half the depth of the tower, z_c — distance between the point of attachment of the cables and the base, D — distance from the centre of the platform to the base, W_p — platform weight, w_T — weight per unit length of the tower, F_b — force from buyancy tanks, z_b — moment arm for F_b, f(t) — wave load, h — moment arm for the wave load; c_1 and c_1' are constants arising from the least-square regression estimation of the horizontal restoring force and c_2, c_2' are the corresponding coefficients for the vertical force.

Therefore, model (6.106) has the form of the Duffing equation with random forcing term characterizing the wave loading (described in the first part of this section). One of the methods presented in Sec. 27.2. can be applied. The most appropriate one is that based on the extension of the state space and the subsequent use of the Itô stochastic equations. To do this, the excitation $f(t)$ must be first represented as the output of a filter with the white noise input. In other words the spectrum of $f(t)$ has to be represented as a rational function of frequency. Since the spectral density is known (e.g. the Pierson-Moskowitz spectrum and the Morison formula), a rational spectral density can be fitted to it by adjusting its parameters. Such an approach has been adopted in paper [332].

39. STOCHASTIC STABILITY OF STRUCTURES

39.1. Stability of a column

Let us consider shortly an important problem of structural mechanics. Let an ideally straight beam of length L be subjected to axial compressing force $P(t)$. If the amplitude of this force is less than the critical static force then a beam will execute only longitudinal vibrations; if, however, the amplitude exceeds this critical value then the transversal vibrations are important. These vibrations can increase infinitely depending on the mutual relation between the beam parameters and the characteristics of the axial force $P(t)$. The problem which occurs is a stability problem for the beam (cf. [307]):

As it is known, the governing equation is as follows

$$EI\frac{\partial^4 u}{\partial x^4} + P(t)\frac{\partial^2 u}{\partial x^2} + m\frac{\partial^2 u}{\partial t^2} + \beta\frac{\partial u}{\partial t} = 0 \tag{6.107}$$

where $u(x,t)$ is the transversal displacement of a beam at point x and in the time t; EI denotes the bending rigidity of a beam, m — is the mass density and β — the damping ratio. The boundary conditions can be taken in the form

$$u = 0, \quad \frac{\partial^2 u}{\partial x^2} = 0, \quad x = 0, \quad x = L \tag{6.108}$$

Taking into account only the first form of transversal vibrations (cf. normal mode approach — Sec. 34.2, formula (6.26)) we represent the solution of (6.107) as

$$u(x,t) = y(t)\sin\frac{\pi x}{L} \tag{6.109}$$

The equation for $y(t)$ is

$$\ddot{y}(t) + \frac{\beta}{m}\dot{y}(t) + \frac{\pi}{mL^2}\left[P_E - P(t)\right]y = 0 \tag{6.110}$$

where $P_E = \frac{\pi^2 EI}{L^2}$ is the Euler critical force.

In reality the axial force $P(t)$ has usually a constant P_o component (average value) and a fluctuating (random) part $P_1(t,\gamma)$, i.e., $P(t) = P_o + P_1(t,\gamma)$. Equation (6.110) takes the form

$$\ddot{Y}(t) + 2\zeta\omega_o\dot{Y}(t) + \left[\omega_o^2 - X(t,\gamma)\right]Y(t) = 0 \tag{6.111}$$

where

$$\omega_o^2 = \frac{\pi^2(P_E - P_o)}{mL^2}, \quad \zeta = \frac{\beta}{2m\omega_o}, \quad X(t,\gamma) = \frac{\pi^2}{mL^2}P_1(t,\gamma) \tag{6.112}$$

Equation (6.111) is a linear differential equation with random coefficient. The stability of a beam dynamics is defined as a boundedness (in the appropriate probabilistic sense — cf. Sec. 22.4) of the solution of equation (6.111).
Let us assume that the axial force variation $P_1(t,\gamma)$ is a white noise with the spectral density g_o; then $X(t,\gamma)$ is a white noise with the intensity $2D$, where

$$D = \frac{\pi^4 g_o}{2m^2L^4}$$

Direct application of the reasoning presented in Sec. 28.3. leads to the following condition for the mean-square stability

$$g_o < \frac{2\beta L^2}{\pi^2}(P_E - P_o) \tag{6.113}$$

In order to obtain the stability condition (with probability one) in the situation where $X(t,\gamma)$ is not a white noise we should use the result of the example discussed in Sec. 28.4.

Let us notice that the problem discussed above — described by the partial differential equation (6.107) — has been reduced to ordinary equation (6.111) by the use of the normal mode approach. However, such an approach is not general enough and above all, non-linear systems can not be treated in this way. Therefore, of great interest are the methods which can be applied directly to the partial stochastic differential equation. Such methods have been elaborated (mainly, through the extension of the Lapunov method) and applied to a wide class of linear and non-linear structural systems (cf. Plant, Infante [352], Kurnik, Tylikowski [337], Tylikowski [379].

39.2. Stability of suspension bridge

The stability of the motion of suspension bridges excited by strong wind constitutes an important problem in structural engineering (Fig. 8).

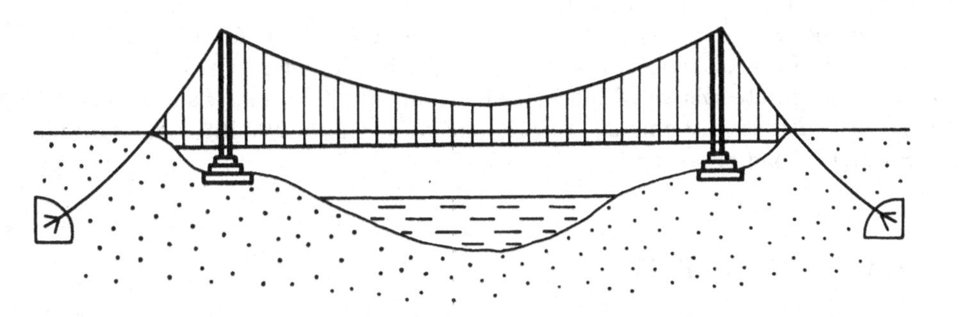

Fig. 8. View of a suspension bridge

It has been shown that a turbulence in wind velocity is able to change the stability conditions for the motion of suspension bridges. When a linear theory of a bridge motion is adopted then the loads generated by the wind can be cathegorized as: buffetting loads (independent of the structural motion) and self-excited loads which in mathematical model manifest themselves as parametric excitation. Since the turbulence-induced load is usually characterized by a stochastic process, we come to the problem of a stochastic stability of suspension bridges. This problem was undertaken by Lin and Ariaratnan in their paper [340]. The simplified differential equation for one uncopled torsional mode $\varphi(t)$ used by the authors has the form

(6.114)
$$\ddot{\varphi}(t) + 2\zeta\omega_o\dot{\varphi}(t) + \omega_o^2\varphi(t) = 2a_1\xi(t,\gamma)\varphi(t) + 2\omega_o a_2\xi(t,\gamma)\dot{\varphi}(t)$$
$$+ a_1\int_{-\infty}^{t}[1 + 2\xi(t,\gamma)]\left[X_{M_\varphi}(t-\tau) - 1\right]\dot{\varphi}(\tau)d\tau$$

where

$$a_1 = \frac{\rho v_o^2 B^2}{I}\frac{\partial C_M}{\partial\varphi}, \quad \omega_o = (\omega_\varphi^2 - a_1)^{\frac{1}{2}},$$
$$a_2 = \frac{a_1 X_\varphi}{\omega_o}, \qquad \zeta = \frac{1}{2\omega_o}(2\omega_\varphi\zeta_\varphi - a_1 X_\varphi)$$

and I is the torsional moment of inertia per unit span length (assumed to be uniform), ω_φ — natural frequency, ζ_φ — ratio of structural damping to the critical damping, ρ — air density, B — width of bridge desk, X_{m_φ} — experimentally determined aerodynamic indicial function, and X_φ and $\frac{\partial C_M}{\partial\varphi}$ are experimentally determined aerodynamic constants; the total horizontal wind velocity v is represented as $v(t) = v_o[1 + \xi(t,\gamma)]$, where v_o plays a role of the mean velocity and $v_o\xi(t,\gamma)$ represents the turbulent fluctuation.

Assuming that $\xi(t,\gamma)$ is a Gaussian white noise the authors formulated equation (6.114) in the form of a system of linear Itô stochastic differential equations, and then the appropriate system of equations for moments. The analysis of this system of equations with use of numerical calculations provided the stability boundaries for the first and second order moments.

40. OTHER PROBLEMS

A.

In applications, stochastic differential equations are introduced as a tool in modelling real dynamical systems subjected to random noises. A solution process provides an information about the response of the system in question. However, most often this information is needed for decisions concerning *safety* or *reliability* of a system. In some cases empirical rules based on the simplest response properties (such as mean and root mean square) can give a necessary estimation of the reliability. In general, a more sophisticated probabilistic analysis is required to provide a solid base for the reliability predictions. The analysis which we have in mind is concerned with characterization of some *random variables* associated with the solution (or response) process $Y(t)$. Most often these random variables are some functionals of the solution process.

The basic reliability characteristics associated with systems described by the response process $Y(t)$ are the following.

a) a number of crossing of a fixed (or time-varying) level u by process $Y(t)$ in time $[0, T]$;
b) duration of intervals between crossings;
c) time spent by process $Y(t)$ above the level u;
d) extremes of process $Y(t)$; especially
   number of maxima,
   number of maxima above a given level,
   supremum of the process on a given time interval;
e) first-passage time, i.e. the time when a process $Y(t)$ reaches a given level for the first time.

The quantities listed above can be determined exactly only in the simplest cases. In general, the analysis associated with the extremes and the first-passage time constitutes an interesting but also difficult branch of stochastic process theory. Recent developments in this direction are presented in the books [6],[22] and in a number of review and original papers (e.g. Bergman, Spencer [311], Krenk [336], Venmarke [381], Shinozuka and Yang [362], Yang [386] and references there in).

Another class of important problems associated with estimation of reliability of mechanical systems subjected to randomly varying excitations includes the modelling and analysis of fatigue process. Because of a very complex nature of fatigue phenomenon as well as the presence of various uncertainties in basic factors provoking fatigue it is now widely accepted that a fatigue of real materials should be studied by use of the stochastic process theory (in particular, by stochastic differential equations). Recently, an increasing amount of attention has been devoted to stochastic modelling of fatigue crack growth (cf. Ditlevsen [318], Sobczyk [367, 368], Sobczyk, Trebicki [369], Spencer, Tang [372], Yang et al. [387]).

B.

The process of using excitation and response data to deduce properties of the dynamic system is usually called *system identification.* In most cases a specific form of system representation with unknown parameters is selected a priori so that the process is one of *parameter identification.*

If the system considered is described by stochastic differential equation then the problem lies in determining the coefficients on the basis of known excitation and solution; this is an *inverse problem* associated with a given stochastic equation. One of the basic difficulties in treatment of such problems is the fact that, in general, inverse problems are not well-posed in the sense of Hadamard (the solution should exist, be unique and depend continuously on the data; otherwise, the problem is ill-posed).

In practical applications to time-invariant vibratory systems the most common parameters are the inertia parameters, the natural and damping ratios. When the excitation and the response of linear system are stationary random processes then (by virtue of the basic formula of the spectral method) the quotient of their densities provides the amplitude (but not the phase) of a system transmittancy. This is often sufficient to estimate the natural frequency and damping ratio.

A possible approach to identification of parameters (and parametric noises) in linear dynamical systems with white noise disturbances is the usage of differential equations for moments. Such a method has recently been elaborated (cf. Wedig [383], Kazimierczyk [333] and applied to engineering systems. A general approach to estimation of parameters in stochastic systems is based on the method of the maximum likelihood (cf. Bellach [310], Kozin [335], Kazimierczyk [334]).

APPENDIX

A.1. CAUCHY FORMULA

The following useful formula (often termed as the Cauchy Formula) holds:

$$\int_a^s \int_a^{s_n} \cdots \int_a^{s_3} \int_a^{s_2} F(s_1) ds_1 ds_2 \ldots ds_{n-1} ds_n =$$

$$= \frac{1}{(n-1)!} \int_a^s (s-t)^{n-1} F(t) dt \tag{A.1}$$

Proof. In order tp prove (A.1) we start from the formula for differentiation of the integral involving a parameter

$$\frac{d}{ds} \int_{\alpha(s)}^{\beta(s)} f(s,t) dt = \int_{\alpha(s)}^{\beta(s)} \frac{\partial f}{\partial s} dt + f(s, \beta(s)) \frac{d\beta}{ds} - f(s, \alpha(s)) \frac{d\alpha}{ds} \tag{a}$$

Let us apply this formula for differentiation of the following function

$$I_n(s) = \int_a^s (s-t)^{n-1} F(t) dt \tag{b}$$

where n is a positive integer and a is a constant.

Let us set $f(s,t) = (s-t)^{n-1} F(t)$ in (a). The result is for $n > 1$:

$$\frac{dI_n}{ds} = (n-s) \int_a^s (s-t)^{n-2} F(t) dt + \left[(s-t)^{n-1} F(t)\right]_{t=s} = (n-1) I_{n-1} \tag{c}$$

and for $n = 1$ we have directly from (b):

$$\frac{dI_1}{ds} = F(s) \tag{d}$$

From recurrence relation (c) we have

$$\frac{d^k I_n}{ds^k} = (n-1)(n-2)\ldots(n-k) I_{n-k}, \qquad n > k \tag{e}$$

which for $k = n-1$ becomes

$$\frac{d^{n-1} I_n}{ds^{n-1}} = (n-1)! I_1(s) \tag{f}$$

Differentiating (f) and using (d) results in the equation

(g) $$\frac{d^n I_n}{ds^n} = (n-1)!F(s)$$

Furthermore, from (b), (e) and (f) it follows that $I_n(s)$ and its $(n-1)$ derivatives all vanish for $s = a$. Therefore, equality (d) and (g) yield

$$I_1(s) = \int_a^s F(s_1)ds_1,$$

$$I_2(s) = \int_a^s I_1(s_2)ds_2 = \int_a^s \int_a^{s_2} F(s_1)ds_1 ds_2$$

$$\vdots$$

(h) $$I_n(s) = (n-1)! \int_a^s \int_a^{s_n} \cdots \int_a^{s_3} \int_a^{s_2} F(s_1)ds_1 ds_2 \ldots ds_{n-1} ds$$

Combining (b) and (h) we obtain the Cauchy formula (A.1).

A.2. GRONWALL-BELLMAN INEQUALITY

Often, especially in analysis of stability the following theorem is useful.

Let us assume that functions $u(t) \geq 0$ and $v(t) \geq 0$ for $t \in [t_o, \infty)$ are continuous and for some constant $C \geq 0$

(*) $$u(t) \leq C + \int_{t_o}^t v(\tau)u(\tau)d\tau;$$

then for $t > t_o$

(A.2) $$u(t) \leq C exp \int_{t_o}^t v(\tau)d\tau$$

Proof.

From inequality (*) we obtain

$$\frac{u(t)}{C + \int_{t_o}^t v(\tau)u(\tau)d\tau} \leq 1$$

and

(a) $$\frac{u(t)v(t)}{C + \int_{t_o}^t v(\tau)u(\tau)d\tau} \leq v(t)$$

Denoting the denominator by $\varphi(t)$ we can read inequality (a) as

(a′)
$$\frac{\dot{\varphi}(t)}{\varphi(t)} \leq v(t), \quad \dot{\varphi}(t) = \frac{d\varphi(t)}{dt}$$

Therefore, integration of (a) yields

$$ln\left[C + \int_{t_0}^{t} v(\tau)u(\tau)d\tau\right] - lnC \leq \int_{t_0}^{t} v(\tau)d\tau$$

Now, by virtue of the hypothesis (*) of the theorem we have

$$u(t) \leq C + \int_{t_0}^{t} v(\tau)u(\tau)d\tau \leq Cexp\int_{t_0}^{t} v(\tau)d\tau$$

which coincides with (A.2).

REFERENCES

CHAPTER I

[1] BHARUCHA-REID A.T., *Elements of the Theory of Markov Processes and their Applications*. McGraw Hill, New York, 1960.

[2] BHARUCHA-REID A.T., *Fixed-point Theorems in Probabilistic Analysis*, Bull. Amer. Soc., **82**, No. 5, p.641, 1976.

[3] BLANC-LAPIERRE A., FORTET R., *Theorie des Functions Aleatoires*, Masson, Paris, 1953.

[4] CHERKASOV I.D., *On the Transformation of the Diffusion Processes to a Wiener Process* (in Russian), Teor. Voroyatnost. Primienienija, **2**, 1957, 384–388

[5] COX D.R., MILLER H.D., *The Theory of Stochastic Processes*. Methuen, 1965.

[6] CRAMER H., LEADBETTER M.R., *Stationary and Related Stochastic Processes*, Wiley, New York, 1967.

[7] CURTAIN R.F., PICHARD A.J., *Infinite Dimensional Linear System Theory*, Lecture Notes in Control and Inform. Sci., **8**, 1978, Springer-Verlag, Berlin.

[8] DITLEVSEN O., *Uncertainty Modelling*, McGraw Hill, 1981.

[9] DOOB J.L., *Stochastic Processes*, Wiley, New York, 1953.

[10] FELLER W,. *Zur Therie der stochastischen Prozesse (Existenz und Eindeutigkeitssätze)*, Math. Ann. **113**, 113-160, 1936.

[11] FELLER W., *The Parabolic Differential Equations and the Associated Semi-groups of Transformations*, Ann.Math., **55**, 468–519, 1952.

[12] FINE L.T., *Theories of Probability: An Examination of Foundations*, Acad. Press, New York, London 1973.

[13] FISZ M., *ProbabilityTheory and Mathematical Statistics*, Acad. Press, New York, 1964 (translated from Polish).

[14] FRIEDMAN A., *Partial Differential Equations of Parabolic Type*, Englewood Cliffs, Prentice Hall, 1964.

[15] GARDINER C.W., *Handbook of Stochastic Methods*, Springer-Verlag, Berlin, 1983.

[16] GIKHMAN I.I., SKOROKHOD A.V., *Introduction to the Theory of Random Processes*, W.B. Saunders, Philadelphia, 1969 (translated from Russian).

[17] HANS O., *Generalized Random Variables*, Trans. First Prague Conf. Inform. Theory, Statistical Decision Functions, Random Processes, Publ. House of Czechoslovak Acad. Sci., Prague 1957, 61–103.

[18] HANS O., *Random Operator Equations*, Proc. Fourth Berkeley Symp. Math. Statistics and Probability, Vol. VII, Univ. of California Press (Ed. J. Neymann), 1961, 435, 1950.

[19] HILLE E., *"Explosive" Solution of Fokker-Planck Equation*, Proc. Intern. Congress of Mathematicians, Cambridge-Mass., Vol. I, p.435, 1950.

[20] HILLE E., PHILLIPS R.S., *Functional Analysis and Semi-Groups*, Amer. Math. Soc., Providence, Rhode Island, 1957.

[21] KRÉE P., SOIZE C., *Mathematics of Random Phenomena*, D. Reidel Publishing Company, Dordrecht, 1983.

[22] LEADBETTER M.R., LINGREN G., ROOTZEN H., *Extremes and Related Properties of Random Sequences and Processes*, Springer-Verlag, New York, Heidelberg, Berlin 1983.

[23] LOÉVE M., *Probability Theory*, Van Nostrand-Reinhold, Princeton, 1963.

[24] MATTHES K., KERSTAN J., MECKE J., *Infinitely Divisible Point Processes*, J. Wiley, New York, 1978.

[25] MORITA T., HARA H., *Solution to the Fokker-Planck Equation with Spatial Coordinate-dependent Moments in Many Dimensions*, Physica **125A**, 1984, 607–618.

[26] PROKHOROV Y.V., *The Methods of Characteristic Functionals*, Proc. Fourth Berkeley Symp. on Math. Stat. and Prob. (Ed. J. Neymann), Vol. II, 403–419, 1961.

[27] PROKHOROV Y.V., ROZANOV Y.A., *Probability Theory*, Spinger-Verlag, Berlin 1969 (translation from Russian).

[28] RICCIARDI L.M., *On the Transformation of Diffusion Process into the Wie ner Process*, J. Math. Anal. Appl., Vol. **54**, 185-199, 1976.

[29] RISKEN H., *The Fokker-Planck Equation; Methods of Solution and Applications*, Springer-Verlag, Berlin, 1984.

[30] ROSENBLATT M., *A Central Limit Theorem and the Strong Mixing Condition*, Proc. Nat. Acad. Sci., USA **42**, 43–47, 1956.

[31] SERFLING, R.J., *Contributuions to Central Limit Theorem for Dependent Random Variables*, Ann. Math. Stat., **4.**, 1968, 1158–1175.

[32] SKOROKHOD A.V., *Constrcutive Methods of Specifying Random Functions*, Usp. Mat. Nauk, Vol. **20**, No. 3, 1965. (in Russian)

[33] SNYDER D.L., *Random Point Processes*, Wiley, New York, 1975.

[34] SOBCZYK K., *Outline of Probability Theory*, Wyd. Polit. Poznańskiej; Poznań, Ed. I (1970), Ed II (1985), (in Polish).

[35] STRATONOVICH R.L., *Topics in the Theory of Random Noise*, Gordon and Breach, New York, 1963 (translation from Russian).

[37] WENCEL A.D., *On Boundary Conditions for Multi-dimensional Diffusion Processes* (in Russian), Teor. Veroyatnost. Primienienija, **4**, No. 2, 1957.

[38] WENCEL A.D., *Lectures on Theory of Random Processes*, Izd. Nauka, Moskow, 1975 (in Russian).

[39] WONG E., *Stochastic Processes in Information and Dynamical Systems*, McGraw Hill, New York, 1971.

[40] UENO T., *The Diffusion Satisfying Wencel's Boundary Conditions and the Markov Processes on the Boundary*, Proc. Japan Acad., Vol. 36, No 9, 1960, 533–538.

CHAPTER II.

[41] CAIROLI R., WALSH J.B., *Stochastic Integrals in the Plane*, Acta Math., **134**, 1975, 111–183.

[42] CURTAIN R.F., FALB P.L., *Itô's Lemma in Infinite Dimensions*, J. Math. Anal. Appl., **31**, 1970, 434–448.

[43] DALETSKII Y., PARAMONOVA S.N., *Stochastic Integrals over Normally Distributed Additive Function of Set*, Dokl. Ak. Nauk USSR, **208**, 1973, 512–515 (in Russian).

[44] FRIEDMANN A., PINSKY M. (Eds), *Stochastic Analysis*, Academic Press, New York, 1978.

[45] GOLDSTEIN J.A. *Second Order Itô Processes*, Nagoya Math. J., **36** (1969), 27–63.

[46] ITÔ K., *Stochastic Integral*, Proc. Jap. Acad., Tokyo, **20**, 519–529, 1944.

[47] ITÔ K., *On a Formula Concerning Stochastic Differentials*, Nagoya Math. J. **3**, 55–65, 1951.

[48] KOTELENEZ P., CURTAIN R.F., *Local Behaviour of Hilbert Space-valued Stochastic Integrals and the Continuity of Mild Solutions of Stochastic Evolution Equations*, Stochastics, **6** (1982), 239–257.

[49] KUNITA H., *Stochastic Integrals Based on Martingales Taking Values in Hilbert Space*, Nagoya Math. J., **39**, 1970, 41-52.

[50] KUO H.H., *Stochastic Integrals in Abstract Wiener Space*, Proc. J. Math., **41**, 469–483.

[51] McKEAN H.P. (jr.) *Stochastic Integrals* , Academic Press, New York, 1969.

[52] McSHANE E.J., *Stochastic Calculus and Stochastic Models*, Academic Press, New York, 1974.

[53] METIVIER M., PELLAUMAIL J., *Stochastic Integration*, Academic Press, New York, 1980.

[54] PONOMARIENKO L.L. *Stochastic Integrals with Respect to Multi-parameter Brownian Motion and Related Equations*, Teor. Veroyatn i Mat. Statist., 1972, No 7, 100–110 (in Russian).

[55] SKOROKHOD A.B., *On One Generalization of Stochastic Integral*, Teoria Verojatn. Prim., **20**, No 2, 223–238, 1975 (in Russian).

[56] STRATONOVICH R.L., *A New Representation of Stochastic Integrals and Equations*, SIAM J. Control, 4 1966, 362–371.

[57] VAHANIYA N.N., KANDELAKI N.P., *A Stochastic Intgeral for Operator-valued Functions*, Teor. Veroyatnosti i Prim., **12** (1967), 582–585, (in Russian).

[58] WILLIAMS D., (Ed.) *Stochastic Integrals*, Lecture Notes in Math., No 851, Springer-Verlag, Berlin, 1981.

[59] WONG E., ZAKAI M., *On the Convergence of Ordinary Integrals to Stochastic Integrals*, Ann. Math. Stat., **36**, 1965, 1560–1564.

[60] WONG E., ZAKAI M., *Martingales and Stochastic Integrals for Processes with Multi-dimensional Time Parameter*, Z. Wahrsch. verw. Gebiete, **29**, 1974, 109–122.

CHAPTER III

[61] ARNOLD L., *Stochastic Differential Equations: Theory and Applications*, Wiley, New York, 1974.

[62] ARNOLD L., *Lapunov Exponents of Nonlinear Stochastic Systems*, in: Ziegler F, Schueller G.I. (Eds), Nonlinear Stochastic Dynamics of Engineering Systems, Springer-Verlag 1988.

[63] ARNOLD L., CURTAIN R.F., KOTELENEZ P., *Nonlinear Stochastic Evolution Equations in Hilbert Space*, Forschungsschwerpunkt Dynamische Systeme, Univ. Bremen. Rep.nr 17, 1980.

[64] ARNOLD L., WIHSTUTZ V., *Stationary Solutions of Linear Systems with Additive and Multiplicative noise*, Stochastics, **7**, 1982, 133–155.

[65] ARNOLD L., WIHSTUTZ V., *Wide Sense Stationary Solutions of Linear Systems with Additive Noise*, SIAM J. Control Optim., **21**, 1983.

[66] ARNOLD L., WIHSTUTZ V., *Lapunov Exponents: A Survey* in: Lapunov Exponents (Eds Arnold L., Wihstutz V.), Lectures Notes in Math., 1186, Springer 1986.

[67] BENSOUSSAN A., TEMAN R., *Equations aux Dérivées Partielles Stochastiques Non Lineares*, Isreal J. Math., **11**, No 1, 1972, 95–129.

[68] BHARUCHA-REID A.T., *Random Integral Equations*, Academic Press, New York, 1972.

[69] BHARUCHA-REID A.T., *On the Theory of Random Equations*, Proc. Symp. Appl. Math., 16th, pp. 40–69, Amer. Math. Soc., Providence, 1964.

[70] BUNKE H.,*Gewöhnliche Differentialgleichungen mit Zufälligen Parametern*, Akademie-Verlag, Berlin, 1972.

[71] CHITASHVILI R.J., LAZRIEVA N.L., *Strong Solutions of Stochastic Differential Equations with Boundary Conditions*, Stochastics, 1981, Vol. 5, pp. 255–309.

[72] CHOJNOWSKA-MICHALIK A., *Stochastic Differential Equations in Hilbert Space*, Prob. Theory-Publ. of Banach Center, Warszawa, 1979, pp. 53–74.

[73] CHOJNOWSKA-MICHALIK A., *On Processes of Ornstein-Uhlenbeck Type in Hilbert Space*, Stochastics, Vol. 21, 1987, 257–286.

[74] CHOW P.L., *Stability of Nonlinear Stochastic Evolution Equations*, J. Math. Anal. Appl., Vol. 89, No 2, 1982, 400–419.

[75] CHOW P.L., *Stochastic Partial Differential Equations in Turbulence Related Problems*, in: Probabilistic Analysis and Related Topics (Ed. Bharucha-Reid A.T.), Vol. 1, Academic Press, New York, pp. 1–43.

[76] CONWAY E.D., *Stochastic Equations with Discontinuous Drift*, Trans. Amer. Math. Soc., **157**, 1, 1971, 235–245.

[77] CURTAIN R.F., *Stochastic Evolution Equations with General White Noise Disturbance*, J. Math. Anal. Appl., **60**, 1977, 570–595.

[78] CURTAIN R.F., *Markov Processes Generated by Linear Stochastic Evolution Equations*, Stochastics, No 1–2, 1981.

[79] CURTAIN R.F., *Stability of Stochastic Partial Differential Equations*, J. Math. Anal. Appl., **79**, 1981, 352–369.

[80] CURTAIN R.F., FALB P.L., *Stochastic Differential Equations in Hilbert Space*, J. Diff. Equs, **10** No. 3, 1971.

[81] DA PRATO G., TUBARO L., *Stochastic Partial Differential Equations and Applications*, Lectures Notes in Math., No 1236, Springer, 1987.

[82] DA PRATO G., KWAPIEŃ S., ZABCZYK J., *Regularity of Solutions of Linear Stochastic Differential Equations in Hilbert Space*, Stochastics, **23** (1), 1–23, 1987.

[83] DAWSON D.A., *Generalized Stochastic Integrals and Equations*, Trans Amer. Math. Soc., **147**, 1970, 473–506.

[84] DAWSON D.A., *Stochastic Evolution Equations*, Math. Biosci., **15** 1972, 287–316.

[85] DAWSON D.A., *Stochastic Evolution Equations and Related Measure Processes*, J. Multiv. Anal., **5**, 1975, 1–52.

[86] DYM, H., *Stationary Measures for the Flow of a Linear Differential Equation Driven by White Noise*, Trans. Amer. Math. Soc., **123**, 1966, 130–164.

[87] FRIEDMAN A., *Stochastic Differential Equations and Applications*, Academic Press, New York, 1975, Vol. 1; 1976, Vol. 2.

[88] GIKHMAN I.I., *On the Theorem of Bogoliubov*, Uhkr. Matiem. Zurnal, **4**, 1952, 215–219 (in Russian).

[89] GIKHMAN I.I., DOROGOVTCEV A.J., *On Stability of Solutions of Stochastic Differential Equations* (in Russian), Ukr. Matiem. Zurnal, **17** No 6, 1965.

[90] GIKHMAN I.I., SKOROKHOD A.V., *Stochastic Differential Equations*, Springer, Berlin, 1972 (translation from Russian edition in 1968).

[91] GIKHMAN I.I., SKOROKHOD A.V., *Stochastic Differential Equations and their Applications* (in Russian), Kijev, Naukovaja Dumka, 1982.

[92] GIRSANOV I.V., *Example of Non-uniqueness of a Itô's Stochastic Equation* (in Russian), Teoria Vieroyatn. Prim., **7**, 1962, 336–342.

[93] HAUSSMANN U.G., *Asymptotic Stability of Linear Itô Equation in Infinite Dimensions*, J. Math. Anal. Appl., **65**, 1978, 219–235.

[94] HORSTHEMKE W., LEFEVER R., *Noise-Induced Transitions*, Springer, Berlin, 1984.

[95] ICHIKAWA A., *Linear Stochastic Evolution Equations in Hilbert Space*, J. Diff. Eqs, **28**, 1978, 266–277.

[96] ICHIKAWA A., *Dynamic Programing Approach to Stochastic Evolution Equations*, SIAM J. Control Optim., **17**, 1979, 152–174.

[97] ICHIKAWA A., *Semi-linear Stochastic Evolution Equations: Boundedness, Stability and Invariant Measures*, Stochastics,, 1984, Vol. 12, 1–39.

[98] IKEDA N., WATANABE S., *Stochastic Differential Equations and Diffusion Processes*, North-Holland, Amsterdam, New York, 1981.

[99] ITO K., NISIO M., *On Stationary Solutions of Stochastic Differential Equations*, J. Math. Kyoto Univ., **4**, 1964, 1–79.

[100] KANAN D., *Random Integrodifferential Equations* in: Probabilistic Analysis and Related Topics (Ed. Bharucha-Reid A.T.), Vol. 1, Academic Press, 1978.

[101] KHASMINSKII R.Z., *On Stochastic Processes Defined by Differential Equations with Small Parameter*, Teoria Vieroyatn. Prim., **11** No 2, 1966 (in Russian).

[102] KHASMINSKII R.Z., *A Limit Theorem for the Solution of Differential Equations with Random Right-hand Side*, Teoria Vieroyatn. Prim., **11** No 3, 1966 (in Russian).

[103] KHASMINSKII R.Z., *Stability of Systems of Differential Equations with Random Disturbances of Parameters*, (in Russian), Nauka, Moscow, 1969.

[104] KLIEMANN W., *Analysis of Non-Linear Stochastic Systems* in: Analysis and Estimation of Stochastic Mechanical Systems (Eds. Schiehlen W., Wedig W.) CISM Courses and Lectures, (no 303), Springer, 1988.

[105] KENECNY F., *On Wong-Zakai Approximation to Stochastic Differential Equations*, J. Multiv. Analysis, **13**, 1983, 605–611.

[106] KOTULSKI Z., *Characteristic Functionals of Stochastic Wave Processes*, PhD. Thesis. Inst. Fund. Techn. Res., Polish Academy of Sciences, Warsszawa 1984 (in Polish).

[107] KOLTULSKI Z., *Equations of Characteristic Functional and Moments of the Stochastic Evolutions with Application*, SIAM J. Appl. Math., Vol. 49, 1989, No 1.

[108] KOTULSKI Z., SOBCZYK K., *Characteristic Functionals of Randomly Excited Physical Systems*, Physica **123A**, 1984, 261–278.

[109] KOZIN F., *Stability of Linear Stochastic Systems* in: Lecture Notes in Math., (No 294), 186–229, Springer-Verlag, 1972.

[110] KOZIN F., *Some Results on Stability of Stochastic Dynamical Systems*, Probabilistic Eng. Mech., **1**, No. 1, March 1986.

[111] KREE P., *Equations Linéaires a Coefficients Aléatories*, Inst. Nazionalle Alta Matematica, Symp. Math., Vol. VIII (1971), Bologna 1971.

[112] KUSHNER H.J., *Stochastic Stability and Control*, Academic Press, New York, 1967.

[113] LEBIEDIEV V.A., *On the Uniqueness Condition of Solution of System of Stochastic Differential Equations*, (in Russian), Teoria Vieroyatn. Prim., **21**, No 2, 1981, 423.

[114] LEIZAROWITZ A., *Estimates and Exact Expressions for Lapunov Exponents of Stochastic Linear Differential Equations*, Stochastics, **24**, No 4, 1988, 335–356.

[115] MAKHNO S.J., *Stochastic Equations of Evolution Type*, Teoria Sluc. Proc., Kijev, No 6, 1978, 101–107 (in Russian).

[116] MARKUS R., *Parabolic Itô Equations*, Trans. AMS, **198**, 1974, 177–190.

[117] NASHED M.Z., SALEHI H., *Measurability of Generalized Inverses of Random Linear Operators*, SIAM J. Math., **25**, 1973, 681–692.

[118] OREY S., *Stationary Solutions for Linear Systems with Additive Noise*, Stochastics, **5**, 1981, 241–252.

[119] PAPNICOLAU G.C., VARADHAN S.R.S., *A Limit Theorem for Strong Mixing in Banach Space and Two Applications to Stochastic Differential Equations*, Comm. Pure Appl. Math., **26**, 1973, 497–524.

[120] PAPNICOLAU G.C., KOHLER W., *Asymptotic Theory of Mixing Stochastic Ordinary Differential Equations*, Comm. Pure Appl. Math., **27**, 1974, 641–668.

[121] PARDOUX E., *Equations aux Derivées Partialles Stochastiques Nonlineares Monotones*, Thése, Univ. Paris XI, 1975.

[122] PARDOUX E., *Stochastic Differential Equations and Filtering of Diffusion Processes*, Stochastics, **3**, 1979, 127–167.

[123] PARDOUX E., TALAY D., *Stability of Linear Differential Systems with Parametric Excitation*, in: Nonlinear Stochastic Dynamic Engineering Systems (Eds Ziegler F., Schueller G.I.) Springer-Verlag 1988, pp. 153–168.

[124] PORTIENKO, N.I., *Generalized Diffusion Processes*, Kijev, Naukovaja Dumka, 1982 (in Russian).

[125] PROTTER P.E., *On the Existence, Uniqueness, Convergence and Explosions of Solutions of Systems of Stochastic Integral Equations*, Ann. Probab., **5**, 1977, No 2, 243–261.

[126] ROZKOSZ A., SLOMIŃSKI L., *On Weak Convergence of Solutions of One-dimensional Stochastic Differential Equations*. (manuskript, 1988).

[127] RUTKOWSKI M. *Strong Solutions of Stochastic Differential Equations Involving Local Time*, Stochastics, Vol. 22, 1987, 201–208.

[128] SCHEIDT VOM J., PURKERT W., *Random Eigenvalue Problems*, Verlag-Akademie, 1983.

[129] SKOROKHOD A.V., *On the Existence and Uniqueness of Solutions of Stochastic Differential Equations*, (in Russian), Sib. Mat. Zurn., **2**, 1961, 129–137.

[130] SKOROKHOD A.V., *Studies in Theory of Random Processes*, Reading Mass., Addison-Wesley, 1965 (translation from Russian published in 1961).

[131] SOBCZYK K., *Methods of Statistical Dynamics* (in Polish), Polish Scientific Publ., Warszawa, 1973.

[132] SOONG T.T., *Random Differential Equations in Sience and Engineering*, Academic Press, New York 1973.

[133] STRAND T.L., *Random Ordinary Differential Equations*, J. Diff. Eqs, **7**, 1970, 538–553.

[134] SUSSMAN H.T., *On the Gap Between Determinsitic and Stochastic Ordinary Differential Equations*, Ann. Probab., 6, 1978, 19–41.

[135] TSOKOS CH. P., PADGETT W.J., *Random Integral Equations with Applications to Life Sciences and Engineering*, Academic Press, New York, 1974.

[136] VIENCEL A.D., FREIDLIN M.I., *Fluctuations in Dynamical Systems Subjected to Small Random Pertubations*, (in Russian), Nauka, Moscow, 1979.

[137] VIOT M., *Solutions Failbles d'Equations aux Derivées Partielles Stochastiques non Lineaires*, Thesis, Univ. Paris VI, 1976.

[138] WATANABE S., *On Stochastic Differential Equations for Multi-dimensional Diffusion Processes with Boundary Conditions*, J. Math. Kyoto Univ., **11**, 1971, 169–180; 545–551.

[139] WONG E., ZAKAI M., *On the Relationship between Ordinary and Stochastic Differential Equations*, Int. J. Eng. Sci., **3**, 1965 213–229.

[140] YEH J., *Existence of Strong Solutions for Stochastic Differential Equations in the Plane*, Pac. J. Math., Vol. 97, No 1, 1981, pp. 217–247.

[141] ZABCZYK J., *On Stability of Finite Dimensional Linear Stochastic Systems*, Proc. Banach Center, Probability Semester, Warsaw, 1976.

[142] ZABCZYK J., *Stationary Distributions for Linear Equations Driven by General Noise*, Bull. Acad. Polon. Sci., Ser. Math., **31**, (3–4), 1983, 197–209.

[143] ZABCZYK J., *Structural Properties and Limit Behaviour of Linear Stochastic Systems in Hilbert Space*, in: Mathem. Control Theory, Banach Center Publ. (Vol. 14), Polish Sci. Publ., Warsaw, 1985.

CHAPTER IV

[144] ABRAMOWITZ M., STEGUN I., *Handbook of Mathematical Functions*, Dover, New York, 1965.

[145] ADOMIAN G., *The Closure in the Hierarchy Equations*, J. Statist. Phys., Vol. 3, 2, 1971.

[146] ADOMIAN G., MALAKIAN K., *Closure Approximation Error in the Mean Solution of Stochastic Differential Equations by the Hierarchy Method*, J. Statist. Phys., Vol. 21, 181–189, 1979.

[147] ARIARATNAM S.T., GRAEFE P., *Linear Systems with Stochastic Coefficients*,

P. I,II,III, Int. J. Control, 1965; P.I: Vol. 1, No 3, p.239; P.II: Vol. 2, No 2, p161, P.III: Vol. 2, No 3, p.205.

[148] ARIARATNAM S.T., TAM D.S.F., *Random Vibration and Stability of a Linear Parametrically Excited Oscillator*, ZAMM, B.59, No 2, p.79, 1979.

[149] ASSAF S.A., ZIRKIE L.D., *Approximate Analysis of Non-linear Stochastic Systems*, Int. J. Control, **23**, 477–492, 1976.

[150] ATALIK T.S., UTKU S., *Stochastic Linearization of Multi-degree of Freedom Non-linear Systems*, Earthquake Engineering and Struct. Dynamics, **4**, 411–420, 1976.

[151] BÉCUS G.A., *Random Generalized Solutions to the Heat Equation*, J. Math. Anal. Appl., **60**, 93–102, 1977.

[152] BELLMAN R., RICHARDSON J.M., *Closure and Preservation of Moment Properties*, J. Math. Anal. Appl., **23**, 639–644, 1968.

[153] BLANKENSHIP G., PAPANICOLAU G.C., *Stability and Control of Systems with Wide-band Noise Disturbances*, SIAM J. Appl. Math., **34**, 437–476, 1978.

[154] BOYCE W.E., *Random Eigenvalue Problems*, in: Probabilistic Methods in Appl. Math., (Ed. Bharucha-Reid A.T.), Academic Press, Vol. 1, 1968.

[155] BRISSAUD A., FRISH U., *Solving Linear Stochastic Differential Equations*, J. Math. Phys., **15**, No 5, 1974.

[156] BRUCKNER A., LIN Y.K., *Application of Complex Stochastic Averaging to Nonlinear Random Vibration Problems*, Int. J. Nonlinear Mech., **22**, 237–250, 1987.

[157] BUCIARELLI L.L., KUO C., *Mean-square Response of Second Order System to Nonstationary Random Excitation*, J. Appl. Mech., **37**, 612–616, 1970.

[158] BURG J.P., *Maximum Entropy Spectral Analysis*, 37th Annual Intern. Meeting Soc. Exploration Geophysists, Oklahoma City, 1967.

[159] CAI G.Q., LIN Y.K., *A New Approximate Solution Technique for Randomly Excited Nonlinear Oscillators*, Florida Atlantic Univ., (Center for Appl. Stoch. Res.), October 1987.

[160] CASCIATI F., FARAVELLI L., *Methods of Nonlinear Stochastic Dynamics for Assessment of Structural Fragility*, Nucl. Eng. and Design, Vol. 90, 341–356, 1985.

[161] CAUGHEY T.K., *Equivalent Linearization Techniques*, J. Acoust. Soc. Amer., **35**, 1706–1711, 1961.

[162] CAUGHEY T.K., *Nonlinear Theory of Random Vibrations*, in: Advances in Applied Mechanics, Vol. 11, pp.209–253, 1971.

[163] CAUGHEY T.K., *On the Response of Non-linear Oscillators to Stochastic Excitation*, Probabilistic Eng. Mech., Vol. 1, No 1, March 1986.

[164] CAUGHEY T.K., STUMPF H.J., *Transient Response of a Dynamic System under Random Excitation*, J. Appl. Mech., **28**, 563–566, 1961.

[165] CODDINGTON E.A., LEVINSON N., *Theory of Ordinary Differential Equations*, McGraw Hill, New York, 1955.

[166] COZZARELLI F.A., HUANG W.N., *Effcets of Random Material Parameters on Nonlinear Steady Creep Solutions*, Int. J. Struct., **7**, 1477–1494, 1971.

[167] CRANDALL S.H., *Zero Crossings and Other Statistical Measures of Random Responses*, J. Acoust. Soc. Amer., **35**, 1693, 1961.

[168] CRANDALL S.H., MARK W.D., *Random Vibration in Mechanical Systems*, Academic Press, New York, 1963.

[169] CRANDALL S.H., *Non-Gaussian Closure for Random Vibration of Nonlinear Oscillators*, Int. J. Non-Linear Mechanics, **15**, 303–313, 1980.

[170] CRANDALL S.H., *Non-Gaussian Closure Techniques for Stationary Random Vibration*, Int. J. Non-Linear Mech., **20** No. 1, 1–8, 1985.

[171] DIMENTBERG M.F., *Nonlinear Stochastic Problems of Mechanical Vibrations* (in Russian), Nauka, Moscow, 1980.

[172] DIMENTBERG M.F., *An Exact Solution of a Certain Non-linear Random Vibration Problem*, Int. J. Non-Linear Mech., **17**, 231–236, 1982.

[173] DOSTUPOV B.G., PUGACHEV V.S., *Equation for the Probability Distribution of Integral of a System of Ordinary Differential Equations Containing Random Parameters*, (in Russian), Automatika i Telemekhanika, Vol. 18, No 7, 1957.

[174] DOWSON D.C., WRAGG A., *Maximum Entropy Distributions Having Prescribed First and Second Moments*, IEEE Trans. Inform. Theory, **19**, 689–693, 1973.

[175] GOANKAR G.H., *Dynamic Systems with Random Initial State*, J. Eng. Math., Vol. 5, No 3, 171–178, 1971.

[176] GOANKAR G.H., *Linear Systems with Non-stationary Random Inputs*, Int. J. Control, Vol. 14, No 1, 1971.

[177] GOPALSAMY K., BHARUCHA-REID A.T., *On a Class of Parabolic Differential Equations Driven by Stochastic Point Process*, J. Appl. Prob., Vol. 12, 98–106, 1975.

[178] GRAHAM R., HAKEN H., *Generalized Thermo-dynamic Potential for Markov Systems in Detailed Balance and Far from Thermal Equilibrium*, Zeitschrift für Physik, Vol. 203, 289–302, 1971.

[179] GRIGORIU M., ARIARATNAM S.T., *Stationary Response of Linear Systems to Non-Gaussian Excitations*, (private communication, August 1987).

[180] HAMMOND J.K., *On the Response of Single and Multi-degree of Freedom Systems to Non-stationary Random Excitation*, J. Sound and Vibr., Vol. 7, pp.393–416, 1968.

[181] HAMPL N.C., SCHUELLER G.I., *Probability Densities of the Response of Nonlinear Structures under Stochastic Dynamic Excitation*, Proc., USA-Austrian Seminar on "Appl. of Stoch. Struct. Mech.", Boca Raton, 1987.

[182] IBRAHIM R.A., *Parametric Random Vibration*, Res. Studies Press — J. Wiley, New York, 1985.

[183] IBRAHIM R.A., SOUNDARARAJAN A., HEO H., *Stochastic Response of Nonlinear Dynamic Systems Based on a Non-Gaussian Closure*, J. Appl. Mech., Vol. 52, p. 965, Dec. 1985.

[184] INGARDEN R.S., URBANIK K., *Quantum Informational Thermodynamics*, Acta Physica Polonica, Vol. 21, No 3, pp. 281–304, 1962.

[185] IWAN W.D., MASON A.B., *Equivalent Linearization for Systems Subjected to Nonstationary Random Excitation*, Int. J. Non-Linear Mech., Vol. 15, pp. 71–82, 1980.

[186] IWANKIEWICZ R., *Response of Vibratory Systems to General Random Impulses*, J. Sound Vibr. **108** (3), 429–446, 1986.

[187] IWANKIEWICZ R., ŚNIADY P., *Vibration of a Beam under a Random Stream of Moving Forces*, J. Struct. Mech., Vol. 12, No 1, 1984.

[188] IWANKIEWICZ R., SOBCZYK K., *Dynamic Response of Linear Structures to Correlated Random Impulses*, J. Sound Vibr., **86** (3), 303–317, 1983.

[189] IYENGAR R.N., DASH P.K., *Study of the Random Vibration of Nonlinear Systems by Gaussian Closure Technique*, J. Appl. Mech., **45**, 393–399, 1978.

[190] JAYNES E.T., *Information Theory and Statistical Mechanics*, Phy. Rev., Vol. 106, 620–630, 1957.

[191] KOTULSKI Z., SOBCZYK K., *Linear Systems and Normality*, J. Statistical Physics, **24**, No 2, 1981.

[192] KOTULSKI Z., SOBCZYK K., *Effects of Parameter Uncertainty on the Response of Vibratory Systems to Random Excitation*, J. Sound Vibr., **119** (1), 159–171, 1987.

[193] KOTULSKI Z., SOBCZYK K., *On the Moment Stability of Vibratory Systems with Random Impulsive Parametric Excitation*, Arch. Mech., Vol. 40, No 4, 1988.

[194] KOZIN F., *On Probability Densities of the Output of Some Random Systems*, J. Appl. Mech., **28**, 161–165, 1961.

[195] KOZIN F., *On approximations to the Densities and Moments of a Class of Stochastic Systems*, Proc. Camb. Phil. Soc., **59**, 463–481, 1963.

[196] KOZIN F., *On Almost Sure Stability of Linear Systems with Random Coefficients*, J. Math. and Phys. 42, No 1, 1963.

[197] KOZIN F., *On Relations between Moment Properties and Almost Sure Lapunov Stability for Linear Stochastic Systems*, J. Math. Anal. Appl., **10** No 2, 1965.

[198] KOZIN F., *The Method of Statistical Linearization for Non-linear Stochastic Vibrations*, in: Nonlinear Stochastic Dynamic Engineering Systems (Eds: F. Ziegler, G.I. Schueller) Springer-Verlag, Berlin, 1988.

[199] KULLBACK S., *Information Theory and Statistics*, J. Wiley, New York, 1959.

[200] LADDES G.S., SAMBANDHAM M., *Error Estimation of Solutions and Mean of Solutions of Stochastic Differential Systems*, J. Math. Phys., **24**, 4, April 1983.

[201] LANGLEY R.S., *A Variational Formulation of the FPK Equations with Application to the First Passage Problem in Random Vibration*, J. Sound Vibr., 123, No 2, June 1988.

[202] LEE J., *Comparison of Closure Approximation Theories in Turbulent Mixing*, Physics Fluid, 9, 363–372, 1966.

[203] LIN Y.K., *Non-stationary Excitation of Linear Systems Treated as Sequence of Random Pulses*, J. Acoust. Soc. Amer., **38**, 453–4601, 1965.

[204] LIN Y.K., *Exact and Approximate Solutions for Response of Nonlinear Systems under Parametric and External White Noise Excitations*, in: Nonlinear Stochastic Dynamic Engineering Systems (eds: F. Ziegler, G.I. Schueller), Springer, Berlin, 1988.

[205] LIN Y.K., CAI G.Q., *Equivalent Stochastic Systems*, Florida Atlantic Univ., (Center for Appl. Stoch. Res.), Sept. 1987.

[206] LUTES L.D., HU S.L.J., *Nonnormal Stochastic Response of Linear Systems*, J. Eng. Mech., ASCE, Febr. 1986, 127–141.

[207] LYONS R.M., *Spectral Analysis of Nonstationary Processes*, J. Roy. Statist. Soc., **B30**, 1–20, 1968.

[208] MADSEN P.H., KRENK S., *Stationary and Transient Response Statistics*, J. Eng. Mech. Div., Proc. ASCE, **108**, No EM4, August 1982.

[209] MAKAROV B.P., *Nonlinear Problems of Statistical Dynamics of Machines and Devices*, Mashinostrojenije, Moskwa, 1983 (in Russian).

[210] MALLOWS C.L., *Linear Systems are Nearly Gaussian*, J. Appl. Probab., **8**, 118–127, 1971.

[211] MATUZAWA T., *On a Class of Stochastic Partial Differential Equations*, J. Math. Anal. Appl., **78**, No 1, 161–172, 1980.

[212] NIEMYTZKI W.W., STIEPANOV W.M., *Qualitative Theory of Differential Equations*, (in Russian), Moscow, 1949: English translation, Princeton 1960).

[213] NGUYEN DONG ANH, *Influence of Exponentially Correlated Stationary Processes on Oscillations of Mechanical Systems with One Degree of Freedom*, Ukrainian Math. J., **38** No 4, 1986 (in Russian).

[214] ORLOWSKI A., SOBCZYK K., *Solitons and Shock Waves under Random External Noise*, Rep. Math. Phys., Vol. 27, No 2, 1989.

[215] PAPOULIS A., *Narrow-band Systems and Gaussianity*, IEEE Trans. Inform. Theory, **IT-18**, 20–23, 1972.

[216] PRISTLEY M.B., *Evolutionary Spectra and Non-stationary Processes*, J. Roy. Statist. Soc., **B27**, p. 204, 1965.

[217] RACICOT R., MOSES F., *Filtered Poisson Process for Random Vibration Problems*, J. Eng. Mech. Div., Amer. Soc. Civil Eng., **98**, 159–176, 1972.

[218] RENGER A., *Equation for Probability Density of Vibratory System Subjected to Continuous and Discrete Stochastic Excitation* (in German), ZAMM, **59**, No 1, 1979.

[219] RICHRADSON J.M., *The Application of Truncated Hierarchy Techniques in the Solution of a Stochastic Linear Differential Equation*, in: Proc. Symp. in Appl. Math., Vol. 16, AMS, New York, 1964, (ed. Bellman R.).

[220] ROBERTS J.B., *On the Harmonic Analysis of Evolutionary Random Vibrations*, J. Sound Vibr., **2**, 336–352, 1965.

[221] ROBERTS J.B., *The Response of Linear Vibratory Systems to Random Impulses*, J. Sound Vibr., **2**, 375–390, 1965.

[222] ROBERTS J.B., *System Response to Random Impulses*, J. Sound Vibr., **24**, 23–34, 1972.

[223] ROBERTS J.B., *Distribution of the Response of Linear Systems to Poisson Distributed Random Pulses*, J. Sound Vibr., **28** (1), 93–103, 1973.

[224] ROBERTS J.B., SPANOS P.D., *Stochastic Averaging: an Approximate Method for Solving Random Vibration Problems*, Int. J. Non-Linear Mech., Vol. 21, No 2, 111–134, 1986.

[225] ROSENBLAT M., *Some Comments on Narrow Band Pass Filters*, Quart. Appl. Math., **18**, 378–393, 1961.

[226] SCHIEHLEN W.O., *Nonstationary Random Vibrations*, in: Random Vibrations and Reliability (Ed. Hennig K.), Akademie-Verlag, Berlin, 1983.

[227] SCHUELLER G.I., BUCHER C.G., *Non-Gaussian Response od Systems under Dynamic Excitation*, in: Stochastic Structural Dynamics; Progress in Theory and Applications (Eds: Ariaratnan S.T., Schueller G.I., Elishakoff I.), Elsevier Appl. Sci., 1988.

228] SCHUSS Z., *Stochastic Differential Equations with Applications*, McGraw Hill, 1981.

[229] SHAPIRO V.E., LOGINOV V.M., *Formulae for Differential and their Use for Solving Stochastic Equations*, Physica, Vol. 91A, 563–574, 1978.

[230] SHINOZUKA M., *Probability of Structural Failure under Random Loading*, J. Eng. Mech. Div., AMCE, **90** (EMS), 147–170, 1964.

[231] SKALMIERSKI B., TYLIKOWSKI A., *Stability of Dynamical Systems*, (in Polish), PWN, Warszawa, 1972.

[232] SOBCZYK K., *Random Vibrations of Statistically Inhomogeneous Elastic Systems*, Proc. Vibr. Probl., **11**, No 4, 1970.

[233] SOBCZYK K., *On the Normal Approximation in Stochastic Dynamics*, in: Random Vibration and Reliability (Ed. K. Hennig), 317–326, Akademie-Verlag, Berlin, 1983.

[234] SOBCZYK K., *Stochastic Wave Propagation*, Elsevier, 1985.

[235] SOBCZYK K., *Stochastic Differential Equations for Applications*, Technical University of Denmark (Dept. Struct. Eng.), Ser. R, No 203, 1985.

[236] SOBCZYK K., *Stochastic Modelling of Fatigue Crack Growth*, in: Probabilistic Methods in Mechanics of Solids and Structures, (Eds. Eggwertz S., Lind N.,) Springer, Berlin, 1985.

[237] SOBCZYK K., TRĘBICKI J., *Maximum Entropy Principle in Stochastic Dynamics*, Probabilistic Eng. Mech., Vol.5, No 3, 1990 .

[238] SOONG T.T., COZARELLI R.A., *Effect of Random Temperature Distribution on Creep in Circular Plate*, Int. J. Non-linear Mech., **2**, 27–38, 1967.

[239] SPANOS P.D., *Stochastic Analysis of Oscillators with Nonlinear Damping*, Int. J. Non-Linear Mech., **13**, 249–259, 1978.

[240] SPANOS P.D., LUTES L.D., *Probability Density of Response to Evolutionary Process*, J. Eng. Mech., **106**, 213–224, 1980.

[241] SPANOS P.D., *Stochastic Linearization in Structural Dynamics*, Appl. Mech. Rev., **34**, 1981.

[242] SPANOS P.D., SOLOMOS, G.P., *Markov Approximation to Transient Vibration*, J. Eng. Mech., **103**, 1134–1150, 1983.

[243] SRI NAMACHCHIVAYA N., HILTON H.H., *Stochastically Perturbed Bifurcations*, in: Nonlinear Stochastic Dynamic Eng. Systems (Eds: Ziegler F., Schueller G.I.), Springer, 1988.

[244] SRINIVASAN S.K., VASUDEVAN R., *Introduction to Random Differential Equations and their Applications*, Elsevier, New York, 1971.

[245] SRINIVASAN S.K., SUBRAMANIAN R., KUMARASWAMY S., *Response of Linear Vibratory Systems to Non-stationary Stochastic Impulses*, J. Sound Vibr., **6**, 159–179, 1967.

[246] SVESHNIKOV A.A., *Applied Methods of Random Function Theory*, (in Russian), Nauka, Moscow, 1968.

[247] SYSKI R., *Stochastic Differential Equations*, in: Modern Nonlinear Equations (Ed. Saaty J.L.), McGraw Hill, New York, 1967.

[248] SZOPA J., *Application of Volterra Stochastic Integral Equations of II-nd Kind to the Analysis of Dynamic Systems of variable Inertia*, J. Techn. Phys., **17**, 423–433, 1976.

[249] TUNG C.C., *Random Response of Highway Bridges to Vehicle Loads*, J. Eng. Mech. Div., ASCE, **93**, 79–94, 1967.

[250] TYLIKOWSKI A., MARKOWSKI W., *Vibration of Nonlinear Single Degree of Freedom System Due to Poissonian Impulse Excitation*, Int. J. Non-Linear Mech., Vol. 21, No 3, 229–238, 1986.

[251] VAN SLOOTEN R.A., SOONG T.T., *Buckling of a Long Axially Compressed Thin Cylindrical Shell with Random Initial Imperfections*, J. Appl. Mech., **39**, 1066–1072, 1972.

[252] VERNE-JONES D., *Stochastic Models for Earthquake Occurance*, J. Roy. Statist. Soc., **B32**, 1–62, 1970.

[253] WEDIG W., *Regions of Instability for a Linear System with Random Parameter Excitation*, in: Lecture Notes in Math., **294**. 160–172, Springer, New York, 1972.

[254] WEDIG W., *Mean Square Stability and Spectrum Identification of Nonlinear Stochastic Systems*, in: Nonlinear Stochastic Dynamic Eng. Systems, (Eds. Ziegler F., Schueller G.I.), Springer, 1988.

[255] WEN Y.K., *Approximate Method for Non-linear Random Vibrations*, J. Eng. Mech. Div., ASCE, **101**, 389–401, 1975.

[256] WEN Y.K., *Equivalent Linearization for Hysteretic Systems under Random Excitation*, J. Appl. Mech., **47**, 150–154, 1980.

[257] WU W.F., LIN Y.K., *Cummulant — Neglect Closure for Nonlinear Oscillators under Random Parametric and External Excitations*, Int. J. Non-Linear Mech., **19**, No 4, 1984.

[258] YONG Y., LIN Y.K., *Exact Stationary Response Solution for Second Order*

Nonlinear Systems under Parametric and External White Noise Excitations, J. Appl. Mech., **109**, 414–418, June 1987.

CHAPTER V

[259] AKHIEZER N. I., *Approximation Theory*, Polish Sci. Publ., Warszawa, 1957 (translation from Russian).

[260] BLUM E.K., *Numerical Analysis and Computation Theory in Practice*, Addison-Wesley, Reading, 1972.

[261] CLARK J,M.C., CAMERON R.J., *The Maximum Rate of Convergence of Discrete Approximations for Stochastic Differential Equations*, in: Stochastic Differential System (Ed. Grigelionos B.), Lect. Notes in Control and Inform. Sci., Vol. 25, Springer, Berlin, 1980.

[262] FRANKLIN J., *Numerical Simulation of Stationary and Non-stationary Gaussian Random Processes*, SIAM Rev., **7**, 68–80, 1965.

[263] GERSCH W., LIU R.S-Z., *Time Series Methods for Synthesis of Random Vibration Systems*, J. Appl. Mech., Vol. 43, 159–165, 1976.

[264] GOROSTIZA L.G., *Rate of Convergence of a Approximate Solution of Stochastic Differential Equations*, Stochastics, Vol. 3, 267–276, 1980.

[265] GREENSIDE H.S., HELFAND E., *Numerical Integration of Stochastic Differential Equations*, Bell System Techn. J., Vol. 60, No 8, 1981.

[266] GRIEGO R.J., HEATH D., RUIZ-MONCAYO A., *Almost Sure Convergence of Uniform Transport Processes to Brownian Motion*, Ann. Math. Statistics, Vol. 42, 1129–1131, 1971.

[267] HELFAND E., *Numerical Integration of Stochastic Differential Equations*, Bell System Techn. J., Vol. 58, No 10, 1979.

[268] IKEDA N., NAKAO S., YAMATO Y., *A Class of Approximation of Brownian Motion*, Publ. RIMS, Kyoto Univ., Vol. 13, 285–300, 1977.

[269] JANSSEN R., *Discretization of the Wiener Process in Difference Methods for Stochastic Differential Equations*, Stochastic Processes and Appl., **18**, 1984, 361–369.

[270] KANTOROVICH L.W., KRYLOV V.Z., *Approximate Methods of Higher Analysis*, (in Russian), Moskwa, 1949.

[271] KLAUDER J.R., PETERSEN W.P., *Numerical Integration of Multiplicative-noise Stochastic Differential Equations*, SIAM J. Numer. Anal., **22**, No 6, 1985.

[272] KLOEDEN P.E., PEARSON R.A., *The Numerical Solution of Stochastic Differential Equations*, J. Austr. Soc., Ser. B20, No 1, 1977.

[273] KOZIN F., *Autoagressive Moving Average Models of Earthquake Records*, Prob. Eng. Mech., Vol. 3, No 2, 1980.

[274] KRASNOSELSKII, G.M., (et al.), *Approximate Solutions of Operator Equations*, Wolters-Nordhoff, Groningen, 1972 (translation from Russian).

[275] KRENER A.J., LOBRY C., *The Complexity of Stochastic Differential Equations*, Stochastics, 4, p.193, 1981.

[276] KRENK S., CLAUSEN J., *On the Calibration of ARMA Processes for Simulation*, in: Reliability and Optimization of Structral Systems (Ed. Thoft-Christensen P.), Lecture Notes in Engineering, Springer, Berlin, 1987.

[277] KRUPOWICZ A., *Numerical Methods of Ordinary Differential Equations*, (in Polish), Polish Sci. Publ., Warszawa 1986.

[278] LAMBERT J.D., *Computational Methods in Ordinary Differential Equations*, J. Wiley, London, 1973.

[279] LISKE H., PLATEN E., *Simulation Studies on Time Discrete Diffusion Approximation*, Math. and Computers in Simulation, 29, (1987), 253–260.

[280] MACKEVIĆIUS V., *On Polygonal Approximation of Brownian Motion in Stochastic Integral*, Stochastics, Vol. 13, 167–175, 1984.

[281] MARUYAMA G., *Continuous Markov Processes and Stochastic Equations*, Rend. Cir. Mat. Palermo, Vol. 4, 1955, 48–90.

[282] MASSERA J.L., SCHAFFER J.J., *Linear Differential Equations and Function Spaces*, Academic Press, New York, 1966.

[283] MIKHLIN S.G., SMOLICKII H.L., *Approximate Methods of Differential and Integral Equations*, (in Russian), Nauka, Moskwa, 1965.

[284] MILSHTEIN G.N., *Approximate Integration of Stochastic Differential Equations*, Theory Prob. Appl., **19**, 1974, 557 (in Russian).

[285] MILSHTEIN G.N., *A Method of Second Order Accuracy Integration of Stochastic Differential Equations*, Theory Prob. Appl., Vol. 23, 2, 1978 (Russian).

[286] NASHED M.Z., ENGL H.W., *Random Generalized Inverses and Approximate Solutions of Random Operator Equations*, in: Approximate Solution of Random Equations (Ed. Bharucha-Reid A.T.), North Holland, New York, 1979.

[287] NEWTON N.J., *An Asymptotically Efficient Difference Formula for Solving Stochastic Differential Equations*, Stochastics, Vol. 19, 1986, 175–206.

[288] ONICESCU O., ISTRATESCU V.I., *Approximation Theorems for Random Functions*, Rendiconti Matem., **8**, 1975, 65–81.

[289] OPPENHEIM A.V., SCHAFER R.W., *Digital Signal Processing*, Prentice Hall, Englewood Cliffs, New York, 1975.

[290] PADGETT W.J., TSOKOS C.P., *Random Solution of Stochastic Integral Equation: Almost Sure and Mean Square Convergence of Successive Approximations*, Int. J. System Sci., **4**, 1973, 605–612.

[291] PAKULA L., *Representation of Stationary Gaussian Process on a Finite Interval*, IEEE Trans. Inform. Theory, **IT-20**, 1976, 231–232.

[292] PARDOUX E., TALAY D., *Discretization and Simulation of Stochastic Differential Equations*, Acta Applicandae Math., **3**, No 1, 1985.

[293] PATTERSON W.M., *Iterative Methods for the Solution of Linear Operator Equation in Hilbert Space — A Survey*, Lecture Notes in Math., 394, Springer,

Berlin, 1974.

[294] PLATEN E., *An Approximation Method for a Class of Itô Processes*, Liet. Matem. Rink., **21**, 1981, 121–133.

[295] RAO N.J., BORWANKAR J.D., RAMAKRISHNA D., *Numerical Solution of Itô Integral Equations*, SIAM J. Control, **12**, 1974, 124–139.

[296] RUMELIN W., *Numerical Treatment of Stochastic Differential Equations*, SIAM J. Numer. Anal., Vol. 19, 3, 1982, 604–613.

[297] SHINOZUKA M., *Simulation of Multivariate and Multidimensional Random Processes*, J. Acoust. Soc. Amer., 49, 1971, 357–367.

[298] SHINOZUKA M., *Monte Carlo Solution of Structural Dynamics*, Computers and Structures, Vol., 2, 1972, 855–874.

[299] SHINOZUKA M., SAMARAS E., *ARMA Model Representations of Random Processes*, Proc. 4th ASCE Conf. on Prob. Mech. and Struct. Reliability, (Ed. Wen Y.K.) 1984, 405–409.

[300] SPANOS P-T.D., *ARMA Algorithms for Ocean Wave Modelling*, J. Energy Resources Technology, Vol. 105, 1983, 300–309.

[301] SPANOS P-T.D., *Numerical Simulation of a Van Der Pol Oscillator*, Comp. and Math. with Appl., Vol., 6, 1980, 135–145.

[302] TALAY D., *Efficient Numerical Schemes for the Approximation of Exptectations of Functionals of the Solutions of Stochastic Differential Equations*, in: Filtering and Control of Random Processes, (Eds. Korezliogliu H., Mazziotto G., Szirglas J.), Lecture Notes in Control and Inform. Sci., 61, Springer 1984.

[303] TALAY D., *Résolution Trajectorielle Analyse Numérique des Équations Differentielles Stochastiques*, Stochastics, 9, 1983, 275–306.

[304] WAGNER W., PLATEN E., *Approximation of Itô Differential Equations*, Report of Zentralinstitut für Math. und Mechanik, Akademie der Wissenschaften DDR, Berlin 1978.

[305] YANG J.N., *On the Normality and Accuracy of Simulated Random Processes*, J. Sound Vibr., Vol. 26, No 3, 1973, 417–427.

CHAPTER VI

[306] AMIN M., ANG A.H.S., *Nonstationary Stochastic Model of Earthquake Motions*, J. Eng. Mech. Div., ASCE, Vol. 94, 1969, 559–583.

[307] ARIARATNAM S.T., *Dynamics Stability of Column under Random Loading*, in: Dynamic Stability of Structures (Ed. Herrmann H.), Pergamon Press, New York, 1967.

[308] AUGUSTI G., BARRATA A., CASCIATI F., *Probabilistic Methods in Structural Engineering*, Chapman and Hall, London 1985.

[309] BATCHELOR G.K., *The Theory of Homogeneous Turbulence*, Cambridge University Press, England, 1959.

[310] BELLACH B., *Parameter Estimators in Linear Stochastic Differential Equations and their Properties*, in: Random Vibrations and Reliability (Ed. Hennig K.). Proc. IUTAM Symp. Frankfurt/Oder, Akademie-Verlag, Berlin, 1983.

[311] BERGMAN L.A., SPENCER B.F., *First Passage Time for Linear Systems with Stochastic Coefficients*, Probabilistic Eng. Mech., Vol. 2, No 1, 1987.

[312] BOLOTIN V.V., *Statistical Methods in Structural Mechanics*, Holden-Day, San Francisco, 1969.

[313] BOLOTIN V.V., *Random Vibrations of Elastic Systems*, Martinus Nijhoff Publ., Hague, 1984.

[314] BOUC R., *Forced Vibration of Mechanical Systems with Hysteresis*, Proc 4th Conference on Nonlinear Oscillation, Prague, Czechoslovakia, 1967.

[315] CIEŚLIKIEWICZ W., *Stochastic Characteristics of Sea Waves in the Vicinity of Still Level*, (in Polish), Ph.D. Thesis, Polish Academy of Sciences, Gdańsk, 1989.

[316] DAHLBERG T., *Comparison of Ride Comfort Criteria for Computer Optimization of Vehicles Travelling on Randomly Profiled Roads*, Vehicle System Dynamics, Vol. 9, 1980, 291–307.

[317] DIMENTBERG M.F., *Nonlinear Vibration of Elastic Panels under Random Loading*, (in Russian), Izv. AN SSSR, Miekhanika i Mashinostr., No. 5, 1962.

[318] DITLEVSEN O., *Random Fatigue Crack Growth — a First Passage Problem*, Eng. Fract. Mech., Vol. 23, 1986, 467–477.

[319] ELISHAKOFF I., VAN ZANTEN A. TH., CRANDALL S.H., *Wide-band Random Aximetric Vibrations of Cylindrical Shells*, ASME J. Appl. Mech., **46**. 1979, 417–422.

[320] ELISHAKOFF I., *Probabilistic Methods in Theory of Structures*, J. Wiley, New York, 1983.

[321] ELISHAKOFF I., LUBLINER E., *Random Vibration of a Structure via Classical and Nonclassical Theories*, in: Probabilistic Methods in Mechanics of Solids and Structures (Eds Eggwertz S., Lind N.C.), Springer-Verlag, Berlin. 1984.

[322] FIEDOROV J.A., *Vibrations of Closed Cylindrical Shell under Random Acoustic Pressure*, (in Russian), In: Zurn., N0 3, 1963.

[323] GERSCH W., NIELSEN N., AKAIKE H., *Maximum Likelihood Estimation of Structural Parameters from Random Vibrational Data*, J. Sound and Vibr., Vol. 31, 1973, 295–308.

[324] GUTENBERG B., RICHTER C. E., *Seismicity of the Earth and Associated Phenomena*, Princeton Univ. Press, 1954.

[325] HAMMOND J.K., HARRISON R.F., *Nonstationary Resonse of Vehicles on Rough Ground — A State Space Approach*, ASME J. Dynamic Systems, Measurement and Control, **103**, 1981, 245–250.

[326] HARRISON R.F., HAMMOND J.K., *Approximate Time Domain Nonstationary Analysis Stochastically Excited Nonlinear Systems with Particular Reference to*

the Motion of Vehicles on Rough Ground, J. Sound Vibr., 105(3), 1986, 361–371.

[327] HOUSNER G.W., *Characteristics of Strong Motion Earthquakes*, Bull. Seism. Soc. Amer., No 1, 1947.

[328] HUANG N.E., LONG S.R., *An Experimental Study of the Surface Elevation Probability Distribution and Statistics of Wind-generated Waves*, J. Fluid Mech., Vol. 101, No 1, 1980.

[329] IRSCHIK H., ZIEGLER F., *Nonstationary Random Vibrations of Yielding Frames*, Nuclear Eng. Design, **90**, 1985, 357–364.

[330] JEFFREYS E.R., PATEL M.H., *Dynamic Analysis Models of Tension-leg Platform*, ASME J. Energy Resources Technology, Vol. 104, Sept. 1982, 217–223.

[331] KANAI K., *Some Empirical Formulas for the Seismic Characteristics of the Ground*, Bull. Earthquake Res. Institute, Univ. Tokyo, Vol. 35, 1957, 309–325.

[332] KANEGAONKAR H.B., HALDAR A., *Non-Gaussian Stochastic Response of Nonlinear Compliant Platforms*, Probabilistic Eng. Mech., Vol. 2, No 1, 1987.

[333] KAZIMIERCZYK P., *Identification of Random Noises in Physical Systems Via Equations for Moments*, Archives of Mechanics, Vol. 37, No 1–2, 1985.

[334] KAZIMIERCZYK P., *Parametric Identification of Dynamical Systems Governed by Markov Processes.* Ph.D Thesis, Inst. Fund. Techn. Research, Polish Academy of Sciences, Warsaw, 1987.

[335] KOZIN F., *Structural Parameter Identification Techniques*, in: Analysis and Estimation of Stochastic Mechanical Systems (Eds. Schiehlen W., Wedig W.), CISM Courses No 303, Springer, Wien-New York, 1988.

[336] KRENK S., *Nonstationary Narrow-band Response and First-passage Probability*, J. Appl. Mech., 46, No 4, 1979.

[337] KURNIK J., TYLIKOWSKI A., *Stochastic Stability and Nonstability of Linear Cylindrical Shells*, Ingenieur-Archiv., **53**, 1983, 363–369.

[338] LIGHTHILL J., *Waves in Fluides*, Cambridge Univ. press., Cambridge, 1978.

[339] LIN Y.K., *Probabilistic Theory of Structural Dynamics*, McGraw Hill, 1967.

[340] LIN Y.K., ARIARATNAM S.T., *Stability of Bridge Motion in Turbulent Winds*, J. Struct. Mech., **8**, 1980, 1–15.

[341] LINDGREN G., RYCHLIK I., *Wave Characteristic Distributions for Gaussian Waves — Wave Length Amplitude and Steepness*, Ocean Engineering, Vol. 9, 1982, 411–432.

[342] LONGUET-HIGGINS M.S., *On the Joint Distribution of the Period and Amplitude of Sea Waves*, J. Geoph. Res., Vol. 80, 1975, 2688–2694.

[343] LOGUET-HIGGINS M.S., *On the Joint Distribution of Wave Periods and Amplitude in Random Wave Field*, Proc. Roy. Soc. London, Vol. A389, 1983, 241–258.

[344] MADSEN H.O., KRENK S., LIND N.C., *Methods of Structural Safety*, Prentice Hall, New York, 1986.

[345] NAESS A., *The Joint Crossing Frequency of Stochastic Processes and its Application to Wave Theory*, J. Applied Ocean Res., Vol. 7, 1985, 35–50.

[346] NARAYANAN S., *Nonlinear and Nonstationary Random Vibration of Hysteretic Systems with Application to Vehicle Dynamics*, in: Nonlinear Stochastic Dynamic Eng. Systems (Eds Ziegler F., Schueller G.I.), Proc. IUTAM Symp., Innsbruck-Austria 1987, Springer-Verlag, Berlin 1988.

[347] NIGAM N.C. *Introduction to Random Vibrations*, M.I.T. Press, Cambridge 1983.

[348] NOORI M., CHOI J-D., DAVOODI H., *Zero and Nonzero Random Vibration Analysis of a New General Hysteresis Model*, Probabilistic Eng. Mech., Vol. 1, No 4, 1986.

[349] OCHI M., *Non-Gaussian Random Processes in Ocean Engineering*, Probab. Eng. Mech., Vol. 1, No 1, 1986.

[350] PENZIEN J., KAUL M.K., BERGE B., *Stochastic Response of Offshore Towers to Random Sea Waves and Strong Motion Earthquakes*, Comp. and Structures, Vol. 2, 1972, 733–756.

[351] PIERSON W.J., MOSKOWITZ L., *A Proposed Spectral Form for Fully Developed Wind Seas Based on the Similarity Theory of S.A. Kitaigordskii*, J. Geophysical Res., Vol. 69, 1964, 5181–5190.

[352] PLAUT R.H., INFANTE E.F., *On the Stability od Some Continuous Systems Subjected to Random Excitation*, J. Appl. Mech., Vol. 92, Ser. E, Sept. 1970, 623–628.

[353] PRICE W.G., BISHOP R.E., *Probabilistic Theory of Ship Dynamics*, Chapman and Hall, London, 1974.

[354] ROBERTS J.B., *A Stochastic Theory of Non-linear Ship Rolling in Irregular Seas*, J. Ship Res., SNAME, **26**, 4, Dec. 1982, 229–245.

[355] ROBSON J.D., DODDS C.J., *Stochastic Road Inputs and Vehicle Response*, Vehicle System Dynamics, Vol. 5, 1975, 1–13.

[356] ROZMARYNOWSKI B., JESIEŃ W., *Numerical Models for Static and Dynamic Calculations of Offshore Platform*, (in Polish), Rozprawy Hydrotechniczne, **31**, 1–2, 1984, 3–14.

[357] RUIZ P., PENZIEN J., *Stochastic Seismic response of Structures*, J. Eng. Mech. Div. ASCE, April 1971, 441–456.

[358] SARPKAYA T., ISAACSON M., *Mechanics of Wave Forces on Offshore Structures*, Van-Nostrand, New York, 1981.

[359] SCHIEHLEN W., *Modelling and Analysis of Nonlinear Multibody Systems*, Vehicle System Dynamics, **15**, 1986, 271–288.

[360] SCHIEHLEN W., *Probabilistic Analysis of Vehicle Vibrations*, Probabilistic Eng. Mech., Vol. 1, No 2, 1986, 99–104.

[361] SCHIEHLEN W., *Modelling Analysis and Estimation of Stochastic Mechanical Systems*, (Eds. Schiehlen W., Wedig W.), CISM Courses No 303, Springer, Wien-New York, 1988.

[362] SHINOZUKA M., YANG J.N., *Peak Structural Response to Nonstationary Random Excitations*, J. Sound Vibr., **16**, 1971, 505–517.

[363] SIGBJØRNSSON R., *Stochastic Theory of Wave Loading Processes*, Engineering Structures, Vol. 1, 1979, 58–64.

[364] SOBCZYK K., *Free Vibrations of Elastic Plate with Random Properties — the Eigenvalue Problem*, J. Sound Vibr., **21**, No 4, 1972.

[365] SOBCZYK K., MACVEAN D.B., *Non-Stationary Random Vibration of Systems Travelling with variable Velocity*, in: Stochastic Problems in Dynamics (Ed. Clarkson B.L.), Pitman, London 1977.

[366] SOBCZYK K., MACVEAN D.B., ROBSON J.D., *Response to Profile-imposed Excitation with Randomly Varying Traversal Velocity*, J. Sound Vibr., **52**(1), 1977, 37–49.

[367] SOBCZYK K., *Modelling of Random Fatigue Crack Growth*, Eng. Fracture Mech., Vol. 24, 1986, 609–623.

[368] SOBCZYK K., *Stochastic Models for Fatigue Damage of Materials*, Adv. Appl. Prob., Vol. 19, 1987, 652–673.

[369] SOBCZYK K., TRĘBICKI J., *Modelling of Random Fatigue by Cumulative Jump Processes*, Eng. Fracture Mech., Vol. 34, Nr. 2, 1989, 477–493.

[370] SPANOS P-T., HANSEN J.E., *Linear Prediction Theory for Digital Simulation of Sea Waves*, J. Energy Resources Technology, Vol. 103, 1981, 243–249.

[371] SPANOS P-T.D., AGARWAL V.K., *Response of a Simple Tension-leg Platform Model to Wave Forces*, J. Energy Res. Technology, Vol. 106, 1984, 437–443.

[372] SPENCER B.F., TANG J., *A Markov Process Model for Fatigue Crack Growth*, J. Eng. Mech., ASCE Vol.114, No 12, Dec. 1988.

[373] STRIEKALOV S., MASSEL S., *Some Problems of Spectral Analysis of Sea Waves*, (in Polish), Arch. Hydrotechniki, Vol. 18, No 4, 1971.

[374] SUZUKI Y. MINAI R., *Application of Stochastic Differential Equations to Seismic Reliability Analysis of Hysteretic Structures*, Probabilistic Eng. Mech., Vol. 3, No 1, 1988.

[375] TAJIMI H., *A Statistical Model of Determining the Maximum Response of a Building Structure During an Earthquake*, Proc. Second World Conf. on Earthquake Eng., Tokyo, Vol. II, 1960, 781–796.

[376] TAYFUN M.A., *Narrow-band Nonlinear Sea Waves*, J. Geophysical Res., **85** (C3), 1980, 1548–1552.

[377] THOFT-CHRISTENSEN P., NIELSEN S.R.K., *Bounds on the Probability of Failure in Random Vibration*, J. Structural Mech., **10**, No 1, 1982, 67–91.

[378] TRIFUNAC M.D., *Response Envelope Spectrum and Interpretation of Strong Earthquake Ground Motion*, Bull. Seism. Amer., Vol. 61, No 2, 1971

[379] TYLIKOWSKI A., *Dynamic Stability of a Nonlinear Cylindrical Shell*, J. Appl. Mech., 1984, 852–856.

[380] VAICAITIS R., YAN C.M., SHINOZUKA M., *Nonlinear Panel Response from Turbulent Boundary Layer*, AIAA Journal, Vol. 10, No 4, 1972.

[381] VANMARKE E.H., *On the Distribution of the First-passage Time for Normal Stationary Random Processes*, J. Appl. Mech., Vol. 42, 1975, 215–220.

[383] WEDIG W., *Fast Algorithms in the Parameter Identification of Dynamic Systems*, in: Random Vibrations and Reliability (Ed. Hennig K.), Proc. IUTAM Symp., Frankfurt/Oder (1982), Springer-Verlag, Berlin, 1983.

[384] WEN Y.K., *Stochastic Response and Damage Analysis of Inelastic Structures*, Probabilistic Eng. Mech., Vol, 1, No. 1, 1986.

[385] WITT M., SOBCZYK K., *Dynamic Response of Laminated Plate to Random Loading*, Intern. J. Solids and Structures, Vol. 16, No 2, 1980.

[387] YANG J.N., SALIVAR G.C., ANNIS C.G., *Statistical Model of Fatigue Crack Growth in a Nickel-based Superalloy*, Eng. Fracture Mech., Vol. 18, 1983, 257–270.

[388] ZEMBATY Z., *On the Reliability of Tower-shaped Structures under Seismic Excitations*, Earthquake Eng. Structural Dynamics, Vol. 15, 1987, 761–775.

[389] ZERVA A., *Seismic Source Mechanisms and Ground Motion Models*, Probabilistic Eng. Mech., Vol. 3, No 2, 1988.

[390] ZIEGLER F., *Random Vibrations: A Spectral Method for Linear and Nonlinear Structures*, Probabilistic Eng. Mech., Vol 2, No 2, 1987.

SUBJECT INDEX

9 780792 303398